# MAPPING THE RURAL PROBLEM IN THE BALTIC COUNTRYSIDE

# Perspectives on Rural Policy and Planning

Series Editors:
**Andrew Gilg**
University of Exeter, UK
**Keith Hoggart**
King's College, London, UK
**Henry Buller**
Cheltenham College of Higher Education, UK
**Owen Furuseth**
University of North Carolina, USA
**Mark Lapping**
University of South Maine, USA

*Other titles in the series*

**Geographies of Rural Cultures and Societies**
Edited by Lewis Holloway and Moya Kneafsey
ISBN 0 7546 3571 6

**Big Places, Big Plans**
Edited by Mark B. Lapping and Owen J. Furuseth
ISBN 0 7546 3586 4

**Young People in Rural Areas of Europe**
Edited by Birgit Jentsch and Mark Shucksmith
ISBN 0 7546 3478 7

**Power and Gender in European Rural Development**
Edited by Henri Goverde, Henk de Haan and Mireia Baylina
ISBN 0 7546 4020 5

**Multifunctional Agriculture**
**A New Paradigm for European Agriculture and Rural Development**
Edited by Guido van Huylenbroeck and Guy Durand
ISBN 0 7546 3576 7

**The Reform of the CAP and Rural Development in Southern Europe**
Edited by Charalambos Kasimis and George Stathakis
ISBN 0 7546 3126 5

# Mapping the Rural Problem in the Baltic Countryside

## Transition Processes in the Rural Areas of Estonia, Latvia and Lithuania

*Edited by*
**ILKKA ALANEN**
*University of Jyväskylä*

**Routledge**
Taylor & Francis Group

LONDON AND NEW YORK

First published 2004 by Ashgate Publishing

Reissued 2018 by Routledge
2 Park Square, Milton Park, Abingdon, Oxon OX14 4RN
605 Third Avenue, New York, NY 10017

First issued in paperback 2021

*Routledge is an imprint of the Taylor & Francis Group, an informa business*

A Library of Congress record exists under LC control number: 2004009030

Notice:
Product or corporate names may be trademarks or registered trademarks, and are used only for identification and explanation without intent to infringe.

Publisher's Note
The publisher has gone to great lengths to ensure the quality of this reprint but points out that some imperfections in the original copies may be apparent.

Disclaimer
The publisher has made every effort to trace copyright holders and welcomes correspondence from those they have been unable to contact.

ISBN 13: 978-0-815-39041-1 (hbk)
ISBN 13: 978-1-351-15328-7 (ebk)
ISBN 13: 978-1-138-35647-4 (pbk)

DOI: 10.4324/9781351153287

# Contents

# List of Figures

# List of Tables

# List of Contributors

**Dr Ilkka Alanen**, Senior Research Fellow at the Academy of Finland (specialized in the study of rural areas in Russia and Eastern Europe), currently serving at the Department of Social Sciences and Philosophy, University of Jyväskylä, Jyväskylä, Finland. As well as having held a professorship, he has been in charge of a number of international research projects. In recent years he has published extensively on rural issues and agricultural reform in post-socialist countries.

**Dr Leo Granberg**, Senior Researcher and Deputy Director of the Finnish Centre for Russian and East European Studies (Aleksanteri Institute), Helsinki, Finland. He has managed several international projects and published extensively on rural issues in Finland and post-socialist countries.

**Dr Leeni Hansson**, Senior Researcher at the Institute of International and Social Studies, Tallinn Pedagogical University, Estonia. She has published sociological studies among other things on family relations and social networks in Estonia.

**Raija-Liisa Kämäräinen**, Master of Social Sciences and doctoral student at the Department of Sociology and Social Psychology, University of Tampere, Tampere, Finland. She has been engaged as researcher in international research projects.

**Dr Irena Krisciukaitiene**, Head of the Agricultural Policy Department at the Lithuanian Institute of Agrarian Economics, Vilnius, Lithuania. In addition to work on agricultural policy in Lithuania, she has published sociological research on rural issues.

**Dr Marjatta Marin**, Professor Emerita of Sociology at the Department of Social Sciences and Philosophy, University of Jyväskylä, Jyväskylä, Finland. She has published extensively on a wide range of issues.

**Dr Jouko Nikula**, Senior Research Fellow at the Finnish Centre for Russian and East European Studies (Aleksanteri Institute), Helsinki, Finland. He has worked on several international projects concerned with class structure and rural issues (especially rural entrepreneurship) in post-socialist countries.

**Dr Rein Ruutsoo**, Professor of Political Science at the Faculty of Social Sciences at the Tallinn Pedagogical University, Tallinn, Estonia. He has published extensively on the history of ideas and civil society.

**Dr Donatas Stanikunas**, Professor of Agricultural Economics and Director of the Lithuanian Institute of Agrarian Economics, Vilnius, Lithuania. His publications deal mainly with Lithuania's agriculture.

**Sandra Sumane**, Master of Social Sciences and doctoral student at the Department of Sociology, University of Latvia, Riga, Latvia. She has worked as a researcher and research assistant on several international projects.

**Dr Talis Tisenkopfs**, Professor of Sociology at the Department of Sociology, University of Latvia, Riga, Latvia. He has published extensively among other things on rural issues in Latvia.

**Dr Romualdas Zemeckis**, Secretary for Research at the Lithuanian Institute of Agrarian Economics, Vilnius, Lithuania. His publications deal mainly with Lithuania's agriculture.

# Foreword

The stream of information concerning post-socialist countries is increasing all the time: it is being produced by the countries themselves for their own uses, by the EU as well as by various international institutions. However, there is still a need for more rigorous academic research with a sound theoretical and critical orientation. The present volume has been written and compiled in this academic spirit.

The articles in this volume deal with different aspects of post-socialist transition in Estonia, Latvia and Lithuania, including agriculture and agricultural companies, industrial enterprises and other kinds of businesses, civil society, as well as some special issues such as poverty and problems in rural communities. Many of the research findings provide a foundation for a critical analysis of the transition policies practised, perhaps even beyond the Baltic countries.

All of the articles except one were written primarily in connection with the research project 'The Decollectivization of Agriculture in the Baltic Countries from a Psychological and Sociological Point of View,' financed by the Academy of Finland. However, since the writing and editing of the articles has taken quite a long time, work from another research project financed by the same institution – focusing on poverty in rural areas – is also reflected herein.

Jouko Peltomäki did the technical editing for this work and also translated one of the articles from Finnish into English. Two shorter articles were translated by Annaleea Quaranta. Most of the texts were checked for grammar and styling by David Kivinen. Karmo Tüür finalized the layout in accordance with the publisher's instructions. I wish to express my thanks to all.

*Ilkka Alanen*
General Editor, Leader of the Research Projects

# Chapter 1

# Rural Problematics
# in the Baltic Countries

Ilkka Alanen

## More Specific and Detailed Information Needed on the Baltic Countries

The Baltic States of Estonia, Latvia and Lithuania remain relatively unknown not only to the general public, but even to scholars in rural studies. Their struggle for independence in the late 1980s and early 1990s did bring them into the focus of world attention for a moment, and their accession to NATO and the European Union in 2004 has certainly brought them back into the Western sphere of influence, but even in post-socialism research there is still a tendency to slot them somewhere in the 'grey zone,' in-between Central Europe and NIC countries, and indeed to lump them together under one heading. It is often overlooked that despite the features that they share in common, these three countries are in fact very different from one another. These differences have their roots in the countryside and in agriculture.

## Common Features

At the beginning of the eighteenth century all three Baltic countries, with the exception of a narrow strip of coastland that belonged to East Prussia, were part of the Russian Empire. However in the provinces where the majority of the people were Estonian, Latvian and Lithuanian speakers, local power was in the hands of an ethnically and linguistically foreign landowner class; mostly German for the Estonians and Latvians, and Polish for the Lithuanians. The breeding ground for national movements and a civil society based on the language and culture of the rural population in the mid-1800s was provided by independent peasant farm production that developed gradually alongside the production by landlords. Because both small-scale production and national culture developed in this struggle against a foreign elite, the national identities took shape mainly on the basis of peasant virtues. Independence wars were waged against the Russians and the Bolsheviks, but also against the domestic landowner class. Independence brought in its wake radical land reforms as well as smallholdings, which had been virtually eliminated by large-scale land ownership. Three independent states were born, each of them a culturally and economically true republic of small farmers. Later, the *coups d'état* by the extreme right made it easier to force the Baltic countries to join the Soviet Union. After German occupation, the Soviet Union collectivized the local agriculture by force, and the wealthiest and most active part of the rural population was accused of being kulaks and deported to Siberia. The countries also lost much of their intelligentsia (either voluntarily or through deportation), and the process of forced collectivization also destroyed the former point of anchorage for national identity. This is why collective farms (kolkhozes) and state farms (sovkhozes) were considered an explicit expression of the occupation force. It also goes some way towards explaining why national movements in all the Baltic countries and especially their humanist leaders embraced a nostalgic trend that idealized petty production in agriculture. Indeed, the primary aim of the restitution of agricultural property was not to prevent the small Russian minority from owning land,[1] but rather to legitimize the original relations of ownership from the times prior to forced collectivization. In some respects, these politics also included a power struggle, although that was not directed against the Russians, but the native 'red barons' (a rural elite composed of nationalist kolkhoz and sovkhoz leaders) who were part of the titular population. Because most of the rural population consisted of titular population, its support was decisive in Estonia and especially in Latvia, where the Russian population was concentrated in the capital city and other major towns. Although the more educated and skilled agricultural employees were certainly not keen on the idea of returning to family farm production, power in the newly independent countries was initially entrusted to an elite who had been committed to the family farm ideology and whose thinking was closely in line with the family farm strategy recommended by the World Bank. In all three countries, decollectivization was characterized by close adherence to these policies.

The Baltic countries continued to make progress during the Soviet era: their level of education rose, they became industrialized and urbanized and their agriculture was mechanized. In a comparison with the other former Soviet republics and the former socialist countries of East Central Europe, Estonia and Latvia have been the best performers among the former Soviet republics and are comparable mainly with Hungary and Slovakia; while Lithuania can be compared to Poland. In terms of agricultural production they all rank among the elite of the socialist camp (see the article by Alanen in this volume).

## Major Country Differences

There are also some noteworthy differences between Estonia, Latvia and Lithuania – which upon closer scrutiny are often found on the same dimensions where the similarities are seen. It is no coincidence that the nationalist movement started in Estonia and only there proceeded to formulate economic programmes (e.g. Panagiotou, 2001). Soviet companies had a less prominent role in the Estonian economy than in Latvia and Lithuania, and Estonian industry was also lighter than

---

[1] This unfounded view is repeated in many of the works by Swinnen et al. (see for example Mathijs and Swinnen, 1997, 18–19).

in its southern neighbours; this allowed the country to retain more of its R&D work within local companies (Paasi, 2000; Nørgaard, 2000, 175–176). Access to Finnish television and the kinship between the languages of Finnish and Estonian meant that Estonia had a window to the west. In addition, a regular ferry service between the Finnish capital of Helsinki and Tallinn since the 1960s provided a steady stream of tourists from Finland. Finnish-Soviet agreements on science and culture, for instance, gave Estonian experts various opportunities to visit Finland. Although this was still limited to a handful of Estonians and although those who did have access were closely monitored, some Estonian innovations even in agriculture, for instance, can be attributed to Finnish influences. At the same time, a network of personal contacts was established between Finns and Estonians that was to become hugely significant later on. For these and other reasons, Estonia was in many ways better prepared for a new way of thinking and for reform programmes than Latvia and Lithuania, which were both linguistically and geographically more isolated.

In terms of rural background, too, the three Baltic States not only share similarities but also differ significantly. Before the Soviet era, Lithuania was less industrialized and Lithuanian agriculture lagged far behind Estonia and Latvia. While the small-scale landowner class in Estonia and Latvia began to differentiate internally between the world wars, the Lithuanian rural structure was much more rigid. The only Baltic country where community structures analogous to the Russian MIR community were preserved until the 1800s was Lithuania. This also explains why some Lithuanian cooperatives were very different from Estonian and Latvian ones between the world wars (cf. Ruutsoo's discussion in this volume). This background, which has its origins deep in the history of Lithuania, was probably reflected during the Soviet era in more widespread private farming and a more peasant way of life (see the article by Alanen).

In the late nineteenth century civil societies in Estonia and Latvia were also more developed than in Lithuania, which had a lower literacy rate. We cannot discount the impact of differences in cultural backgrounds either: very much influenced by Germany, Estonia and Latvia belong to the sphere of Protestant culture, while Lithuania is a Catholic country, heavily influenced by Poland. Bearing in mind Lithuania's past as a major political power, it is understandable that Lithuanian culture is characterized by state-centrism to a greater extent than the two other Baltic countries. This also helps to explain why Lithuanians often display a greater tendency to reforms organized from the top down (see the article by Ruutsoo). Perhaps the state-centrism of the process of rural de-collectivization in Lithuania can also be explained by reference to historical facts. However, although legislation on de-collectivization is very similar in Estonia and Latvia, the outcomes of these processes are very different. Even though all three countries have an abundance of raw material for the wood processing industry, this industry is by far the most developed in Estonia; Latvia ranks second in this comparison, while Lithuania is the least developed (see the articles by Nikula in this volume). In most cases the rank-order of the countries can be predicted from an historical point of view. However, not every explanation can be found in history: it is important to remember that in practice, explanations are generally mediated by structural factors. In many cases the two dimensions are intertwined and the immediate structural factor only assumes its full meaning when examined against its historical background (such as the extent of plot farming by private households). Anyone who is inclined to look upon the Baltic countries as a single block should study the historically sensitive works of Lieven (1994) and Ruutsoo (2002).

## The Importance of Research on the Baltic Countryside

A brief historical overview should suffice to make it clear that agriculture and the countryside are crucial to our understanding of all the Baltic countries. In these three countries, more than in any others, an in-depth understanding of the countryside is essential to deepening our knowledge of the nations as a whole. The three Baltic countries provide a fruitful research frame. Paradoxically, the many features that these countries share in common also highlight the significance of the historical differences that set them apart. Because similar policies produce different results in different countries, it is clear that those policies must be tailored according to local needs and circumstances; witness the agricultural reforms in Estonia, Latvia and Lithuania, which have failed to meet the expectations of both national elites and international institutions. Any attempts to explain the success of agricultural reforms by reference to their adherence to certain patterns (individualization of agriculture, the rapidity of the process, synergy, legal characteristics of different types of corporation, liberalization of the national economy) are bound to fail. Also the failures and partial successes of de-collectivization in the Baltic countries prove that good rural policies cannot be based upon dictation from above to below. Instead, those policies have to be adapted to the circumstances prevailing in each country. In addition, the rural population must be involved both in making the decisions on those reforms and in their execution.

## References

Lieven, A., (1993), *The Baltic Revolution: Estonia, Latvia, Lithuania and the Path to Independence*, Yale University Press, Newhaven.

Mathijs, E. and Swinnen, J. F. M. (1997), *The Economics of Agricultural Decollectivization in East Central Europe and the Former Soviet Union*, Policy Research Group Working Paper No 9 (Revision of Working Paper No 1), Policy Research Group, Department of Agricultural Economics, Katholieke Universiteit Leuven.

Nørgaard, O. (2000), *Economic Institutions and Democratic Reform: A Comparative Analysis of Post-Communist Countries, Cheltenham*, Edward Elgar, Cheltenham.

Paasi, M, (2000), *Restructuring the Innovation Capacity of the Business Sector in Estonia*, European Bank for Reconstruction and Development, Blackwell, Oxford.

Panagiotou, R. A, (2001), 'Estonia's Success: Prescription or Legacy?' *Communist and Post-Communist Studies*, vol. 32, no. 2, pp. 261–277.

Ruutsoo, R, (2002), *Civil Society and Nation Building in Estonia and the Baltic States*, Acta Universitatis Lapponienses 49, University of Lapland, Rovaniemi.

# Chapter 2

# The Transformation of Agricultural Systems in the Baltic Countries – A Critique of the World Bank's Concept

Ilkka Alanen

## Introduction

It is only quite recently that the effects of the political choices made during agricultural decollectivization in East-Central Europe have begun to take clearer shape; the publication of new agricultural and population censuses has also very much facilitated the analysis of their outcomes. At the time of writing, however, only some of the results of agricultural censuses were available, and they were confined to Estonia and Latvia. This article is therefore mainly based on ordinary public statistics, earlier research, as well as interview data collected in the course of my previous research projects.[1] Since the results of agricultural decollectivization have fallen well short of expectations, I hope this present article will inspire new critical debate while we wait for new data to come out. The Baltic countries may have a special role in this debate, since their national elites have – at least at the theoretical level – quite faithfully followed the decollectivization strategy promoted by the World Bank, IMF, OECD, EBRD and a number of other international organizations. Despite their differences, the paths pursued by these countries clearly indicate the problems involved in the line advocated by international organizations, even when there are some successes, such as in Estonia. My argument is that the failures are mainly due to two untenable premises: first, politics have been formulated and carried out overwhelmingly according to the top-down principle, with little or no regard for the opinions of the agricultural population, and second, they have been too firmly anchored to the family farm system as the ideal organization of agricultural production. Instead of maximal utilization, the doctrinal policy of decollectivization followed in the

---

[1]   The interview data were collected in a research project ('The Decollectivization of Agriculture in the Baltic Countries from a Psychological and Sociological Point of View' in 1998–2000) that I was responsible for conducting in three municipalities/villages in the Baltic countries. Even though I have not been able to draw directly on these data in the present article due to their general nature, they have had a key role in the critical evaluation of all the data I utilize in the text.

transformation of the agricultural system has led to vast amounts of material and human resources being lost and ruined.

## Decollectivization and the World Bank

The general principles lying behind the reconstruction of agricultural systems in Estonia, Latvia and Lithuania can be found, first, in official World Bank sources that include recommendations issued to the new countries created after the collapse of the former Soviet Union (WB, 1992, 1993b, 1993c and 1993d); and second, in an unofficial Internet publication written by three leading World Bank researchers, Zvi Lerman, Csaba Csáki and Gershon Feder ('Agriculture in Transition', 2001). Both of the sources are marred by similar contradictions that probably arise from the authors' differing opinions as well as political pressures. Not surprisingly, the World Bank sources are by far the more contradictory and harder to decipher, owing to the official nature of the documents and the semi-official nature of the researchers' reports. However, the Internet book,[2] even though it does repeat some of the same contradictions, is far less ambiguous. It gives a far more coherent and intelligible account of the main arguments behind the policy of the World Bank and its collaborative institutions. The same set of concepts is in fact also used by a group of researchers, including Johan F.M. Swinnen, that is instrumental in drafting EU agricultural policy. The World Bank researchers mentioned above and Swinnen and colleagues work in close collaboration with each other (see Lerman et al., 2001, Introduction, 9). In addition, they share the same narrow focus on just one science, Agricultural Economics, its theory tradition and methodology. Furthermore, neither of the groups comment on the often challenging and critical research carried out in the fields of sociology and anthropology. Nevertheless, there are also some interesting differences between the analyses and conclusions of these research groups. Despite their narrow theoretical and methodological focus,[3] they have probably had a greater impact on agricultural transition policies than all other social scientists taken together.

The predilection for family farming at the World Bank and among its researchers is based both upon the argument of the excellence of family farming, and upon the strategic idea of moving towards family farming through the ideal intermediate steps of certain corporate forms of large-scale production.

---

[2]   The arguments found in this unfinished manuscript (2001) also appear in the authors' other articles, notably in those by Lerman. I prefer to use the Internet book because the argumentation is more concise and condensed and because it can be easily accessed for verification.
[3]   This theme would warrant an article of its own, but we cannot dwell upon it any longer here.

## Predilection for Family Farming

Although the Baltic reform was supposed to be neutral with regard to enterprise type,[4] 'Food and Agricultural Policy Reforms in the Former USSR' (WB, 1992, 70–77) that focused solely on making recommendations concerning agricultural transition, gave a definite preference to family farms or individual farms;[5] these are taken to refer primarily to non-collective enterprises (Lerman et al., 2001, Chapter 2, 1). However, the concept of collective farms includes not only conventional kolkhozes, sovkhozes and other similar East European large-scale cooperative farms, but also all kinds of stock-holding companies or cooperative farms based on employee ownership (Ibid., Chapter 3, 2–3 and 10–11). Ultimately, individual or family farming is the opposite of corporate farming (Ibid., Chapter 3, 12–13). According to Lerman et al., the schizophrenia between a neutral policy and the predilection for family farming was resolved on a scientific basis:

> Theoretically, it is individual farms that are expected to achieve highest levels of productivity and efficiency due to personal involvement and direct accountability of family members. Corporate farms are inherently disadvantaged by various monitoring, transaction, and agency costs, which are unnecessary in family farms and are unavoidable in corporate structures with hired labor and professional managers[6] (Lerman et al., Chapter 3, 18).

---

[4] Neutrality is highlighted in those World Bank recommendation documents (1993b, 1993c and 1993d) that deal with the overall transition policy of each Baltic State – to such an extent that the Estonian document states the following: 'Again, it is critically important that legal and administrative arrangements facilitate this process, and that the process is not subjected to administrative partiality or predilection, such as a strong desire to establish and maintain "family farms."' The document then goes on to suggest that 'some farms with a large indivisible infrastructure (such as large, centrally located barns) should be maintained as corporate farms with private sector investors and managers' (WB, 1993b, 111). If the actual policies of international organizations and the Baltic governments had complied with these recommendations, this article would be more or less without foundation. However, it is perfectly clear to anyone who systematically studies the international organizations' reports that they are fully committed to establishing and maintaining family farms (see also Spoor and Visser, 2001).

[5] There was lively and varied debate on the forms of small-scale production in agriculture in the 1970s and 1980s (Alanen, 1991 and 1992). Against this background the formulations of the World Bank are inexcusably vague. Family farms, individual farms or peasant farms are just some of the concepts that are treated synonymously in World Bank publications, and in most cases the reader needs to rely on context to determine the actual meaning of each term.

[6] In the contradictory fashion that characterises most publications dealing with the subject, the text goes on to state: 'To offset these costs, corporate farms have to achieve a substantially greater reduction in operating costs. Only corporate farms that undergo significant internal restructuring of operations and management are theoretically expected to be competitive with individual farms by measures of productivity and efficiency.' This raises the question as to how is it 'theoretically' possible to overcome the drawbacks of corporate farms, which had just been considered 'theoretically ... indispensable in corporate structures.'

This theory explains why 'large farms are a rarity' in the west (Western Europe and North America) (Ibid., Chapter 1, 30). Corporate entrepreneurship is only suitable in specific cases in agriculture, such as pig and poultry farming, which are also practised in large 'factories' in the west (Ibid.). Despite the authors' conflicting and neutral formulations, the general drift is clearly towards adopting family farming, 'a transformation from collective to individual agriculture as an ultimate goal' (Ibid., Chapter 1, 37). Upon closer inspection, the goal is not just the creation of individual, family farms, but the establishment of technologically advanced family farms complete with the associated infrastructure and institutions (road, utility, etc. networks, as well as credit, education and counselling services) that are *de facto* at the core of the western family farm system. Hence, this alternative involves the adoption of another system, not just coming together in an agglomeration of individual farm enterprises. It is notable that while World Bank documents do not recommend the creation of any particular system, they do offer plenty of advice on how to meet the financing needs of the new enterprises, for instance.

All the Baltic countries decided to adopt the family farm system as their transition goal, even though the legislation in Estonia and Latvia was technically quite neutral. It was crucial that the legislation enabled and even endorsed the preservation of technological units as functional wholes.[7] Perhaps the most important obstacle to the establishment of corporate farms in Estonia and Latvia was the ban on land ownership by corporate farms. The ban was subsequently lifted in both countries, but it did clearly deter the establishment of corporate farms during the decollectivization of non-land assets. The risk was that such farms would become totally dependent on outside supplies, particularly the cattle feed sold by local family farmers. Public opinion obviously played a significant role, too, above all the negative attitude of government ministers, but so too did the critical view taken by the media on the continuation of large-scale production by kolkhozes and sovkhozes in any other legal form. Soviet farm chairmen who came out in defence of large-scale production were often considered 'traitors' (Abele, 1995, 4). However, according to my interviews, the commitment and willingness of ordinary people, particularly the agricultural middle-class, to continue large-

---

[7]    According to the author of the OECD Country Report on Latvia, the law passed in 1991 'clearly stated that corporate farms created out of the old collective and state farms were transitional structures that existed only as long as it was necessary to privatise all their non-land assets' (1996b, 79). However, the same law also made it clear that you could only remove those movable assets whose removal would not affect the operation of a technologically integrated unit (Abele, 1995). In Estonia, support for the handover of non-land assets to family farms grew steadily prior to the reform, but the final legislation nonetheless retained the existing technological units (Tamm, 2001). In Estonia, too, the legislation facilitated the establishment of cooperative enterprises. The purpose of the law was to create service enterprises on family farms in accordance with the policy promoted by the World Bank, but in Estonia this opportunity actually contributed to the preservation of large production combinates and, in a few cases, the entire agricultural part of the kolkhoz – even if industrial and construction activities, for example, were privatized separately from this whole (Tamm, 2001, 434–435).

scale farming (or their reluctance to establish family farms) was even more important than the opinions of farm chairmen.

It was perhaps because of this opposition that legislation on decollectivization and its implementation in Lithuania were aimed directly at the abolition of large-scale production. At the first stage – often for political reasons and with little regard for the expertise of kolkhoz and sovkhoz leaders – the former Soviet farms were, however, split into smaller limited liability companies, partnerships.

*Corporate Agricultural Enterprises as Tools for a Family Farming System*

According to World Bank recommendations, corporate ownership should only be a temporary stage in agricultural production, or at the very least the rules should be such that they do not to prevent individual shareholders from leaving and transferring their shares to themselves in the form of physical land and non-land assets in order to establish a family farm (WB, 1992, 76). Instead of productive agricultural enterprises, former kolkhozes and sovkhozes could be turned into 'all types of voluntary and commercial cooperatives (service, supply, marketing, production) (Ibid., 77).' Cooperatives with direct agricultural production are, as a rule, considered the least suitable in the preference hierarchy of the World Bank researchers. Since corporate enterprises are primarily understood as temporary arrangements on the path to a family farming system (Ibid., 75–77), it is recommended that the chief legal form of agriculture should be limited liability companies, particularly partnerships. All of these facilitate the realization of shareholders' shares, and in partnerships, the physical separation of the share (Lerman et al., 2001, Chapter 3, 10). Whereas the 'more conservative' farm 'reconstruction options' (WB, 1992, 74) involve the establisment of closed enterprises, in which the realization of a shareholders' shares is in fact much more difficult, even if it might be possible in theory. According to the World Bank, they constitute a representation of the 'resistance' of (former Soviet) 'farm managers' and 'rural nomenklatura' (Ibid., 74). The establishment of this type of enterprise is not recommended. In the event that such a negative development has occurred, the reconstruction is not performed voluntarily through shareholder activity and the market, but by dividing the former enterprise mechanically into new corporate enterprises based on its current division into departments. Later, the 'rural nomenklatura' participated in the preservation of the system by helping to prevent bankruptcies, for instance by forgiving old debts and by granting new loans – in the vein of the old Soviet tradition (Lerman et al., 2001, Chapter 3 28–33 and 41–42).

Lerman et al. raise another, directly opposite theoretical argument to explain why it is not a good idea to have closed stock-holding companies and cooperative corporate enterprises, namely the 'syndrome of labor-managed firm' (Ibid., Chapter 3, 39), of which the cooperative represents most disastrous form not only in the former socialist countries but 'everywhere in the world.' 'All production cooperatives based on member labor suffer from shirking and free riding' (Ibid., Chapter 1, 33). This theory is based not on the assumption that the managers are conservative, but rather that the employees are in a position to force their managers to practise inefficient agriculture. This is why such enterprises employ unnecessarily large numbers of people, and their work discipline on the whole is poor. In this situation the enterprise's function effectively changes from an independent agricultural unit to a support organization for the employees' household plots. For the employees, the large enterprise provides a 'safety umbrella' in difficult social circumstances. In addition, the researchers assume that any pensioners and people employed elsewhere who own shares prefer large enterprises because they lack confidence in new ones (Ibid., Chapter 3, 40). The logical solution to this problem would be to separate ownership from management and to impose strict market legislation, including a concrete threat of bankruptcy. Thereby the agricultural enterprise would bear an increasing resemblance to non-agricultural, well-functioning corporate firms. According to Lerman et al., the separation of ownership from management explains why the heavily restructured corporate farms in East-Central Europe, which are smaller than those of the socialist era, have succeeded much better in adapting to market conditions than those in the CIS countries. However, 'the ultimate goal' is still family farming, and from a theoretical viewpoint corporate farms in East-Central Europe are also just transitional arrangements.

The family farm theory and the corporate farm theory thus constitute two sides of the same agricultural theory. In the same way as the classical romantic (or populist) tradition from Eduard David (1903 and 1922) to A. V. Chayanov (1966), the World Bank and its researchers take the view that the natural development of agriculture into family farming is ultimately based on its special relationship to nature, which differs markedly from that of industry (Lerman et al., Chapter 1, 30–32). They believe that agriculture in the western world has developed into a family farm system since the free market economy has imposed no artificial restrictions on its development, and therefore 'among other performance criteria, we evaluate the transition countries against the benchmark of individual farming' (Ibid., Chapter 3, 4).

## Good Management of Large-Scale Production or Specialization of Small-Scale Production?

There are two recent articles on agricultural productivity that provide useful support to the present research design. Following the established World Bank line, Lerman, Kislev, Kriss and Biton (2003) say that production has increased the most 'in Armenia and Georgia that resolutely switched from large-scale Soviet agriculture to small-scale individual farming' because 'by specializing in high-value and labour-intensive products, the individual sector could follow low-input farming practices particularly avoiding reliance on machinery and equipment' (Ibid., 1015). This is the way family farms have developed in Europe over the past one hundred years plus (Hussain and Tribe, 1981). However, the authors still believe that even the tiniest plot farms of under half a hectare could follow the former path of development in Europe. Thus, both household plots and family farms would specialize in vegetable and potato farming in crop production, and in meat, milk and in part egg production in livestock production. Looking at these research results, it is immediately apparent that the Baltic country, which most closely resembles Armenia and Georgia, characterized by Lerman (2001a, 110) as the 'two CIS stars,' is Lithuania. According to the calculations of Lerman, Kislev et al., Armenia and Georgia are the top transition countries in terms of increased labour productivity, with Lithuania following in close pursuit as the best Baltic country. Lithuania's high ranking should come as no surprise if the World Bank's theory of family farms and limited liability companies (as a transitional stage to family farming) is valid. Labour productivity in Latvia, on the other hand, has developed much more slowly, and that is something the authors find hard to explain. In principle, Latvia should be among the top performers from the World Bank viewpoint, since its agricultural production is the most individualized of the three Baltic countries. However, the authors content themselves with the statement that 'privatization did not help Latvia' (Ibid., 1016). Estonia is another country that fits in poorly with their theory, since labour productivity in Estonian agriculture has increased at almost the same rate as in Lithuania, despite the fact that large-scale production based on closed stock-holding companies is dominant in Estonia. The authors make no comment on this at all.

Karen Macours and Johan F.M. Swinnen (2002) found another explanation for differences in labour productivity. Compared with the typical viewpoint of the World Bank, theirs is a rather curious theoretical perspective. Drawing on time series data from the first five years of transition, the authors propose three basic models: the CVA model is typically found in China, Vietnam and Albania; the RUB model in Russia, Ukraine and Belarus; and the CSH model in the Czech Republic, Slovakia and Hungary. In the CVA countries, both production and productivity are on the increase,[8] whereas in the CSH countries production is declining while productivity is increasing. In the RUB countries, then, both production and productivity are on the decrease. Other countries not mentioned here are more or less 'hybrids' of these three basic models. Among the Baltic States, Estonia closely resembles the CSH model, while Latvia and Lithuania share many features in common with the RUB model.

Both Lerman, Kislev et al. and Macours and Swinnen explain the differences between the RUB and CSH models by reference to the 'paper share' theory: private ownership by corporate farm part-owners (employees and pensioners) exists only on paper, since the owners cannot transfer their share (of the land and non-land assets) in any form into such purposes as financing the start-up of a small business, nor sell or rent it to outsiders. However, according to the calculations of Macours and Swinnen (and compared with the results of Lerman, Kislev et al.), not only Latvia but also Lithuania has shown, somewhat surprisingly, reduced labour productivity. Poor productivity in Lithuania or Latvia cannot be explained by reference to the 'paper share' theory that is applicable to the RUB countries, since 'strong use rights on agricultural production factors, including land, to individuals' (Macours and Swinnen, 2002, 387) characteristic of the CSH model (Ibid., 386)

---

[8] I call this model the 'development country model.' The example of China had a profound effect on World Bank policies in Europe, but the results have failed to live up to expectations (WB, 2002a, 26).

exist as 'key conditions for productivity improvement' in all the Baltic countries. Even if one considers the 'larger fall in relative prices and large disruption, which reflect more important price and trade distortions in the Baltics' (Ibid., 375 and 387), caused by the collapse of the Soviet Union, the situation in Latvia and Lithuania still requires further explanation – Estonia being almost a CSH country – especially as all the Baltic countries have pursued 'good,' i.e. completely liberal and radical, privatization and economic policies.

Latvia's and Lithuania's poor performance is attributed by the authors to these countries' restitution policies, 'which resulted in an important shift to individual farming' (Ibid., 386). The productivity problem is thus linked to the handing over of land to people who are unlikely ever to work in agriculture; often these people have lived in towns almost all their lives and do not know the first thing about farming. The landed property is further fragmented as it has to be split up among several beneficiaries. With much of the land falling into the hands of people who had neither any experience of nor any motivation for farming, it would have made more sense to apportion the land between the workers of Soviet farms (see e.g. Mathijs and Swinnen, 1997, 18–19). However, the theory presented by Macours and Swinnen is not applicable to all the Baltic countries, since Estonia followed the same policy of land restitution, yet the country nearly reached the CSH level.

The main difference between Lerman, Kislev et al. (the traditional World Bank line) and Macours and Swinnen, however, lies in the conclusion of the latter that 'the extent to which management differs effectively from the pre-reform management, rather than the shift to individual farming, is crucial for improving productivity. [...] The key difference lies in the extent to which farm management of these large-scale farms, and its incentives, have restructured' (Macours and Swinnen, 2002, 387). According to Macours and Swinnen, the productivity problem is not always caused by lack of individualization, but individualization can – at least if you use the restitution method – even become the cause of these problems – as was the case in Latvia and Lithuania. However, in RUB countries, the main reason behind lowered productivity still is the distortion of corporate farm management by the paper share system.

Why would two empirical measurements of labour productivity produce diametrically opposed results, as in the case of Lithuania? The results are explained by differences in the methods of measurement. Whereas Lerman, Kislev et al. calculate labour productivity on the basis of a combination of several factors (arable land/worker, farm machinery hp/worker, livestock standard head/worker, fertilizer/arable land), Macours and Swinnen settle for a much simpler calculation of the ratio of GAO to agricultural labour force. Complicated measurements under transition conditions involve numerous problems. Lerman, Kislev et al. themselves acknowledge that the results of their measurements should be interpreted with caution because of market imperfections. Unfortunately, they fail to provide an account of their exact method of calculating productivity as well as estimating subsistence farming and the informal economy. The present article shows that subsistence farming is very common among small-scale producers, especially in Latvia and Lithuania, and, in all likelihood, a larger than average proportion of these people are also involved in the informal economy (cf. OECD, 2003, 63–64).

If their calculations only cover chemical fertilizers (the sale of which is fully reflected in the statistics) but ignore household manure, for instance, then the productivity of plot farms may appear to increase because of the method of measurement used. Since the calculations of Lerman, Kislev et al. are inconsistent with the statistical and research data and with the impressions I gained from my field interviews (which are analyzed in more detail later in this article), I must assume that the method used by Lerman, Kislev et al. to measure labour productivity creates an unwarranted bias that favours the countries where plot farming is the most widespread. In the early stages of transition it makes sense to use simple and straightforward indicators, particularly in the measurement of labour productivity. This is why in this article I have used a few simple time series and their combinations (the ratio of GAO to labour force, milk yield per cow and specialization in dairy farming) as indicators of productivity and production efficiency.

Lerman, Kislev et al. consider it more important to express their confidence in the ability of small farmers to adapt than address the question of where farmers can find a market for their potatoes and vegetables, for instance, or how can they specialize in dairy farming if they cannot afford to build or renovate a cowshed – not to mention the problems of milk marketing.

Even though Macours and Swinnen are somewhat more realistic in their thinking, and even though they have correctly identified the question regarding the conditions under which good management is possible, their theory is too general to account for the differences found between the transformation of the agricultural system in the different Baltic countries and indeed in the other countries of East-Central Europe. I would also have liked to see a more differentiated approach to the problem of 'initial conditions.' The scarcity of large-scale production in Poland and the former Yugoslavia are certainly not the only differences between the farm structures of these countries. As I will show, Estonia and Lithuania represent two extreme cases with regard to the ratio of large- to small-scale production in the former Soviet Union. Since there was only a minimal need to develop new infrastructure and institutions (road, sewage, water and electricity networks, as well as credit, education and counselling systems) for small-scale production during the transformation of the agricultural system in the Baltic countries, small-scale production was separated from its previous symbiotic relationship with large-scale production and faced inevitable problems, even though the plot farming tradition, with its modest tools, houses and production buildings, meant that large numbers could continue with some sort of small-scale production. However, since the small farming tradition in the Baltic countries was different, it could provide an answer to a number of questions. Why would the post-socialist 'path dependence' debate (Stark and Bruszt, 1998) not be relevant in agriculture, too?

Second, Macours and Swinnen fail to see that the paper share system is by no means the only possible cause for poor management conditions. In the light of the facts put forward in the present article, there is good reason to believe that the

Lithuanian partnership system[9] will also produce below-average management conditions.

Third, Macours and Swinnen focus too heavily in their approach upon the government. They fail to grasp the fundamental difference between the centralized, top-down decollectivization of non-land assets in Lithuania and the decentralized, locally managed process in Estonia and Latvia. In Estonia and Latvia, the local agricultural population was offered the opportunity to take part in and influence the course of events; in Lithuania they were excluded from decision-making, perhaps because they were seen by the government as their political opponents – forced collectivization was followed by forced decollectivization. A great deal of local professional knowledge was lost in the process.

Fourth, the positive attitude shown by Macours and Swinnen to rapid, radical agricultural (and other transition) reforms (a view they share in common with World Bank researchers) is, on closer inspection, more problematic than their article would give to understand.

Fifth, Macours and Swinnen (together with the World Bank researchers) spend too much effort trying to explain agricultural changes via labour mobility to other industries, but they are unable to present anything other than indirect empirical evidence (such as the wage differences between agriculture and other production types in Lerman, Kislev et al.).

## Dairy Farming from an Empirical Perspective

Dairy farming is highly characteristic of Estonia, Latvia and Lithuania. There are a number of reasons for this. First, within the Soviet division of labour, the Baltic countries specialized in livestock production, and dairy farming was by far the most important part of livestock production (more on this later). The resources for agricultural production, including infrastructure (production buildings), machinery, cattle as well as education were designed with a view to dairy farming. Secondly, the natural conditions in the Baltic States favoured specialization in dairy farming, and continue to do so today. Thirdly, given these circumstances, dairy farming has provided the foundation for the development of family farming, particularly in North Western Europe, allowing even the smallest farms largely to rely on family

---

[9]   The authors are well aware of the intrinsic nature of the Lithuanian reform, as well as of the replacement of the leaders of former collective farms by other people during the implementation phase. In an earlier article Swinnen pays close attention to this, but from a different perspective. Swinnen believes that the partnership legal form was used by the Sajudis government (formed on the basis of the Lithuanian Popular Front) as a means of reducing the transaction costs linked to the withdrawal of assets by the shareholders of corporate farms. For the same reason Mathijs and Swinnen (1997, 16–17) criticize the subsequent ex-communist government for changing the structure and composition of the reform institutions (that had previously been centralized to the government) once they had won the elections and gained power in 1993. The lack of knowledge among the new leaders could also have been one of the reasons for the initial failures.

labour. It has also been considered competitive when compared with large-scale production, which has to use wage labour in this highly labour intensive business.

The statistical data presented in the tables of this article have been collected from various sources, including both public statistics and individual research publications. Where different sources present conflicting data,[10] I have selected the data so as to make comparisons between the Baltic countries as easy as possible. I have also tried to keep each time series technically consistent[11] with the other time series. Since international statistics are largely based on national statistics, I have used national statistics as far as possible.

## The Initial Conditions for Transition

International as well as national agricultural macro-level starting points are, of course, essential to the success of transition.

**Table 2.1 Some macro-level indicators before transition**

| Indicator | Estonia | Latvia | Lithuania | Year |
|---|---|---|---|---|
| GDP per Capita (USD) (1) | 4,646 | 4,522 | 2,002 | 1987–1991 |
| Employment (% of population) (2) | | | | 1990 |
| Agriculture, forestry and fishing | 21 | 17 | 18 | |
| – Industry | 29 | 28 | 30 | |
| – Construction | 8 | 10 | 12 | |
| – Services | 42 | 45 | 40 | |
| Share of Rural Population (%) (3) | 29 | 30 | 35 | 1980–1989 |
| Share of the Soviet Population (%) (4) | 0.5 | 0.9 | 1.5 | 1986–1990 |
| Share of the Soviet Grain Production (%) (4) | 0.5 | 0.8 | 1.5 | 1986–1990 |
| Share of the Soviet Milk Production (%) (4) | 1.2 | 1.8 | 3.0 | 1986–1990 |
| Agricultural Output (SUR per worker) (5) | 12,800 | 9,600 | 9,700 | 1965–1990 |
| Agriculture (% of GDP) (3) | 20.9 | 18.3 | 25.8 | 1980–1989 |

*Sources:* (1) annual average statistics, Lerman et al., 2001, Chapter 1, 39; (2) OECD, 2000, 236; (3) annual average statistics, Lerman et al., 2001, Chapter 1, 38; (4) annual average statistics, WB, 1992, 194; (5) Lerman, et al., 2001, Chapter 1, 26.

---

[10]   Since the statistical principles have been changed over time on several occasions, I have decided to use comparative data only. For example, both Estonia and Lithuania changed their methods of compiling statistics on the employed population in 2000, and the new statistics were no longer comparable with the older ones. Since I do not always have access to complete time series based on the new statistical principles, I use old statistics that are only available up to 1999. However, I will always use the latest information as long as comparability is not affected.

[11]   By technical consistency I refer to a situation where inconsistency between two sets of statistics excludes one (or both) sets, for instance.

At the outset of transition the industrial structure in the three countries was strikingly similar, even though the share of services was slightly larger in the Latvian economy, making it more modern than the others. In terms of production efficiency, however, the Soviet republics differed widely from one another. The Estonian economy was the most efficient in the entire Soviet Union, with Latvia following close on the heels of its northern neighbour. In Lithuania, GDP per capita was well above the average for all Soviet republics (by around 30 per cent), but still some 50 per cent lower than the figures for Estonia and Latvia and some 30 per cent lower than in Russia. GDP per capita in Estonia and Latvia was in fact comparable to the figures in Hungary and Slovakia, the wealthiest countries in East-Central Europe, but still about 20 per cent behind the Czech Republic. Among East-Central European countries, Poland recorded efficiency figures comparable to those of Lithuania (GDP statistics, see Lerman et al., 2001, Chapter 1, 39).

At the end of the Soviet era, all the Baltic States were among the most industrialized republics, and in agriculture they ranked among the Soviet elite. Estonia, which again was the most efficient, was 54 per cent more productive than Russia (in fourth place) in 1983; even Latvia, which came last in a Baltic comparison, recorded productivity figures that were 27 per cent higher than for Russia (Lerman, Kislev et al., 2003, 1004).[12] Due to differences in efficiency between agriculture and industry, Latvia is the most industrialized country, and Lithuania is a predominantly agricultural country, whereas in Estonia there are no substantial differences between the efficiency of agriculture and industry.

Lithuania was slightly more rural than the other countries, probably because of its small-scale agricultural production (see Table 2.3).

The Baltic States were more highly specialized in animal husbandry than any other European transition country (Lerman, Kislev et al., 2003, 1003). Partly because of their emphasis on dairy production, but also because of their highly developed pig farming, another area of specialization for them was meat production. Grain production in each republic corresponded with their shares of the total Soviet population, but milk production was twice as high as their share of the population. Relative to population size Estonia, Latvia and Lithuania were by far the most important net exporters of milk to other republics (WB, 1992, 230). Their milk (and pork) production was heavily dependent on cattle feed and protein deliveries from other republics. This factor alone (in addition to the dependence of these republics on the import of machinery and fertilizers) explains the agricultural depression in the early 1990s, before the launch of decollectivization proper (Tamm, 2001; WB, 1993c, 89; OECD, 1996c, 66).

---

[12]  Agriculture is too dominant in the picture that is drawn of Estonia in Table 2.1. Unlike some other sources which suggest that Lithuania has by far the largest agricultural population among the Baltic States, the method used by the OECD fails to take into account the popularity of private plot farming (especially among pensioners) in Lithuania (Lerman, Kislev et al., 2003, 1013). The figures above include fishing and forestry, which were more important sources of livelihood in Estonia than in Latvia and especially Lithuania, where they were very low.

Today, the Baltic States are generally considered to form part of the group of East-Central European countries. Before transition, Estonia and Latvia ranked among the top performers in this group, while Lithuania was roughly midway down the table. However, the dependence of Baltic industry and agriculture on other Soviet republics, together with the absence of many key economic and social institutions characteristic of nation-states, put them in a more disadvantageous position. Against this background it understandable that the economic recession of the transition stage was deeper and longer in the Baltic countries than in the more developed East-Central European countries.

Looking at individual dairy farming indicators (Table 2.2), there is very little to tell apart the three Baltic countries: Estonia was most efficient in all respects, usually followed by Lithuania, with Latvia holding up the rear.

**Table 2.2 Agricultural indicators in Estonia, Latvia and Lithuania**

|  | Estonia | Latvia | Lithuania | Year |
|---|---|---|---|---|
| Agricultural Land per Collective Farm (ha) (1) | 4,490 | 4,041 | 3,094 | 1980–89 |
| Workers per Collective Farm (1) | 412 | 412 | 338 | 1980–89 |
| Share of Livestock Production in Total Agricultural GDP (%) (2) | 71 | 70 | 67 | 1987–89 |
| Feed Consumption (oat units) per Milk Kg (3) | 0.98 | 1,18 | 1,1 | 1986–90 |
| Milk Production (kg per cow) (4) | 4,230 | 3,636 | 3,808 | 1989 |
| Farm Machinery (hp per worker) (5) | 38.8 | 38 | 35.4 | 1980–85 |
| Cattle (standard heads per worker) (5) | 7 | 6.5 | 7 | 1980–85 |

*Sources:* (1) annual average statistics, Lerman et al., 2001, Chapter 1, 41; (2) annual average statistics, Lerman et al., 2001, Chapter 1, 38; (3) annual average statistics, WB, 1992, 229; (4) WB, 1993d, 177; (5) annual average statistics, Lerman, Kislev et al., 2003, 1004.

Kolkhozes and sovkhozes in each Baltic country differed from their counterparts in the neighbouring countries both in terms of acreage and number of employees. Estonia had the largest farms and Lithuania the smallest; Latvia was in the middle, but closer to Estonia than Lithuania. However, this appears to have had little effect on the decollectivization process. On the other hand it seems that household plot farming, the nature and economic significance of which was clearly different in each country, assumed much greater importance.

**Table 2.3 The importance of household plot farming in Estonia, Latvia and Lithuania**

|  | Estonia | Latvia | Lithuania | Year |
|---|---|---|---|---|
| Arable Land (%) (1) | 4.5 | 5.2 | 10.5 | 1989 |
| Gross Crop Production (%) (1) | 33.8 | 26.0 | 30.8 | 1989 |
| Gross Livestock Production (%) (1) | 15.9 | 23.9 | 30.5 | 1989 |
| Gross Agricultural Production (%) (1) | 21.3 | 24.5 | 30.6 | 1989 |
| Plot Farmers with Dairy Cows (%) (2) | 25 | 49 | 75 | 1987 |
| Share of All Dairy Cows (%) (3) | 14.6 | 26.8 |  | 1990 |
| Share of Milk Production (%) (4) | 17.0 | 31.3 | 41.3 | 1990 |
| Milk Buyer (%) (5) |  |  |  |  |
| – Kolkhoz Market Customers | 1 | 0 | 0 | 1988 |
| – State and Cooperatives | 0 | 0 | 85 | 1988 |
| – Kolkhoz | 99 | 100 | 15 | 1988 |
| Income from Plot (SUR per family per per month) (6) | 62 | 77 | 134 | 1988 |
| Average Wage (SUR per month) |  |  |  |  |
| – Agriculture (7) | 235.2 | 235.1 | 216.2 | 1988 |
| – Industry (7) | 268.0 | 248.2 | 242.4 | 1988 |

*Sources:* (1) WB, 1992, 193; (2) Wegren, 1991, 125. Wegren's data only comprise kolkhoz workers. There were also plot farms on sovkhoz-controlled land; (3) AFE, 1994, 1997 and 2000; AgrLa, 1997, 31–35; (4) AFE, 1994, 1997 and 2000; SYLa, 1996, 195; WB, 1993d, 178; (5) Wegren, 1991, 132. Wegren's data only comprise kolkhoz workers. There were also plot farms on sovkhoz-controlled land; (6) Wegren, 1991, 134. Wegren's data only comprise kolkhoz workers. There were also plot farms on sovkhoz-controlled land; (7) WB, 1993b, 327; WB, 1993c, 276; WB, 1993d, 373.

It is quite clear from the table above that Estonia and Lithuania represented two extremes in household plot farming. In Lithuania, private plot farming had a much more important role than in Estonia both as a source of livelihood and in terms of agricultural output. Even in the broader Soviet context, Lithuanian households had one of the highest incomes from plot farming, while in Estonia the figures were among the lowest (Wegren, 1991, 134). Even so wage levels in Lithuania were among the highest in the Soviet Union. This suggests that the wide differences between the countries can probably be attributed in large part to cultural reasons. The three countries concentrated on rather different areas of plot farming: The Estonians emphasized crop farming, while the Lithuanians focused more on animal husbandry, and dairy farming in particular (Ibid., 132). Latvia, where dairy farming was predominant on household plot farms, came somewhat closer to Lithuania.

Results from working time research also indicate that dairy farming by plot farmers in Lithuania was more binding and time intensive (Ibid., 123). You could say that agricultural workers in Estonia placed more value on enjoying their leisure time, including holidays and hobbies, than their southern colleagues. Lithuanians in particular were inclined to the traditional peasant way of life. It also emerged clearly from my interviews that employees with a higher level of education were less interested in plot farming. Rein Ruutsoo's article in this volume also reveals that participation in association activities was much more popular in the Estonian countryside than in Lithuania.

Lithuanian plot farmers were not only more experienced and more willing to take up animal husbandry, but they had also managed to keep their livestock buildings and other production buildings more often than colleagues in Latvia and Estonia. In addition, they had managed more often to preserve the tools required for small-scale farming (including their own horses). The World Bank assumed (WB, 1993b, 1993c and 1993d) that plot farming would constitute the basis for a new kind of individual farming. In the World Bank's view it was clear that Lithuania would offer the best opportunities for establishing new farms, while the conditions would be the worst in Estonia. Even in Latvia, plot farming was slightly more important than in Estonia. In fact, the table above also suggests a third argument. Instead of marketing their milk through the local kolkhoz or sovkhoz in the manner of Estonians and Latvians, Lithuanian household plot farmers signed direct marketing contracts with the government and co-operatives. The contracts signed with co-operatives (the number of which is not indicated by our source) in particular would suggest that smallholders were more independent in Lithuania. However, according to a study (Alanen, 2001, 79–80) conducted in Estonia, the government and Soviet farms also competed with each other to some extent for the plot farmers's production at the end of the Soviet era.

## A Comparative Overview of the Transformation of Agricultural Systems

### Speed, Synchronization and the Success in Transferring Resources

The overall starting-points for the privatization of non-land assets were roughly the same in Estonia, Latvia and Lithuania. First, the assets of kolkhozes and sovkhozes were used for restituting the collectivized non-land assets, and the remaining part (which in Latvia should have been at least 50 per cent of the assets, but often fell short of that mark) was privatized in favour of former and current employees based on various combinations of working time and wages. In Lithuania, decision-making was in the hands of a centralized, national organ, whereas in Estonia and Latvia, local reform commissions, typically composed of the management of local kolkhozes and sovkhozes, made all decisions. In both cases the decollectivization process involved widespread anarchy and lawlessness, but in Estonia and Latvia the reform committees had in principle the best expertise at their disposal, and practical matters such as the legal and practical monitoring of the decollectivization process were easier to organize at the local level, too. Of the three Baltic countries, Lithuania's choice of limited liability companies and particularly partnership as the transitional form to family farming answers well to the description of the strategy recommended by the World Bank and its researchers. Some partnerships were established in Estonia and Latvia, too, but their number was small, and based on interviews conducted in the late 1990s hardly any of them continue to survive. From the legal perspective, some cooperative enterprises established in Estonia played an important role in the preservation of large-scale production during the first stage of decollectivization (Tamm, 2001). However, the closed stock-holding company, which was ranked by the World Bank as the second-worst alternative after cooperatives, became the prevalent form of ownership of corporate agricultural enterprises in Estonia and Latvia.

In all three countries agricultural land was transferred to new owners using two methods, the most important of which was restitution. Restitution was supplemented by giving land (even to people who were not entitled to restitution) for private use. In Latvia, the process was completed within a relatively short space of time, whereas in Estonia and Lithuania restitution has still not been completed because of complicated and conflicting legislation and administrative problems. Even though numerous studies have stressed the importance of transferring land to private ownership, my interviews have shown that people entitled to restitution have generally succeeded in acquiring land for farming before acquiring legal land ownership – it has been relatively easy to rent land. In many cases the shortcomings of land rental legislation are a greater problem than land ownership. Renting small areas of land from a large number of landowners is complicated and time-consuming, and short-term rental agreements make investment very difficult.

The information available (Table 2.4) on the number of privatized individual farms (excluding plot farms, since they are on rented land), unprivatized farms (i.e. kolkhozes and sovkhozes) and privatized corporate farms (open or closed companies, partnerships, cooperative agricultural enterprises) indicates that the decollectivization of non-land assets took the longest time (1991–1993) and the smallest steps in Estonia, but it was completed with determination during 1993. In Latvia and Lithuania, decollectivization was implemented in full force in 1991– 1992, that is, immediately after the countries had regained independence in August 1991.[13] We shall look at the pros and cons of radical and gradual reforms in more detail later on.

Many researchers (WB, 2000a, 27) believe that one of the main reasons for the failure of the reform was the poor synchronization of the actual land reform and the decollectivization of non-land assets; I have suggested the same myself (see Alanen, Nikula and Ruutsoo, 2001). Latvia was the most successful with synchronization, making great progress with restitution as early as 1992; Lithuania, too, made good progress in 1993. In Estonia, by contrast, the number of family farms did not grow significantly until after 1997 (Table 2.4, Columns 3, 6 and 8).

---

[13] Some collective farms remained in both countries.

**Table 2.4  The number of large-scale production units, corporations and family farm units 1990–2001 [1]**

| Year | Estonia | | | Latvia[2] | | | Lithuania | |
|---|---|---|---|---|---|---|---|---|
| | Corp's (number) | Comp's (number) | Farms (1,000s) | Collective & state farms [3] (number) | Comp's (number) | Farms (1,000s) | Corp's (number) | Farms (1,000s) |
| 1980–89 | 293 (av.) | | | 570 (av.) | | | 1,057 (av.) | |
| 1990 | ... | | | ... | | 7,5 | 1,212 | 0,1 |
| 1991 | 340 | 30 | 2,3 | 823 | | 17,5 | 1,219 | 2,3 |
| 1992 | 501 | 396 | 7,0 | 304 | | 52,3 | 4,279 | 5,1 |
| 1993 | 719 | 589 | 8,4 | 265 | | 57,5 | 3,484 | 71,5 |
| 1994 | 1,146 | 1,013 | 10,2 | 103 | 1,101 | 64,3 | 2,880 | 112 |
| 1995 | 1,013 | 983 | 13,5 | 95 | 656 | 74,1 | 2,611 | 112 |
| 1996 | 925 | 874 | 19,8 | 92 | 617 | 94,9 | 2,328 | 135 |
| 1997 | 898 | | 22,7 | 81 | 474 | 94,9 | 2,004 | 166 |
| 1998 | 834 | | 34,7 | | | | 1,503 | 196 |
| 1999 | 771 | | 41,5 | | | | 1,244 | 200 |
| 2000 | 705 | | 51,5 | | | | 1,063 | |
| 2001 | 1,003 | | | 75 | 477 | 37,6 | 561 | |

[1] Estonian and Latvian data are for the end of the year, Lithuanian data for the beginning of the year.

[2] No systematic data are available on the total number of corporate farms in Latvia. Subsequent state and municipal farms are listed under collective and state farms. Privatized corporations are presented separately.

[3] Includes sovkhozes, subsequent state and municipal owned farms, as well as kolkhozes.

*Sources:* AgrE, 1995; Ministry of Agriculture, 2002; SYLa, 2000; Sepp, 1997; ACE I, 2002; Bratka, 1998; Pirksts and Rozenberga, 1997; OECD, 1996b; Meyers and Kazlauskiene, 1998; ACLa, 2003; SYLis, Agriculture and Rural Development Plan 2000–2006, 2000; WB, 1996; Lerman et al., 2001.

Initially, production declined sharply in all the Baltic countries, but later on the trends slowed down to a more moderate decline (Annex Table A-1). By late 1994, when the immediate results of the non-land asset reforms in all the countries were already visible, Gross Agricultural Output (GAO) had fallen to 58 per cent of the 1990 level in Estonia, to 50 per cent in Latvia and to 54 per cent in Lithuania (Annex Tables A-1A, B and C, Column 1). As far as our further analysis is concerned, there were three interesting differences between the countries. First, although the differences are not dramatic, the Estonian reform would appear to have been less destructive than the Lithuanian and, in particular, the Latvian reform. Second, the subsequent trends in Estonia and Latvia have principally shown a downward trend, while Lithuania displayed a period of strong growth in 1994–1997. Third, the strong growth in Lithuania was followed by a sharp decline in 1998–2000, which is why production in Lithuania in 2002 is still no more than 65 per cent of the production levels recorded in 1990. However, the drop in Lithuania was significantly smaller than the 46 per cent drop in Estonia and the 41 per cent drop in Latvia compared to 1990.

The farm structure in the Baltic countries developed in different directions and at different speeds. In Estonia, corporate farms accounted for slightly less than 50 per cent of GAO immediately after the main reform year, but in Latvia and Lithuania[14] their share of GAO continued to decline to about 20 per cent in the late 1990s (Annex Table A-1, Column 6).

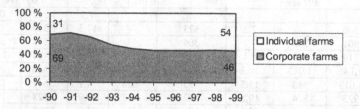

A. Estonia

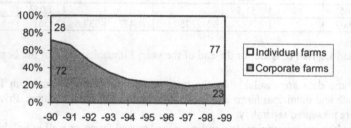

B. Latvia

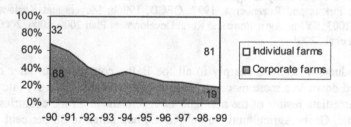

C. Lithuania

**Figure 2.1 Distribution of Gross Agricultural Output (GAO) between corporate and individual farms (%) 1990–1999**

*Sources:* SYEs; SYLas; Poviliunas and Batuleviciute, 1997 (1990–1995), SYLis (1996–1999).

_____

[14] The data provided by Statistics Lithuania on production by corporate farms as a proportion of GAO in 1991–1993 differ from the OECD (1996c, 82) statistics I have used (Annex Table A-1C, column 6). I have used OECD statistics because they also indicate the GAO share of plot and family farms, to which I refer later on in this article.

Large-scale production has retained its previous role in Estonian agriculture, whereas Latvia and Lithuania have turned into de facto small-scale production countries.

Successful transition to a family farm system requires that a significant part of the labour force as well as of the land and non-land assets of Soviet farms is transferred to family farms (or individual farms in general). The rapid implementation of Latvian (1992) and Lithuanian (1993) land restitution simultaneously with the decollectivization of non-land assets led to an increase in the total agricultural labour force: in addition to the labour coming from large-scale farms, individual farms also saw large numbers joining them from other industries (Annex Table A-3). The family farms created in Latvia and Lithuania were roughly of the same size, while in Estonia they were larger, partly on account of the slow progress of land restitution.[15] In Estonia, the decline of the agricultural labour force has been continuous, and the acceleration of land restitution in 1997 did not change this general trend. The decline of the labour force was most pronounced during the reconstruction of large-scale enterprises (1992–1993) and in the years immediately following privatization (1994–1995) (Annex Tables A-1 and A-3).

Changes in the cow population are a good indicator of the decollectivization of non-land assets, since cow statistics are readily available and cows are easily transferable (unlike buildings) and they are, in principle, accessible to all farmers (unlike machinery). Changes in the cow population also indicate the fate of cowshed complexes, the most important technological units of typical Soviet farms. Figure 2.2 shows both the redistribution of the cow population and the absolute development of the cow population.

As expected, the redistribution of cows to individual farms reached its peak during the decollectivization of non-land assets (though the changes occurring during the calendar year are not included in the statistics until the beginning of the following year, 1 January). This is visible in the decline of the cow population on corporate farms; in Latvia the decline from 1990 to 1995 was 84 per cent, in Lithuania 67 per cent, but in Estonia only 44 per cent (Annex Table A-2, Column 4). However, these cows were not transferred to individual farms only. Figure 2.2 clearly shows that by far the greatest reduction in the cow population was seen in Latvia, where it was almost halved (a decline of 45 per cent) in five years. In Estonia and Lithuania, the decline was more moderate and progressed at nearly the same rate, 25 per cent and 27 per cent (Annex Table A-2, Column 3). At a more

---

[15]   In Latvia, the average size of a family farm was 11.8 ha after the 1993 restitution and 11.7 ha in 1995. In Lithuania, it was 8.9 ha in 1993 and 8.6 ha in 1995 (SYLa, 1995, 217; AgrfLa, 1997, 10; Meyers and Kazlauskiene, 1998, 97; Agriculture and Rural Development Plan 2000–2006, 2000, 138). Meanwhile in Estonia, the figure in 1993 was 25.4 ha and in 1995 23.1 ha (AFE, 1993, 15; AFE, 1994–1996, 14). Comparative studies of farm sizes tend to overlook the fact that farm sizes in Latvia and particularly in Estonia also include a fair amount of forestland. Bearing this in mind, the difference between the average size of family farms in Lithuania and Latvia is almost negligible. Mathijs and Swinnen (1997, 356) may be right in highlighting the relatively large average farm size in Latvia from the viewpoint of restitution, but in the context of the Baltic countries Latvia did not significantly differ from Lithuania in the early stages of the agricultural reform.

general level, this also suggests that the percentage of other dairy farming-related production resources (such as the production buildings of corporate farms and their fixed technical equipment) discarded or destroyed was the highest in Latvia.

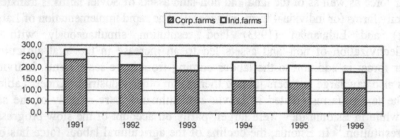

A. Estonia

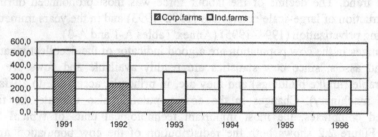

B. Latvia

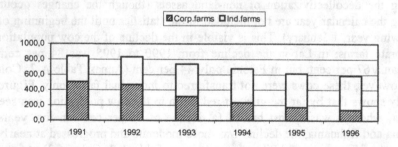

C. Lithuania

**Figure 2.2 Distribution of cows (in 1,000s) between corporate and individual farms 1991–1996**

*Sources:* AFE, 1994 and 1997; Bratka, 1998; Pirksts and Rozenberga, 1997; Poviliunas and Batuleviciute, 1997; SYEs; SYLas; SELis.

This massive loss of resources in Latvia during the phase that was most decisive of all to the fate of the agricultural system – the decollectivization of non-land assets – can also be seen in Table 2.5. Compared to Estonia and Lithuania, the reductions in sown area and total agricultural production are particularly striking.

**Table 2.5  Cow population and milk production compared with 1990, and some other indicators 1990–1995**

|  | Estonia 1995 | Latvia 1995 | Lithuania 1995 |
|---|---|---|---|
| 1. Cow index 1995 (1990=100) | 75 | 55 | 73 |
| 2. Milk production index 1995 (1990=100) | 59 | 55 | 58 |
| 3. Sown area index 1995 (1990=100) | 76 | 57 | 79 |
| 4. Corporate farms in GAO* | 61 | 24 | 31 |
| 5. GAO* index (1990=100) | 58 | 47 | 61 |
| 6. Employees in agriculture index (1990=100) | 41 | 80 | 118 |

\* GAO=Gross Agricultural Output.

*Sources:* AFE, 1994 and 1997; Pirksts and Rozenberga, 1997; SYLis; SYEs; SYLas; Bratka, 1998; Poviliunas and Batuleviciute, 1997.

The destruction of resources in Latvia indicates the failure of the family farm project. As long as the empirical focus remains solely on cows, on total sown area and overall production figures, one might well conclude that Lithuania has had the most success in the transfer of resources. However, we also have to consider losses that tend to remain hidden: the dramatic decline of the cow population on corporate farms shows that a major part of the cowsheds and other Soviet farm fixed assets related to dairy farming were out of production in Lithuania as well – and judging by my interviews, most of them were destroyed. However, the destruction of these technological units is the inevitable price that has to be paid for the transfer of resources determined by the family farm process. According to the WB assessment (2000a, viii) of the past years of transition, both Lithuania and Latvia were 'reformed radically and rapidly rather than gradually,' which is why one might expect 'short term adjustment difficulties' – but by the end of the 1990s they will rank among the countries making the 'best progress.' In most respects, the creation of the family farm system appears to have been a success in Lithuania. GAO continued to decline until 1994 (Annex Table A-1C, Column 1), but showed a 7 per cent increase in 1995 and over the next three years (1995–1997) went up in all by 22 per cent. In comparison, the first leg of Latvia's journey showed little promise, since there was no corresponding growth after the redistribution of production resources. In Estonia, on the other hand, the transfer of resources to family farms was generally a failure, with the majority of the cow population remaining at the corporate farms. On the other hand, relatively few technological entities and other production resources were destroyed, and the decline in GAO was not as great as in Latvia.

If Lithuania is the best-performing example of the new family farm system hypothesized by World Bank researchers, why is Latvia lagging so far behind? Will the most radical and rapid reform yield a double reward at the end of the decade? What about Estonia, which has remained largely dominated by large-scale production, without suffering any major resource losses? Most corporate enterprises in Estonia remained closed employee-owned companies, some were even established as co-operatives. According to the World Bank theory, however, they are threatened in the future by the 'syndrome of the labour-managed firm.' Similarly, the slow process in Estonia, the privatization of non-land assets and the privatization of land may still cause problems in the future. Indeed, the first five years of decollectivization in the Baltic countries produced an interesting research design from the perspective of the World Bank theory.

*Labour: Productivity and the Causes of Productivity Differences*

In Estonia the 'syndrome of the labor-managed firm' did not prevent the sharp decrease in the number employees in relation to GAO, since Estonian labour productivity – when measured in terms of the ratio between the labour force and GAO – increased (roughly) above the average Soviet level in the first year after the decollectivization of non-land assets (1994).

Meanwhile, labour productivity in Latvia and Lithuania fell sharply during the years of decollectivization and remained below the level of the Soviet era until at least 1999.[16] In all three countries the development of productivity is closely linked to the decollectivization of non-land assets. In Latvia and Lithuania, labour-intensive small-farms represented the reverse side of the disappearance of large-scale production. In Estonia, the relatively more numerous corporate enterprises were able effectively to adapt their labour force in the years immediately following privatization (1994–1997) – even though the number of family farms (and perhaps also their labour) continued to increase in Estonia for some time (Table 2.4).

In Latvia, the number of family farms continued to increase for a longer period (Table 2.4), but based on the cow population it may be assumed that individual farms soon ran into increasing difficulties. Since the number of cows started to decrease as early as 1993 (see Annex Table A-2B, Column 6), and since the total number of agricultural population soon returned to the levels recorded in the Soviet era following a sharp increase in 1992, dropping by about 20 per cent by 1995, we may assume that labour-intensive dairy farming on small-scale farms was being replaced by more extensive crop production. This sheds further light on our understanding of the highly destructive nature of the Latvian transition: the recession in large-scale production was followed by a recession in small-scale production.

---

[16] In 2000 labour statistics criteria in the Baltic countries changed so radically that, in the absence of of time series based on the new statistical criteria, I have used 1999 as the cut-off year.

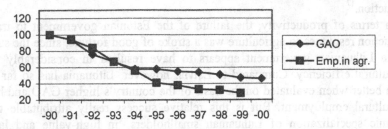

Figure A. Estonia

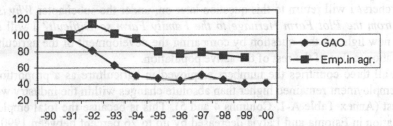

Figure B. Latvia

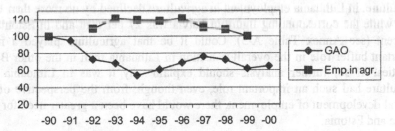

Figure C. Lithuania

**Figure 2.3 Development of Gross Agricultural Output (GAO) and the number of employees in agriculture 1990–2000**

*Sources:* Lerman, 2001b; SYEs; SYLas; SYLis.

In Lithuania, the number of corporate farms began steadily to decrease from 1992 onwards. As the number of family farms was now growing (Table 2.4), the resources available to individual farms, at least when measured in terms of the number of cows, continued to increase until 1998 (Annex Table A-2C, Column 6) and provided better employment for their new labour. The Lithuanian development model differs from the Latvian model in terms of its dynamics as well. Lithuania may have been more successful than its neighbours in transferring resources from Soviet farms to individual farms, but even so its productivity dropped below the Latvian level in 1992–1994 as a result of the process. However, this dip was only temporary, and thanks to the growth boom in 1995–1997, it rapidly caught up with Latvia. This was due to the increased efficiency in both crop and livestock production.[17]

In terms of productivity, the failure of the Estonian government to transfer production resources to agriculture was a stroke of good fortune, since the success of the Lithuanian government appears to have resulted in considerably lower agricultural efficiency. Compared to Latvia, however, Lithuania has so far fared much better when evaluated on the basis of the country's higher GAO and higher agricultural employment. But is this relative success really attributable to the dynamic specialization of Lithuanian smallholders 'in high-value and labour-intensive products [...] avoiding reliance on machinery and equipment,' as one would expect if they followed the development path advocated by the World Bank researchers? I will return to this question later on, under the subchapter *Why is the Path from the Plot Farm Heritage to the Family Farm so Difficult?* I will shed some new light on this question by comparing the development of the agricultural population to that of the rest of the active population.

In all three countries the numbers employed in agriculture as a proportion of total employment remained higher than absolute changes within the industry would suggest (Annex Table A-1, Columns 4 and 5). This is because the total employed population in Estonia and Latvia decreased by up to 26 percent between 1990 and 1999, and in Lithuania by 10 per cent. Part[18] of this decline can be attributed to agriculture: in Lithuania employment in agriculture declined by no more than 1 per cent, while the corresponding figure in Latvia was 27 per cent and in Estonia 68 per cent (see Annex Table A-3). Could it be that agriculture played a more important buffer role in the overall transition in Lithuania than in the other Baltic societies? If so, further analysis should explain why it was in Lithuania that agriculture had such an important role, even though, from the perspective of the general development of employment, there would have been a greater need for it in Latvia and Estonia.

Both the World Bank researchers and Macours and Swinnen assume that country differences in productivity are caused by the fact that agriculture can act as a buffer against the income problems resulting from the overall societal transformation by providing jobs to people who would otherwise be unemployed. This results in higher productivity in countries where the overall economic development has been most favourable, because part of the agricultural labour force in these countries is able to move to other sectors of the economy (as well as to migrate to cities). Crucial to this thinking, however, is the development of

---

[17] Even if the relative share of production by corporate farms has declined, increased productivity is not linked to farm type. This is evident from the crop and animal production figures found in the Statistical Yearbook of Lithuania for 1998.

[18] With regard to Estonia and Latvia, the explanation lies in part in the fact that emigration to the other former Soviet republics was much greater at the time. Other people filled many of the jobs vacated by emigrants. This makes it difficult accurately to assess the effect of emigration on the employment figures.

employment, not the overall development of the national economy[19] as such. In Estonia, GDP growth was somewhat more rapid than in the other Baltic countries, but total employment in Estonia decreased in every year of the transition, until it reached roughly the same level as in Latvia, the least developed Baltic country (Annex Tables A-1 and A-3). Even though Lithuanian agriculture should have been best placed to increase its productivity, given the overall employment situation in the country, it is possible that agriculture was in fact more in the role of a buffer for the transition than in the other countries.

I will be showing later on that Lithuanian agriculture may well have had a major role as a buffer, but that its role is more readily understandable against the background of its Soviet heritage and the legal enterprise type (partnership) chosen for corporate farms in connection with privatization rather than against the overall development of the Lithuanian national economy, e.g. employment or GDP. Furthermore, this explanation enables a more concrete interpretation for the consequences of the fast and radical reform in Lithuania. On the other hand, the problems caused by the outflow of labour from agriculture may therefore be most acutely felt in structural unemployment (unfortunately no accurate statistical data are available for the early phases of the transition), or perhaps in the transfer of people from the labour force using various socio-political measures (see the Estonian case in the article by Hansson and Marin in this volume).

*Intensive and Extensive Specialization Trends in Agricultural Production*

During the Soviet era, the agricultural sectors in Estonia and Latvia were more heavily specialized in livestock production than in Lithuania: in the former two countries it accounted for 66 per cent of GAO, in Lithuania for about 10 per cent less. Within the agricultural sector, Estonian Soviet farms were particularly specialized in dairy farming. In Latvia and Lithuania, household plot farms accounted for a much higher percentage of milk production (Table 2.3). By 1999, specialization in livestock production in Estonia and Lithuania had decreased by about 10 per cent, and in Latvia by 20 per cent (Annex Table A-2, Column 1).

A comparison of the development of the dairy cow population on corporate farms (Annex Table A-2, Column 7) and the share of livestock production (Annex Table A-2, Column 5) indicates that Estonia's path kept diverging from those of Latvia and Lithuania. Estonian corporate farms continued to focus – after a short drop in the critical year of the reform – on milk production, in which they had already been highly specialized during the Soviet era. Estonia was the only country to display this counter-trend, which emphasized livestock production and specialization in the main direction of the development. Meanwhile in Lithuania, the proportion of livestock production on Lithuanian corporate farms did not

---

[19] This view is supported by data in the OECD labour market survey (2003, 162). In Estonia, 60 per cent of the people moving out of agriculture and forestry during the first ten years of the transition had found a job in another sector of the economy, while this figure was only 40 per cent in Latvia and Lithuania. Still, moving out of agriculture and forestry has been difficult in all three countries.

decrease, but the focus of livestock production on these farms shifted from dairy farming to other types of production. In 2001, corporate farms still produced 70.2 per cent of the eggs and 44.35 per cent of the meat, but only 8.8 per cent of the milk in Lithuania (SYLi, 2002, 398). Unfortunately, there are no statistical data available on the share of livestock production on Latvian corporate farms, but based on indirect evidence (Annex Table A-2B) the trends in Latvia were roughly the same as in Lithuania.

The overall strengthening of crop production may be considered, at least tentatively, a sign of extensification in Baltic agriculture. However, this trend is due solely to the growth of small-scale production. With reference to overall production focus, crop and livestock production on individual farms remains, on average, at the same level as during the Soviet era (Annex Table A-2, Column 8.). Although accurate GAO data are not available, it is highly likely that Latvia is no different in this respect. I will analyze small-scale production in greater detail in the following four subchapters, but before that, by way of an interim summary of what I have said so far, I would like to take a stand on the following 'lesson' of agricultural transition brought up by the World Bank.

According to the World Bank report (2000a, 27), 'The greatest -reform progress has been made by those countries that are reforming in very large steps, despite the great difficulties that these efforts are causing in the short term.' In the light of the empirical facts presented here, the thesis above cannot be universally applicable because the most radical (fastest, synergetic) decollectivization process in the Baltic countries, i.e. the decollectivization of Latvian agriculture, was also without question the most destructive one. The more gradual and slow process of reform in Estonia – which has also been more successful in preserving key technological units – has facilitated a greater increase in productivity and specialization in high-value and labour-intensive products, despite the presumed 'syndrome of labour-managed firm.' One may also ask whether Estonian agriculture in the end was fortunate to lag so far behind in the restitution of landed property, if that encouraged people to establish farms with poor future prospects in Latvia.

At least in Estonia and Latvia, then, things have moved in more or less the opposite direction to what the World Bank researchers expected in their theories. I will move on now to explore the topic in more depth by making a distinction between two types of individual farms, namely plot farms and family farms.

## The Difference Between Plot and Family Farming as a Tool for Further Analysis

Family farms, peasant farms and private farms are all classified in official statistics under the same heading of individual farms on the grounds that the land (or part of it) is in private ownership. They have been entered in a farm register on this criterion. Household plot farms, however, usually only have the right to use the land. Most of the early family farms created through restitution were bigger than household plot farms, but not always. Family farms constituted a new category established as a result of the transformation of the agricultural system, and their creation indicates the progress of the family farm project. Household plot farms, on the other hand, are an extension of the individual household plot tradition typical of the Soviet era, even though plot farmers typically received more land during the transition.

Household plot farmers, including those who were not entitled to land restitution, also received new land.[20] This applied to pensioners as well. In Lithuania, current agricultural workers as well as pensioned kolkhoz and sovkhoz workers were given three hectares of land (Hejbowicz, 2001, 94). All in all, plot farmers had by 1995 received in relative terms the largest share of land ownership or usage rights granted to individual farms in Latvia, where they held 30 per cent, followed by Estonia with 22 per cent and Lithuania with 21per cent. This explains why household plot farms were the largest (3.6 ha of arable land) in Latvia compared to Lithuania (2.5 ha) and Estonia (1.5 ha).

**Table 2.6 Household plot farming characteristics in 1995**

|  | Estonia | Latvia | Lithuania |
|---|---|---|---|
| 1. Share (%) distributed land | 22 | 30 | 21 |
| 2. Number (1,000s) | 187 | 246 | 397 |
| 3. Agricultural land on average | 1.7 | 3.6 | 2.5 |
| 4. Share (%) of total GAO | 33 | 40 | 49 |
| 5. GAO of household plot index |  |  |  |
| – Estonia (1991=100) | 103 |  |  |
| – Latvia (1990=100) |  | 90 |  |
| 6. Share (%) of total livestock production | 28 | – | 50 (1994) |
| 7. Share (%) of total milk production | 30 | 48 | 60* |
| 8. Share (%) of total dairy cows | 26 | 47 | 51 (1993) |
| 9. Dairy cow index (1990=100) | 137 | 96 | – |

* In 1993 plot farmers' share of total milk production in Lithuania was 53 per cent.

*Sources:*
1. Mathijs and Swinnen (1997).
2. Meyers and Kazlauskiene (1998).
3. Meyers and Kazlauskiene (1998).
4.–5. AFE (1997); AgrLa (1997); SYLa (1996).
6. AFE (1997); Bratka (1998); OECD (1996c).
7. AFE (1997); Pirksts and Rozenberga (1997); Poviliunas and Batuleviciute (1997).
8. AFE (1997); Pirksts and Rozenberga (1997); Girnius (1994).
9. AFE (1997); Pirksts and Rozenberga (1997).

---

[20] Not all land recipients had a farming background. The Latvians even had their own legal term for a new farm type: subsidiary farms. Land was given to rural people who generally worked outside agriculture. These farms seldom had the small-scale production infrastructure or the farming experience demonstrated by those plot farmers who had worked on collective farms. Once these plot types had been established, their economic activity appears to have decreased rapidly (Bratka and Dzepuka, 1997, 128).

Although household plot farmers received significantly more land in Latvia than in the two other countries, this did not appear to have an effect on production figures (Table 2.3): in 1995 Lithuania was still the top plot farming country, while plots were least popular in Estonia. Therefore, we can once again draw a conclusion that supports path dependency thinking: the decisive factor in household plot farming was not the amount of land available, but the non-material heritage of the Soviet era.

During the Soviet era, private small-scale production on household plot farms relied on large-scale production in a number of ways. The support provided by Soviet farms was comparable to the external support that Western small-scale producers receive in the form of infrastructure (roads, utility networks, etc.), as well as support from government institutions, cooperatives, associations and organizations. When the provision of machinery, various other production resources, product marketing and professional counselling services that had been based on the symbiosis of large- and small-scale production began to dry up (see later) with transition, not only did the establishment of new farms become more difficult, but existing household plot farmers also had somehow to compensate for the losses or accommodate their production to the new circumstances. The plot farm heritage (skills, buildings, tools) did mean that the owner(s) could continue production at a reduced level, but this is not even remotely comparable to the technological, skill, marketing and management heritage that is handed down on a typical Western family farm to the younger generation in the form of financial and social heritage.

As we will be using the distinction between plot and family farming in our further analyses, it should be noted that once a household plot farm is registered as a family farm at the end of the restitution process, or once a land user acquires ownership to the land for which they have previously had rights of use, they will be recorded as a family farmer in the classification. (Since the restitution process was most rapid in Latvia, this country has the largest proportion of former plot farms registered as family farms). Therefore, the socio-historical arguments that make the comparison of these two legal forms meaningful gradually lose their significance as well.

This is why most of the analyses that apply this distinction in the present article focus on the first years following decollectivization.

*Individual Farms Opt for a Mixed Economy Instead of Specialization*

A comparison of how much household plot farms, family farms and corporate farms contributed to total production and milk production during the first years (1990–1996) of transition (see Annex Figures A-1 and A-2) shows that Estonian corporate farms were disproportionately dominated by dairy farming. In Latvia, and especially in Lithuania, on the other hand, dairy farming was dominant on plot farms, but on family farms it was even slightly dominant only in Lithuania.

If we consider the total share of animal production on individual farms (instead of just dairy farming) without differentiating between plot and family farms, that is more or less equally widespread in Estonia and Lithuania, and there have been no changes in the figures since the Soviet era (Annex Tables A-2A and B, Columns 8). Besides cows, Estonian farms also have more pigs, chicken, bull calves, etc. than Lithuanian farms.

However, in both countries individual farms are somewhat more specialized in crop production than the GAO shares would seem to suggest.[21] For Estonia, I will provide differentiated data on the extent of crop cultivation by acreage on both plot and family farms. This will allow us to analyze what exactly this specialization in crop production means. The focus is on the most recent information available, i.e. for 1996–1999 (AFE, 2000, 22–23). On plot farms, crop production continued to be dominated by fodder plants for animal feeding, while the production of potatoes and root vegetables appeared to be on the decrease rather than on the increase. The same holds good for family farms, with one exception: about half of the sown area was used for growing cereals, whereas on plot farms cereals accounted for only about 6 per cent of the sown area. The basic production structure remained more or less the same on both farm types throughout the period under review. Estonian family farms continued to specialize somewhat more on crop production, which in practice usually meant expanding into cereal farming. The situation in Latvia does not conflict with the data for Estonia, with the exception that the production of cereals is relatively more widespread on Latvian plot farms (which were on average larger than those in the neighbouring countries) than on their Estonian counterparts (AgrfLa, 1997, 12; AgrfLa, 2002, 13–14).

Although systematic statistical data are not available from all countries, it can quite safely be concluded that in each case both plot and family farms are essentially continuing the private farm tradition of the Soviet era. Crop production is obviously more important to family farms, since they have just received more land to use, and also since their cattle sheds – typically dating back to the Soviet household plot farming days – are very small. On average, both farm types were in the 1990s also characterized by the preservation of the mixed economy almost in its original form, with no clear statistical evidence of the dynamics of specialization.

## The Community of Faith between Corporate and Plot Farming

The differences in the paths of development followed by the Baltic countries cannot, however, be explained simply by reference to the plot farm heritage. This is particularly apparent when it comes to dairy farming. In Lithuania, the number of cows on individual farms continued to increase up until 1998, whereas in Latvia the growth came to a halt in 1993 and in Estonia (somewhat surprisingly in view of its weakest plot farming tradition) as late as 1994 (Annex Tables A-2A and B, Column 6). Did the Estonian cow population start to decline a little later (1995) because Estonian Soviet farms were privatized later than Latvian farms, and because there were so many large-scale successors to the Soviet farms that the Estonian plot farmers were able to acquire inexpensive services from them after decollectivization? The difference in the rate of decline of the dairy cow population

---

[21]   AFE, 1997; AFE, 2000; SYLi, 1997, 363, SYLi, 2002, 372.

both in Estonia and Latvia could against this background also be explained by the different rates of progress in breaking the Soviet symbiosis.

Our comparisons above of the rates of decline in the number of cows on corporate farms with the overall trend showed that the total cow population decreased most rapidly in Latvia. During the decollectivization of non-land assets Latvian individual farms were unable to accommodate as many cows available from corporate farms as their counterparts in Estonia or Lithuania (Annex Table A-2, Columns 3 and 4, and Figure 2.2). Indeed, this is why the decollectivization of non-land assets resulted in greater destruction in Latvia than in Estonia, despite the better synchronization with the land reform and the stronger plot farming tradition in Latvia. According to several sources, mobile machinery and technology was often sold separately from the cowsheds and other production complexes in Latvia; sometimes even the cows were sold separately from the cowshed complex. This is why many newly established corporate farms were unable to start up their operation, or why they soon faced bankruptcy or continued to operate on a very modest scale (OECD, 1996b, 79; Abele, 1995). In Latvia, the Gross Agricultural Output (GAO) of agriculture declined by 53 per cent from 1990 to 1995, and production by corporate farms as a proportion of total GAO fell from 72 per cent to 24 per cent, while the figure in Estonia stood at 46 per cent (Annex Table 2.1A and B, Columns 1 and 6). This is why the OECD report (1996b, 81) not only states that 'individual farms [...] are now the cornerstones of the county's agricultural sector,' but also adds that 'some 40 per cent of agricultural is probably operated by farms that are not oriented to market production.'

If the main reason behind the relatively moderate loss of the cow population in Estonia was the preservation of large-scale production, Lithuania's success was partly due to the country's strong smallholder tradition. This, however, is still not the full explanation, since Lithuanian small-scale production can only be understood in the context of large-scale production. The second distinctive feature of the Lithuanian large-scale production system was the partnership enterprise.

*Management Environment on Lithuanian Partnership Farms*

In a partnership system, shareowners can withdraw the capital they have invested in the enterprise at will 'in kind or cash' (Klimaðauskas and Kasnauskiene, 1995). Since share ownership in Lithuania was determined on the basis of working years and age (Ibid., 326), shareowners were often pensioners who continued to cultivate their household plot (often with the help of their children working in the area). This system served as a new way of integrating plot and corporate farms. For the shareowner who did not work on the corporate farm but who cultivated a household plot instead, the corporate farm was only useful as a producer of support services. If the shareowner withdrew his/her capital from the enterprise it could cause a chain reaction where only the fastest shareowners could hope to receive assets of any concrete value. Therefore, corporate farm managers had to focus their management efforts on meeting the needs of plot farmers.

In all three countries the interviewees often described how the quantity and quality of the feed for dairy farm cattle deteriorated during the transition period.

The amount of proteins, vitamins and minerals was reduced, and the cows were often treated very poorly: the daily rhythm of milking was disturbed due to increased indifference, drinking and the overall deterioration of work discipline, there was shouting and more and more abuse of the cows. All this had an adverse effect on milk yield and quality, which in turn had a major impact on producer prices. Milk yield per cow is an excellent indicator of stress levels on corporate farms striving to survive, since systematic and comparative milk production statistics are readily available.

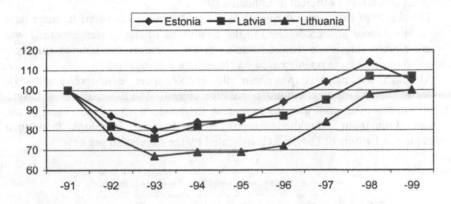

Figure 2.4 Milk yield per cow index on corporate farms 1991–1999 (1991=100)

*Sources:* AFE, 1994, 1997, 2000; SYLas; SYLis.

Time series trends for 1991[22] to 1999 from Estonian, Latvian and Lithuanian corporate farms indicate that the milk yield per cow decreased most in Lithuania, where it also remained depressed for the longest time.

Edvardas Rudys from the Lithuanian Institute of Agrarian Economics has studied the adaptation of Lithuanian agricultural employees and enterprises to the market economy. Rudys (1998, 87) has the following interpretation of a set of empirical statistical data on Lithuanian corporate farms from 1995 and 1996:[23]

> Members of companies understood that on break-up of the companies a considerable part of the property is lost and they will receive only old big buildings for their shares. That is why they try to do their best to prolong the life of the company for at least several years. Such an attitude is caused because household plots of members of companies are tilled for half price or free of charge, also yields are gathered and employees of the company receive some earnings. It is necessary to mention that the expenses for the cultivation of household plots or gathering of yield are very high and make up half or more of the total annual labour costs of the company. Of course such a situation considerably reduces the interest of employees in the improvement of their

---

22  Data by different farm types are only available for Estonia from 1991 onwards.
23  Only minimum changes have been to the original quotation with a view to grammatical correctness.

work and causes serious contradictions between property owners and the employees, but also one of the reasons accelerating the bankruptcy of agricultural companies.

Analysis of the activities of the strongest agricultural companies showed that in their cases, too, the so-called mechanism of expenses bound with the former public farm was in force.

Rydus also points out that the cost mechanism of these agricultural enterprises is similar to the one on Soviet farms, which he says also explains why large family farms are much more efficient in Lithuania (Ibid., 87).

The ability of Lithuanian corporate farms of the partnership-type to adapt their operations (labour and production) to the conditions of the market economy was therefore much inferior to that of Estonian farms – or even the Latvian corporate farm systems that were mainly based on closed stock companies.

Although the partnership system did enable more extensive small-scale production and a larger agricultural population compared to Estonia or even Latvia, it did not, however, contribute to increasing the productivity of individual farms. Instead, Lithuanian individual farmers suffered more hardships than their colleagues in Estonia or Latvia, as is indicated by the milk yield per cow.

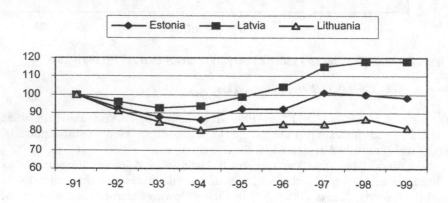

**Figure 2.5 Milk yield per cow index on individual farms 1991–1999 (1991=100)**

*Sources:* AFE, 1994, 1997, 2000; SYLas; SYLis.

Absolute yield figures will help to provide a clearer overall view of milk yield trends in each country. At the same time as the average milk yield per cow increased from 4,164 kilograms per year (1990) to 4,660 kilograms per year (1999) in Estonia, and from 3,437 kilograms to 3,754 kilograms in Latvia, the figure decreased from 3,743 kilograms to 3,228 kilograms in Lithuania.[24]

---

[24]   Estonia: WB, 1993b, 177; AF, 1999; Latvia: SYLa, 1996, 200; SYLa, 2000, 135; Lithuania: SYLi, 1994–1995, 310; SYLi, 2002, 392.

The empirical evidence presented above strongly supports the view of Macours and Swinnen, according to whom 'the extent to which management differs effectively from the pre-reform management, rather than the shift to individual farming, is crucial for improving productivity [...]. The key difference lies in the extent to which farm management of these large-scale farms, and its incentives, have restructured' (Macours and Swinnen, 2002, 387). However, our empirical evidence expands on the empirical content of the conclusion by Macours and Swinnen through the incorporation of different legal forms of corporate farms. Thus, in the form of ownership (partnership corporate farms) preferred by World Bank researchers, not only did the operating environment turn out to be detrimental to the management 'incentives' of the corporate farms, but it was also an exceptionally poor solution for small-scale agricultural producers.

## A Family Farm or Plot Farm Project?

The plot farming heritage of the Soviet era was evident in the number of agricultural enterprises that were still engaged in dairy farming at the beginning of the 21st century. According to agricultural census data for 2001, 26 per cent of all active Estonian farms and 53 per cent of all active Latvian farms practised dairy farming.[25] No data are currently available for Lithuania, but it is likely that the share of farms practising dairy farming is higher in Lithuania than in Latvia because of the strong milk production tradition on plot farms. This was clearly reflected in the small number of cattle on farms. The share of farms with one or two heads of cattle remained remarkably high. In Estonia, the proportion was still 75 per cent, in Latvia 79 per cent and in Lithuania as high as 84 per cent.[26] For the majority of dairy farmers in all the Baltic countries, then, agriculture has not changed – at least in dairy farming – towards family farms.

The reverse of the plot farming tradition is corporate farming. In 2001, 32.4 per cent of all cows in Estonia were on farms with a minimum of 300 heads, and 56.5 per cent on farms with a minimum of 100 heads. The corresponding figures for Latvia were 6.2 per cent and 12.8 per cent. No data are available for Lithuania. In all three countries, nearly all of the large farms were corporate farms. In 2001, corporate farms accounted for 53.9 per cent of the total cow population in Estonia, for 12.9 per cent in Latvia and 8.4 per cent in Lithuania. Since the proportion of corporate farms is a relatively accurate indicator of the degree of centralization in milk production, the figures indicate that milk production was by far the most centralized in Estonia, and somewhat more centralized in Latvia than in Lithuania.[27]

But to what extent have people established enterprises that on the basis of their external characteristics might be considered to be technologically advanced family farms in the zone between plot farming and large-scale farming? Since we have

---

[25] Percentages of farms reported to be active in the Estonian and Latvian agricultural censuses.

[26] ACE I, 2002, 109; ACLa, 2003, 223–224; SYLi, 2002, 395.

[27] ACE I, 2002, 109; ACLa, 2003, 223–224; SYLi, 2002, 395.

comparative data on the number of dairy cows in different countries, we can assume – based on the Western system and also considering the Baltic farms still under construction – that a technologically advanced farm would have 10–29 cows.[28] Such a farm would probably not only ensure the livelihood of the family, but also be manageable by its members The figures below[29] indicate that Estonia has the largest number of dairy farms that fall in the category of developed family farms in proportion to the total number of dairy farms, with Lithuania again showing the lowest number.

Number of dairy farms with 10–29 cows and their proportion of all farms in 2001:

|  | Estonia | Latvia | Lithuania |
| --- | --- | --- | --- |
| Dairy farms | 797 | 1,340 | 1,014 |
| Share | 4.5 per cent | 1.8 per cent | 0.5 per cent |

In Estonia, small-scale production has already become differentiated into subsistence farming that carries on the Soviet tradition and more dynamic, market-dependent and technologically developed family farming, but in Lithuania this process of differentiation is only just beginning.

We can also draw the paradoxical conclusion that the family farm project has been most successful in a country where large-scale production was most successful in maintaining its position, too. And in reverse, the development towards technologically developed family farms is most difficult in Lithuania, the country where the individualization of agriculture had progressed the furthest.

The share of family farms is as high as it is in Estonia because the few peasant farms that got a good head start at the end of the Soviet era now constitute a larger proportion of all farms, since the total number of dairy farms in Estonia is relatively smaller than in the two other countries. Other possible explanations include the same reasons that account for the preservation of enterprises based on technological large-scale units: people had more time to reflect and to make the necessary social arrangements, and the economic circumstances were slightly more stable during the decollectivization of the non-land assets of Soviet farms (more on this later).

Specialization in crop farming does not require major initial investments, since land has been cheap and available for rent at a reasonable cost. Moreover, the heavy machinery acquired through privatization is more suitable – at least in the intermediate stage – for the farming of cereal and industrial plants than animal production. Therefore, it would be logical to assume that more developed family farms had been established for farming cereal and industrial plants (e.g. flax and sugar beet). Unfortunately, we do not have a simple indicator such as the number of cows (that can be proportioned to the technology and family labour required) for

---

[28] The ideal herd size is growing all the time. According to Krisciukaitiene, Stanikunas and Zemeckis (see their article in this volume), 'EU experience suggests that for reasons of stability the ideal size of dairy farms is from 30 to 60 cows.'

[29] ACE I, 2002, 109; ACLa, 2003, 223–224; SYLi, 2002, 395.

analyzing them, but we would need to address this question in a separate empirical study. However, based on the data shown in the tables and figures in this article, as well as the information available on the specialization on crop production on legally constituted family farms, it is unlikely that such new information would significantly change the overall picture of the potential number of technologically advanced family farms. Although family farms show slightly above average specialization in crop production, the principal rule appears to be that most of them are characterized by a technologically and financially modest and unspecialized mixture of crop and animal production.

In Lithuania, much more than in Latvia and especially Estonia, the forced top-down family farm project has turned against itself by actually turning into a plot farm project. Some further explanations for the country differences can also to be found in civil society traditions and cultural traditions (see Ruutsoo's article in this volume).

Baltic accession to the EU will have a major effect on these countries' farm structure through the introduction of more stringent milk quality requirements alone. Estonia adopted a stricter set of quality requirements for purchasing raw milk in July 2002 (Ministry of Agriculture, 2002, 15). According to the latest statistics (AgrE, 2003, 73), the Estonian cow population declined by about 10 per cent in a single year, with the decline coming entirely through individual farms. This marked a huge leap forward in the concentration of dairy industry on corporate farms. While the share of cows held by corporate farms stood at 60.2 per cent at the beginning of the year, the figure increased to 65.9 per cent by the end of the year. In 2002, milk production was even more concentrated: corporate farms accounted for 68.7 per cent, since the milk yield per cow on individual farms was significantly lower (4,533 kilograms) than on corporate farms (5,500 kilograms) (Ibid., 78).

The statistics do not provide information on the development of the cow population and milk yield on farms of different sizes. However, the Estonian Ministry of Agriculture (2002, 15) has in previous year estimated that the more stringent quality requirements would cause difficulties on farms with 1–5 heads in particular, since they will not be able to make the required investments. According to the agricultural census, less than one-third of the 3,000 or so farms with 3–9 heads had even such basic equipment as milk coolers (ACE II, 2002, 82), yet such farms still accounted for 19.6 per cent of the cow population (Ibid., 15). Since production on farms with 1–2 heads, which made up three-quarters of all dairy farms in Estonia, was probably geared mainly for own consumption, the Ministry estimate appears to be well justified. As the pressure of the more stringent EU quality requirements is focused not on the (most primitive) plot farms, but primarily on technologically in advanced family farms, a large proportion of family farms in the Baltic States will be destroyed unless support is made available for their critical investment needs, and small-scale producers will be divided even more sharply into a large group of elementary subsistence farms and a small number of technologically adequate commercial farms.

Conventional agricultural policy has little or no effect on the conditions of primitive technology[30] (see also the article by Krisciukaitiene, Stanikunas and Zemeckis in this volume). Based on my interview data, this group represents the poorest segment of the rural population, and a reasonable livelihood for these people can first and foremost be ensured through social policy, particularly special programmes designed for overcoming poverty.

## *Why is the Path from the Plot Farm Heritage to the Family Farm so Difficult?*

The plot farm heritage is particularly important for former Soviet household plots. World Bank (1993b, 101; 1993c, 91; 1993d, 177) country studies from 1993 pinned a lot of hope on plot farms established by Soviet farm workers, since the authors believed that plot farming had served to preserve 'technical and entrepreneurial farming skills.' Former Soviet household plot farmers also received more land during the transition by restitution or by land distribution. Dairy farming has traditionally been the most important area of specialization through which smallish farms have been able to increase their production volume and value added. Indeed, this is one of the development paths specifically mentioned by Lerman, Kislev et al. (2003), who believe that household plot farmers can increase their productivity by following this path. However, the development of dairy farming on plot farms involves major economic challenges, most notably the construction of sufficiently large and modern cowsheds.

According to agricultural censuses carried out in 2001, only 5 per cent of the 2,367 Estonian cowsheds specifically designed for dairy farming had been built after 1991 (ACE II, 2002, 123 and 130). In Latvia, only 13 per cent of the 8,625 cowsheds had been built in or after 1989, 36 per cent in 1940–1988 and 43 per cent before 1940. Cowsheds built in 1989–1994, a period that effectively coincides with the peasant farm phase (1989–1992), account for about 10 per cent. Very few new cowsheds, only 3.6 per cent of the total number, have been built after decollectivization (1995–2001), and likewise only very few have been renovated. Most of the former kolkhoz and sovkhoz cowsheds built in 1961–1998 are now mainly owned by farms with the largest farmed acreage (ACLa, 2003, 283–285).

When the cowshed statistics drawn from agricultural censuses are compared with the number of farms with dairy cattle in Estonia and Latvia, it is apparent that only a small proportion of those farms could have had a cowshed specifically designed for dairy cows (for a definition, see ACE II, 2002, 158). Most of the cowsheds on individual farms today are very small and old, representing as they do the cowshed heritage of the Soviet household plot farms. After decollectivization, the construction of a modern, Western-standard cowshed is only possible in exceptional cases (e.g. through the sale of timber acquired through restitution),

---

[30] Of the 13,306 dairy farms in Estonia, only 1,861 had a milking machine and 941 a milk cooler. It is likely that very few smallholders owned the most basic of all farm machines, i.e. a tractor, since less than 20 per cent of all farms had a tractor, while only 16 per cent had a soil cultivation machine and the same proportion a sower (ACE II, 2001, 14, 82 and 84).

since the income provided by small-scale agriculture or wage labour is wholly inadequate for such purposes.

In terms of the goal of the family farm project, dairy cows were probably the ideal resource of Soviet farms: they were easy to transfer to small farms and there were plenty of them around for all prospective buyers (who would primarily be using privatization and restitution vouchers). However, cows needed to have some sort of a cowshed, but only some families had inherited one from the Soviet era, and the number of plot farms with cowsheds varied widely from country to country. The number of farmers keeping cows on household plot farms during the Soviet era gives a good idea of this opportunity, since on the basis of Estonian and Latvian agricultural censuses (see above), the number of dairy farms as a proportion of all farms was still the same in both countries in 2001. In Lithuania, 75 per cent of former kolkhoz workers have kept dairy cows, in Latvia this figure was 49 and in Estonia 25 per cent (see Table 2.3). According to my interviews, other rural people also owned some cowsheds.

If the cowsheds and technology remained unchanged or were only supplemented with individual pieces of machinery acquired in conjunction with the privatization of large-scale farms, the consequences for farm production were as follows: only a limited number of heads of cattle could be added, since the majority of the buildings were too small, too dilapidated and outdated, nor did they have the required machinery (milk coolers, feed mixers, etc.). The problems of plot farms can be illuminated by comparing them with the fate of peasant farms. Only the owners of a few of the thousand peasant farms established during the Soviet era or at the very beginning of independence (1989–1992), who had been able to establish such strong farms with the help of financial support from the government that their operation no longer required the support of a large-scale farm, were able to develop their farms into modern family farms. Government support for the construction of family farms dried up in all the Baltic countries shortly after they had regained independence, other less successful peasant farms faced the same problems as plot farmers (Alanen, 1999). Based on a case study in Southern Estonia, owners of the best peasant farms had also developed a good sense of how to upgrade their farm technically. This is why they were better able to make rational use of the chances to acquire machinery, equipment, cattle, spare parts, building materials, etc. during the decollectivization of non-land assets. Meanwhile, traditional plot farmers and other prospective family farmers were largely acting without any clear plan (Alanen, 2001).

The plot farms also required labour for animal husbandry. Many farmers could start up or expand into dairy production with one or two cows only if they could rely on their children or other close relatives to take care of the animals. The interview material suggest that old people living on plot farms were mainly aided by their children who had worked on former Soviet farms, especially as large numbers of jobs were lost when the privatized corporate farms reduced or suspended operations or rationalized their production. However, the average cowshed was not large enough to accommodate very many additional cows.

Outside services were also available after the privatization of Soviet farms, but most plot farms did not have the money to pay for those services. In Estonia and

Latvia, many new enterprises redeemed the shares of their stockowners by providing machinery services. Often one half of the sum would be paid in cash and the other half by deducting the amount (i.e. EEK or LTL) from the nominal value of the ownership share. This for its part helped to reduce the transition shock. According to my interviewees, plot farmers considered share ownership otherwise worthless since corporate farms were for a long time unable to pay out dividends to their owners. Meanwhile in Lithuania, the partnership system meant that plot farmers were in a stronger position than in Estonia and Latvia. Still, the buffer effect of corporate farms was only temporary in all three countries.

Former Soviet household plot producers faced great difficulties as soon as the crucial external factor, i.e. the privatized corporate farms, gradually ceased to function as their infrastructure and institutional support structure as a result of the transformation of the agricultural system. Since the new corporate farms were forced to cease operating more often in Latvia and Lithuania than in Estonia, the problems of plot farmers increased accordingly.

Even if an isolated plot or family farm were able to resolve the problems with the newly arranged labour and technology, it still had to contend with additional obstacles in marketing. Some were able to expand their production to five or six heads using the old buildings. A farm of this size can only market its produce through the dairy, but collecting milk from such small units is not economical for the dairy, and the lack of cold storage equipment means that there are bound to be problems with quality as well. Therefore, even if they were able to sell their produce to the dairy (which had become increasingly difficult), people were not getting paid very much. Stricter milk quality requirements thus appear to have become a new insurmountable obstacle to the gradual development towards a technically developed family farm.

Given the difficulties involved on the path towards a family farm, 75 to 84 per cent of all dairy farmers (see above) in the three Baltic countries have chosen to limit themselves to 1–2 cows for family consumption[31] and sell any excess milk to neighbours at better prices than they would get from the dairy. This is one way of adapting production levels to the size of the market (the family and possibly neighbours).

The results of the Latvian agricultural census illustrate most dramatically that this development path, heavily restricted by the Soviet heritage, is everyday reality for the majority of individual farms. In 2001, only 11.8 per cent of all private farms sold more than half of their produce on the open market, while 59.9 per cent did not sell any of it (ACLa, 2003, 79). Agricultural census data from Lithuania remain unpublished, but it is clear that subsistence farming is widespread in Lithuania, too. According to Poviliunas and Batuleviciute (1997, 66), 73 per cent of the agricultural output of Lithuanian individual farms in 1996 was consumed in the form of own production. Own consumption is also the only reasonable explanation for the large proportion of tiny dairy farms in Estonia. True, a large part of production was reserved for own consumption during the Soviet era as well, but at that time people had a steady income from wage labour or pensions. Today, some

---

[31]  ACE I, 2002, 109; ACLa, 2003; SYLi, 2002, 395.

plot producers live essentially outside the monetary economy (for more details on poverty, see *Rural Poverty*).

Drawing on extensive empirical datasets, I am now in a position to complete my critical evaluation of the argument made by Lerman, Kislev et al. (2003), who suggest that individual farms will develop 'by specializing in high-value and labor-intensive products.' This argument is not supported practically at all by the empirical evidence for the Baltic States. Instead, most agricultural enterprises remain technically trapped by the old Soviet heritage. Furthermore, the productivity of Lithuanian agriculture, which according to Lerman, Kislev et al. should rank among the best-performing transition countries, lends little support to the argument. On the contrary, the empirical findings (see also Table 7.3 in the article by Krisciukaitiene, Stanikunas and Zemeckis in this volume) not only cast a shadow on the methodological reliability of their productivity calculations, but also undermine the credibility of the family farm theory propounded by the World Bank.

## External Influences on the Agricultural System

Compared to other countries in East-Central Europe, all the Baltic States suffered an additional blow upon independence when their agricultural division of labour and relations with the other former Soviet republics were weakened and largely severed. Before they regained independence in 1990–1991, agricultural output showed similar trends in development in all countries (Annex Table A-1, Column 1).

The external problems were aggravated over the next two years (1992–1993). Inflation, particularly in 1992, reached similar peaks of hyperinflation in all three countries – even if the figures were considerably higher in Lithuania in 1993 than in its neighbours (Annex Table A-1, Column 8). PSE (percentage of value transfers to the agricultural producer, see Annex Table A-1, Column 7), an indicator of price relations between agricultural producers and other sectors, was heavily negative for agriculture in 1992, the year of hyperinflation, signifying that other sectors of the economy were 'taxing' agriculture at a ratio varying from –97 in Estonia to –124 in Lithuania. In terms of PSE, the country differences were the greatest in 1991. PSE was heavily positive in Estonia at +54 per cent, and even more so in Latvia at +83 per cent, while Lithuanian agriculture was severely 'taxed' at –262 per cent. Amazingly, Lithuanian agricultural production declined less than in the other Baltic countries in that period. And even though the circumstances in Latvia were slightly more favourable than in Estonia (cf. also Davidova and Buckwell, 2000, 15), Estonian agricultural production declined most sharply.

The new transition debate emphasizes that the development of institutions must coincide with the key phases of transition. Earlier transition reports suggested that Estonia was ahead of the other Baltic countries in agriculture as well, but current statistics (see EBRD, 2002, 78) indicate that the countries are in fact very close to one another. In all three countries the farmers interviewed complained about the lack of the same resources, above all the inefficiency of the agricultural credit system.

Gross Domestic Product (GDP), which describes the overall development of the national economy, is an external factor that exerts an indirect effect on agriculture (it is mediated to agriculture through such factors as consumer demand and employment development) (Annex Table A-1, Column 2). However, GDP and GAO are largely unsynchronized, and it is often easier to find an explanation for special characteristics in the development of GAO within agriculture (cf. Lithuania's growth boom in 1995–1997). Another indicator that is clearly based on the agricultural reform and that is even better than GAO in describing the relative independence of agricultural development, is agricultural productivity.

It would clearly be wrong to claim that external factors had no impact upon the development path that agriculture would take in Estonia, Latvia and Lithuania, but those factors certainly were not decisive in determining that path. The decisive factor in this respect was the decollectivization of non-land assets in the early 1990s – which means that the development model in each country must be understood as the result of the decollectivization process and the Soviet heritage, its initial condition.

It will not be possible to understand this differentiated development without considering the effects of the agricultural population on its own destiny.

### The Effect of Soviet Farm Workers on the Privatization of Non-Land Assets

In all three Baltic countries, Soviet farm workers would have wanted (though undoubtedly for different reasons and with different emphases) to see the large-scale system remain in place. This was by no means peculiar to the Baltic countries: similar response results were also reported in other former socialist countries in both Eastern Europe and the former Soviet Union (see e.g. Swinnen and Mathijs, 1997). World Bank reports indicate that it, too, was well aware of these results: 'Surveys of collective farm members indicate that most members do not wish to strike out on their own' – this is a direct quote from the recommendation document for Latvian agricultural reform. In addition, readers are cautioned against considering the phenomenon 'a sign of entrepreneurial (or moral) deficiency,' because the conditions for establishing family farms were difficult at the time (WB, 1993c, 93). For World Bank researchers and many others, the reluctance to start up family farming was more or less 'a conservative factor in the restructuring process' (Swinnen and Mathijs, 1997, 361) that should be overcome. The governments of all the Baltic countries that had committed themselves to the family farm project were also well aware of the survey results. In Lithuania, the decisive 'resolution of the Supreme Soviet of Republic of Lithuania [30 July 1991] was based on the conviction that privatization was a necessity and peasants were often coerced into accepting this without being offered any alternatives' (Hejbowicz, 2001, 97). Although the Lithuanian government took an extreme line, other governments did not respect the opinion of the agricultural population either, let alone adopt their view as the basis for the agricultural reform. Both government representatives and researchers from international institutions also overlooked the critical analyses of local experts in the former social countries – in the Baltic countries, the critics were labelled as communists. However, there were also voices of caution within the World Bank. M. Karen Brooks wrote that 'the new cooperatives [...] may be the necessary institutions of the transition,' and warned that 'forced decollectivization should not be pushed on rural people' (1991, 5).

However, a more differentiated analysis indicates that at least in the Baltic countries, people were not opposed to the changes as such. Even though they wanted to maintain some large-scale production, kolkhozes and sovkhozes were generally (and according to my interviews particularly in Estonia and Latvia) considered manifestations of the Russian occupation, which is why they were resented. They were also widely criticized for their inefficiency. In their defence of large-scale production, people actually meant more or less reformed large-scale production units, and in this sense the mood was favourable to their reforms. Besides, part of the people, especially the elderly and uneducated, wanted to preserve the large-scale farm mainly because it would have been extremely difficult, if not impossible for them to take up the cultivation of a restituted family farm. The people who were most committed to large-scale production were trained specialists (agronomists, zoologists, veterinarians, etc.) and elite workers (such as machine operators), i.e. the Soviet farm middle-class. The value of their education, professional status, income and modern way of life all depended on a work organization based on large-scale production. It was the middle-class that also produced the 'citizen activists' of the decollectivization process (Alanen, 1999 and 2001). People were justified in their concern about their future, and with regard to popular opinion it would have been important to keep all options and their combinations open. However, a creative and effective reform would not only have required disengagement from the pure family farm doctrine, but also much more varied support measures (systematic information, counselling, financial support and other institutional arrangements) than international organizations and national governments were willing or able to provide. If they had wanted to, the World Bank and other international organizations whose recommendations had a major influence on the course of the process, could have recommended that the population being subjected to the reform be integrated in the decision-making process.

The model of decentralized decollectivization adopted by the Estonian and Latvian governments, which required the approval of the general assembly of those entitled to the privatization of non-land assets, provided a channel of influence for the subjects of the reform, whereas in Lithuania this channel was deliberately closed. However, according to my interviews the decollectivization process was not particularly transparent in Estonia or Latvia either. Information was sporadic (and sometimes used simply for personal benefit) and legal control of the process was neglected. Neither were the Estonian and Latvian reform laws neutral enough in terms of the selection of enterprise type. Corporate farms were not entitled to land ownership at the time that new companies were established in either of the countries; this often had a decisive discouraging effect on those who were thinking of setting up an enterprise based on a technological entity.

Judging by my interviews, decollectivization involved a great deal of uncertainty, pressuring, abuse of power, secrecy, anarchy and rivalry between different actors and alliances in all the Baltic countries (cf. also Alanen, 2001; Alanen, 2002). The field interviews also indicated that the actual decollectivization processes varied widely from one former Soviet farm to another in all countries (including Lithuania) – and further that elite farms were sometimes broken down into nonviable pieces, while in an opposite case an average Soviet farm might develop into one of the most successful agricultural enterprises in the country. However, if we look at the bigger picture, the Estonians were the most successful in preserving technological entities – which became apparent in the continuity of large-scale production. If their reform was very similar to that carried out in Latvia in legal terms, what was the secret of Estonia's success?

In my opinion, the main reason for Estonia's success lies in its middle class that was educationally and socially strong (large and modern) enough to break the hegemonist, anti-large-scale production view of the national government and media at the local level, to come up with ideas for large-scale agricultural enterprises based on the Soviet farm and its technological units, and at the same time capable of forming shareholder alliances to finance the purchase of major lots. The power of the middle class was based not just on numbers, even though Estonia probably has the largest middle class in the Baltic countries (there are no reports of comparative studies on this subject). The empirical data presented by Rein Ruutsoo indicate that Estonia had a more active tradition of civil society than the other Baltic States. They had more activists, i.e. people who engaged in the activities of several organizations at the same time. It is only natural that conflicts also arise more often as a result of active participation (see Ruutsoo's article in this volume).

In many cases, the decollectivization of a Soviet farm was led by a member of the former leadership (together with its trustees). However, even a former leader needed the support, whether active or passive, of the middle class, because it appears that not a single farm survived the process without antagonism. An open struggle between different ideologies and interest groups was more characteristic of the Estonian process than the patrimonial direction of passive people, and in many cases the Soviet middle class took over the control of the process from the old kolkhoz or sovkhoz leadership (Alanen, 2001). In Estonia, there were several 'little revolutions' where the old leadership was ousted.

The main reason for the destruction of large-scale production in Latvia was the exact opposite of the above: the inability of people (the middle class) to take sufficient control over the decollectivization process at the local level. Although Latvian agriculture was quite highly developed compared to other transition countries, it was still the weakest Baltic case and lagged clearly behind Estonia (see Table 2.2 above). According to the field interviews, the strength of the middle class depended on the economic resources of the farm: a wealthy farm was able to hire a larger number of highly trained professionals (a sought-after resource).

In addition to the strength of the Soviet middle class, another important factor was probably the overall political situation. Rein Ruutsoo (see his article in this volume) thinks that the bloodshed and the occupation of certain strategic targets by Soviet troops in Lithuania and Latvia in the year preceding independence radicalized the population and governments of these countries. Estonia escaped this. This may explain why Latvia and Lithuania set about decollectivizing Soviet farms with such speed and determination as soon as the countries had regained their independence, whereas Estonian parliament entered into a complex and long-winded debate, prolonging the preparation of special legislation. In Latvia the decollectivization of non-land agricultural assets took place in 1992, the year of hyperinflation and very low producer prices (negative PSE value) (Annex Table A-1B, Columns 7 and 8), while in Estonia it was postponed to a somewhat more tranquil phase. Latvian agricultural employees had little time to gather information, make their personal decisions and form the shareholder alliances that were needed for the purchase of major lots. Hyperinflation made it difficult to identify economically viable solutions, and the extraordinarily high PSE indicated that no commercial agriculture venture could be profitable at the time of decision-making.

In Estonia, there were less external pressures in 1993, when most of the decollectivization of non-land assets took place than in 1992, although even then temporary pressures did have a profound effect on people's behaviour.[32] Gradual implementation was a definite advantage, too. According to the interviews, decision-making was influenced by successes and failures on other farms, both home and abroad. Some farms even obtained consulting assistance from the former GDR. Empirical evidence (some of which was acquired from Latvia) tended to support the preservation of large-scale agriculture, instead of transferring all farm resources to family farms. The interviewees used the fate of one of Estonia's former elite farms as a warning example. Agriculture in this area dominated by former Soviet farms collapsed as a result of radical, family farm system oriented decollectivization of land assets.

In Lithuania, the development resulted in a combination of top-down coercion, mechanical solutions and management problems subsequently caused by the partnership system.

However, the final result of these processes cannot be understood in any of these countries without considering the former symbiosis of large- and small-scale production that disintegrated and dissolved after the decollectivization of non-land assets at different paces and in different ways, depending on its initial conditions and the nature of the overall agricultural reform.

---

[32] In a detailed case study of the decollectivization of kolkhozes in southern Estonia, we found that the former kolkhoz leadership was particularly disinclined to acquire cowsheds and other animal facilities due to the difficult economic circumstances. Therefore, the reform committee managed to get an owner for most of them only after much persuasion. However, complexes such as these have subsequently developed into the backbone of Estonian agriculture.

## Rural Poverty

Due to the nature of Soviet farms, their privatization involved much more than just agricultural transition. An 'Assessment of Rural Poverty' published by IFAD (2002, xiv) states in a very critical tone that the agricultural reform to a 'market-based agricultural economy' in itself 'requires much more than the privatization of the land and other assets.' This is true, since if you set the goal of moving over to a family farm system, you should have built up the supporting infrastructure required by the enterprises and the vital formal institutions at the same time. But still it must be added that it should also have covered all the other tasks (an array of economic, administrative, social, educational and cultural functions) that were the responsibility of kolkhozes and sovkhozes in the Soviet countryside. It was not just a question of the transformation of the agricultural enterprises. If international research is concerned with the fact that agricultural enterprises especially in Russia and other CIS countries (see Lerman et al., 2001) still provide public services, the explanation for that lies, at least in part, in overly limited reform programmes. Particularly in Estonia and Latvia, new corporate farms disengage themselves very quickly from their public obligations, yet the extent and nature of rural poverty can only be understood from this broader perspective. Therefore, it would be important to know just how successful the creation of non-agricultural enterprises on former Soviet farms was. Much more systematic rural research is still needed along such lines. In 1994, 76 per cent of the economically active population in the Estonian countryside received their main income from other sources than agriculture, in Latvia the figure was 66 per cent and in Lithuania just 48 per cent (Alanen, 1998, 50). Based on these statistics and the interviews with entrepreneurs in Estonia, a non-land assets reform of Soviet farms where large-scale production was preserved turned out to be the best way of preserving both agricultural and non-agricultural resources (and providing employment in the latter).[33] Therefore, the extent of elementary plot farming in Lithuania could also be explained by decollectivization that led to rural economic activities being curtailed.

From this overall perspective then, rural poverty is in good harmony with what I have said in this article about the economic success of agricultural decollectivization – even though poverty measurement results are of course affected by the level and coverage of the social security system (EBRD, 2002, 90–91). According to a recent assessment by the EBRD (2002, 91), 31 per cent of the rural population in Estonia and 51 per cent in Latvia live in absolute poverty when measured on the basis of their income level (Ibid., 91). Information from Lithuania was not available. According to a slightly older World Bank study from the 1990s (WB, 2000b, 97), the proportion of poor people was 42 per cent in the Latvian countryside and 52 per cent in Lithuania, and according to risk assessments based on personal profiles (education, age, etc.) the risk of falling into poverty is 1.4-fold

---

[33] Although Soviet farms were very important in terms of non-agricultural production too, there were also a few independent industrial enterprises and services in the countryside. Separate legislation was passed to address their privatization (cf. the articles by Jouko Nikula in this volume).

among the Latvian rural population when compared to the urban population, and 1.7-fold among the Lithuanian rural population. Information from Estonia is missing (Ibid., 76).

The picture is complemented by household budget survey information on household consumption. Among the Latvian rural population, food made up 49 per cent of total consumption, among the Lithuanian rural population the figure was 54 per cent. The importance of subsistence production is highlighted by the fact that home-produced food and other products accounted for 20 per cent of total consumption in Latvia (in 2000), while home-produced food accounted for 37 per cent of all food consumption. In Lithuania (2002), the figures were 32 per cent and 53 per cent, respectively. In both countries, slightly more than one-third of the rural population's disposable income came from income transfers during the corresponding period (OECD, 2003, 75–76). We also have comparable information on food consumption for rural households in all the Baltic countries from 1997. At that time, home-produced food accounted for 19 per cent of total household consumption in Estonia, for 17 per cent in Latvia and for 32 per cent in Lithuania (OECD, 2000, 241–242). According to all research results, food consumption in the countryside is several tens of percentages higher than among the urban population, and the higher the amount of food consumption as a proportion of total consumption and the higher the share of home-produced food, the poorer the household (OECD, 2000, 154 and 241–242; OECD, 2003, 75–76). In other words, poverty and home produced food are closely linked to each other. Since poverty is most widespread in Latvia and especially in Lithuania, there is no reason to doubt the following causal relationship: The more disruptive the failure of the agricultural reform, the greater the role of plot farming, and the more severe the problem of poverty in the countryside. The family farm project thus turned not only into a plot farm project, but also or at the same time into a poverty production project.

A more general poverty assessment project has already been carried out in Latvia, which according to one project representative 'has revealed the alarming depth of poverty, most notably rural poverty.' The 'frequency and depth of poverty' in the countryside 'reflect the difficult adjustment associated with the transition to the market economy' (Kim, 2001, 182–183). Indeed, as Csaba Csáki et al. write in the World Bank Rural Development Strategy (2000a, 1), 'the transition process in agriculture is far more complex than originally envisaged by both the countries themselves and the international community, including the Bank.' Unfortunately, this report does not really offer anything new or innovative, but merely recycles old ideas that have largely proved untenable. However, in order to eradicate poverty, the same World Bank that was busy presenting recommendations for agricultural reform with the authority of an important lending institution, is now repairing the consequences of the failed reforms by developing and financing a poverty eradication programme of its own (e.g. WB/IBRD, 2002).

Based on our case studies (see Alanen et al., 2001 and Kämäräinen's article in this volume), the problems of poverty are in many ways intertwined with the paralysis of civil society, the weakness of local government (and social programmes), the anomic state of local communities, the grey or even black economy, etc. Most of the mistakes already made at least in the accession countries will ultimately have to be carried by the EU and its Member States. Overcoming the problems will not only be expensive, but it will also be a long and winding road, since – as the findings of the EU project reported in this book show (see the article by Tisenkopfs and Sumane in this volume) – helping those who have fallen into poverty back into an independent life is a challenging task indeed.

## The Three Baltic Development Models and the Concept of the World Bank

The new agricultural production systems evolving in Estonia, Latvia and Lithuania and essentially their new rural structures are thus primarily an outcome of national agricultural decollectivization policies.

In Estonia, the development has been characterized by the prominent creative role of local people, the preservation of robust large-scale production, but also by the development of a large number of potential technologically advanced family farm enterprises, increased productivity in agricultural work and other industries, and the ability of large-scale farms to specialize in high value-added production and effectively adapt their business to the conditions of a capitalist market economy at the same time. The key to Estonia's relative success has been its ability to make better use than either of its Baltic neighbours of the heritage of Soviet farms. Since the Soviet middle class had a decisive role in this process, I will call the Estonian model 'the middle class large-scale production project.'

The Latvian model, too, was conceived as a result of local decisions on the decollectivization of non-land assets and the rapid restitution of agricultural land. The process was faster and more radical than in Estonia, but it was perhaps because of the speed of it all that the process turned into chaos, beyond the control of local decision-makers. As far as existing resources were concerned, the direct consequences of the radical reform in Latvia were more destructive than in any other Baltic country. It resulted in a decline of agricultural productivity and a radical decrease in the degree of specialization. The end result was an ineffective industry structure based on small-scale production. Compared to Estonia, a far smaller number of Latvian enterprises were potential technologically advanced family farms. Due to the potential influence of local people, I will call the Latvian model 'the unintended small-scale production project.'

The decollectivization of non-land assets in Lithuanian agriculture differed clearly from both the processes described above. Decision-making was concentrated in the hands of the government and where necessary the reform was carried out by force (by displacing the opponents). Partnership, a variation of the limited liability company preferred by the World Bank, was employed as a transitional stage in the family farm project. Since this legal form made corporate enterprises dependent on the interests of local owners as successors to the Soviet plot farming heritage, it made the management environment of corporate farms extremely challenging, and thus prevented the effective adaptation of corporate farms to the requirements of capitalist market economy. Even though the Lithuanian model came closest to the ideal solution recommended by the World

Bank that was supposed to produce an effective family farm system, the reform resulted in a smaller number of potential technologically advanced family farms than in any other Baltic country, the greatest number of elementary household plot farms mainly used for family subsistence farming, a marked reduction in productivity levels and the deepest rural poverty. I will call the decollectivization of non-land assets implemented by the Lithuanian government 'the enforced World Bank project.'

All the Baltic countries were in principle capable of achieving the high productivity pattern of CSH countries introduced by Macours and Swinnen (2002): in terms of their 'initial condition,' all of them were developed enough, and Estonia was also capable of realizing the pattern despite the transition shock caused by independence from the Soviet Union (a disadvantage shared by all the Baltic States, but not by East Central European countries). One thing that only Estonia shared in common with all the CSH countries was that they remained essentially dependent on large-scale production. Meanwhile, in Latvia and Lithuania, agriculture became dominated by small-scale production in the fashion of CVA countries or countries such as Armenia or Georgia (where the economy collapsed due to wars and natural disasters), but in their case the outcome was much worse in terms of production and productivity. A family (or rather plot) farm project might well be a good solution for developing countries (CVA countries) or countries that have undergone near total social collapse (such as Armenia and Georgia).

The principal cause of the failures was the doctrinal and romantic (see the articles by Alanen and Ruutsoo in this volume) view of history and the family nature of agricultural entrepreneurship. Although it is true that corporate farms are not dominant in the agriculture of either North America or Western Europe, a closer examination of history shows that agricultural systems in both these cases have always had several different types of farms operating side by side, and that their relative numbers have varied both over time and production sectors. The relative proportion of different farm types has depended on such factors as overall economic policy (cf. British free trade and state customs policy in Continental Europe), land reforms (cf. land reforms in many Eastern and Central European countries prior to World War II), natural conditions and the special characteristics of agricultural policy. In California, large-scale production has gained dominance even in milk production due to the combined effect of (state) agricultural policy and natural conditions. Milk production has traditionally been a bastion of family farming, which it still is the US Midwest. (For details on the development of farm types and its general causes, see Alanen, 1991 and Alanen, 1992).

For reasons of principle alone, then, no enterprise type should be prioritized in isolation from its historical background or environment.

It is particularly unfruitful to focus on the more than 100-year-old dispute over the superiority of small- or large-scale production as the key issue – which is exactly what the list of recommendations from the World Bank and its researchers primarily implies. Right now, the latest statistical data on both crop and animal production from Estonia (AgrE, 2003) speak strongly in favour of large-scale production (better harvests, higher yields, centralization of production). However, there is no way we can tell for sure whether a situation that at one moment appears to be promising for one type is in fact due to a specific phase in the development of the agricultural system or some temporary circumstance (cf. milk quality requirements). And how will EU regulations eventually impact future developments? All in all, the results of comparisons concerning the efficiency of different production factors on different types of farms are so contradictory that they warrant conclusions in favour of both large-scale production and the benefits of family farming (Fernandez, 2002). Instead of a single doctrine, i.e. the family farm ideology, we should also beware of another trap, the large-scale production ideology. In more theoretical terms, this is because capitalist commodity economy does not develop in any one direction, that of centralization, but it always allows for interruptions and reorganization. Besides, political choice comes into the equation as well: government policy has always had a significant effect in terms of strengthening some agricultural enterprise types and weakening others (Alanen, 1991 and 1992).

According to Lerman et al. (2001, Introduction, 7–8), 'sociologists and anthropologists working on transition issues have a tendency to stress the dramatic deterioration in the standard of living and the provision of social services throughout the region.' In other words, World Bank researchers admit that major damage has indeed been caused, but at the same time they ask 'has there been an alternative?' Their own answer is: 'the damage to social conditions was an unavoidable price.'

The question as to what kind of damage – including human suffering – is acceptable as the unavoidable price of progress, for me, brings to mind the justifications presented for forced collectivization. As far as the transformation of the agricultural system was concerned, that too was quite unnecessary. The empirical evidence presented in this article from the three Baltic countries indicates that political decision-makers have in fact made a number of choices, some of which were good, but some of which are less so. In reality, the problem of good agricultural transition policy has nothing at all to do with the old dispute as to which enterprise type is the ideal form of agricultural production (family farm vs. corporate farm or small-scale production vs. large-scale production). Instead, the question we must ask is how do we make the best possible use of the human and material resources of an agricultural system and cause as little damage as possible so as to conform to the requirements of capitalist market economy. In the Baltic countries, Estonian agricultural employees were more successful in this regard than their Latvian colleagues; this is why Estonian agriculture is in better shape than elsewhere and why there is less rural poverty. If the government or the press had been allowed to determine the path of development of Estonia, Estonia would hardly be regarded as a member of the CSH transition pattern group, as it is now.

Making the best possible use of existing material and human resources favours, at the initial stages of transition, large-scale production, but it does exclude the possibility of family farming either, as demonstrated by the case of Estonia. Nor does it exclude the continuous restructuring of new large-scale production units, as again exemplified by the CSH pattern countries. Granting strong ownership rights to the new owners might be necessary for the effective adaptation of agricultural enterprises (as shown by the negative example in the RUB countries), but making the redemption of shares invested in an enterprise too easy is not desirable either, as illustrated by the example of Lithuanian partnerships. In the future, the pace at which agriculture in different countries will continue to diversify into different types of farms based on areas of specialization, natural conditions, historical traditions and other factors will depend on the policy (institutions, food supply chain, etc.) of each country (and ever more decisively on EU policy). It is impossible to predict what the agricultural structure of individual transition countries will look like in twenty years' time. However, it is possible that the large-scale production tradition preserved in Estonia and the CSH pattern countries in general may yield some positive innovations in the field of production efficiency, as Grabher and Stark (1998) expect to see more widely in post-socialist countries. For example, could Estonians establish a milk production system where large-scale production enterprises specializing in dairy production are not engaged in crop production? A few Estonian enterprises are seriously considering this option.

The World Bank's decollectivization concept was also too narrow. Although it borrowed some terminology from institutional economics (e.g. 'transaction costs'), it did not incorporate the problem of informal institution characteristic of this branch of economics that emphasizes all-inclusiveness – to say nothing of the fundamental questions involved in values, norms and civil society (see e.g. Alanen, 2002) that are analyzed in a more differentiated way by sociology and anthropology.

I am sure that World Bank researchers follow the debate in other social sciences and not just the narrow discipline of economics. However, they make hardly any reference to the findings of other disciplines in their publications. Transition policy should have been developed and evaluated in continuous dialogue not only with other social scientists, but also with such traditionally humanist disciplines as history. Many problems in the transformation of the agricultural system would have appeared in a completely different light if they had been placed in their historical context.

A third major problem arises from the fact that part of the content of agricultural decollectivization has been left out of the analysis: this process, after all, was essentially about a restructuring of rural society in its entirety. Once the socialist large-scale farms had been dissolved, someone had to take on all the tasks previously performed by the large-scale farms, but which were cut during the transformation due to the nature of the transition. If plot farmers no longer get the support they had previously received from large-scale farms, and if they have no alternative work to go to, that clearly is an essential social problem. The rise of poverty should therefore come as no surprise to experts specializing in agricultural transition. And if the local government does not have the resources to provide the required social and infrastructure services, one can hardly criticize the managers of the successor of the large-scale farm for conservatism (see e.g. Lerman et al., 2001) if the enterprise is caught in a situation where it has to maintain the local infrastructure (e.g. road or drainage networks vital to the enterprise itself). In addition, farm employees need these services that were previously provided by the Soviet farm in order to be good, productive employees. With regard to the conceptualization of the restructuring of rural society, the set of concepts offered by economics is not adequate. This, if anything, should be a common concern for all disciplines focusing on the countryside, and a key task for the World Bank is to bring scholars from all disciplines in contact with each other.

## References

Abele, D. (1995), *Privatization in Latvian Agriculture and Employee Property Rights*, http://www.cbs.dk/centres/cees/publications/pfpb_project/book/olegs2.PDF

ACE I (2002), *2001 Agricultural Census I, General Data, Crop Production, Livestock*, Statistical Office of Estonia, Tallinn.

ACE II (2002), *2001 Agricultural Census II, Agricultural Machinery and Equipment, Storages, Animal Husbandry Facilities*, Statistical Office of Estonia, Tallinn.

ACLa (2003), *Results of the 2001 Agricultural Census in Latvia*, Central Statistical Bureau of Latvia, Riga.

AFE (1994/1997/2000), *Agriculture, Forestry, Fishing* (1994), Statistical Office of Estonia, Tallinn; *Agriculture in Figures 1994–1996* (1997), Statistical Office of Estonia, Tallinn; *Agriculture in Figures 1999* (2000), Statistical Office of Estonia, Tallinn.

AgrE (1995/2003), *Agriculture 1994* (1995), Statistical Office of Estonia, Tallinn; *Agriculture 2002* (2003), Statistical Office of Estonia, Tallinn.

AgrfLa (1997), *Agricultural Farms in Latvia 1996*, Central Statistical Bureau of Latvia, Riga.

AgrfLa (2002), *Agricultural Farms in Latvia 2001*, Central Statistical Bureau of Latvia, Riga.

*Agriculture and Rural Development Plan 2000–2006* (2000), the Ministry of Agriculture, the Republic of Lithuania, Mimio, Vilnius.

AgrLa (1997), *Agriculture in Latvia*, Central Statistical Bureau of Latvia, Riga.

Alanen, I. (1991), *Miten teoretisoida maatalouden pientuotantoa*, Jyväskylä Studies in Education and Social Research 81, Jyväskylän yliopisto, Jyväskylä.

Alanen, I. (1992), *On the Conceptualization of Petty Production in Agriculture*, paper presented at the 8th World Congress for Rural Sociology, Pennsylvania State University, University Park, USA, 11–16 August 1992.

Alanen, I. (1998), 'Petty Production in Baltic Agriculture: Estonian and Lithuanian Models,' in M. Kivinen (ed.), *The Kalamari Union: Middle Class in East and West*, Ashgate, Aldershot, pp. 39–70.

Alanen, I. (1999), 'Agricultural Policy and the Struggle over the Destiny of Collective Farms in Estonia,' *Sociologica Ruralis*, vol. 39, no. 3, pp. 431–458.

Alanen, I. (2001), 'The Dissolution of Kanepi Kolkhoz,' in I. Alanen, J. Nikula, H. Põder and R. Ruutsoo (eds), *Decollectivisation, Destruction and Disillusionment – A Community Study in Southern Estonia*, Ashgate, Aldershot, pp. 63–276.

Alanen, I. (2002), 'Soviet Community Spirit and The Fight Over the Rural Future of the Baltic Countries,' *Eastern European Countryside*, vol. 8, pp. 15–29.

Alanen, I., Nikula, J., Põder H. and Ruutsoo, R. (2001) (eds), *Decollectivisation, Destruction and Disillusionment – A Community Study in Southern Estonia*, Ashgate, Aldershot.

Alanen, I., Nikula, J. and Ruutsoo, R. (2001), 'The Significance of the Kanepi Study,' in I. Alanen, J. Nikula, H. Põder and R. Ruutsoo (eds), *Decollectivisation, Destruction and Disillusionment – A Community Study in Southern Estonia*, Ashgate, Aldershot, pp. 389–406.

Bratka, V. (1998), 'Farm Business Analysis Based on Latvian FADN,' in *Farm Economy in Baltic States and EU: Data Bases, Accounting and Planning, The Eighth Finnish-Baltic Seminar of Agricultural Economists*, Kuressaare, Estonia 1998, Working Papers 12/98, Finnish Agricultural Economics Research Institute, Helsinki, pp. 27–35.

Bratka, V. and Dzepuka, I. (1997), 'FADN as a Decision Basis for Structural Adjustment of Latvian Agriculture,' in *Structural Adjustment of National Agriculture and Food Industries within the Framework of Integration in the EU, The Seventh Finnish–Baltic Seminar of Agricultural Economists*, Vilnius, Lithuania 1997, Working Papers 13/97, Finnish Agricultural Economics Research Institute, Helsinki, pp. 128–133.

Brooks, M. K. (1991), 'Decollectivization and the Agricultural Transition in Eastern and Central Europe,' in *Policy, Research, and External Affairs Working Papers, Agricultural Policies*, WPS 793, The World Bank, Washington, DC.

Chayanov, A. V. (1966), *Peasant Farm Organization*, in D. Thorner, B. Kerblay and R. E. F. Smith (eds), *A. V. Chayanov On the Theory of Peasant Economy*, Richard D. Irwin, Homewood.

David, E. (1903), *Sozialismus und Landwirtschaft*, Verlag der Sozialistischen Monatshefte, Berlin.

David, E. (1922), *Sozialismus und Landwirtschaft. Der Zweite umgearbeitete und vervollständigte Auflage*, Verlag Quelle & Meyer, Leipzig.

Davidova, S. and Buckwell, A. (2000), 'Transformation of CEEC Agriculture and Integration with the EU: Progress and Issues,' in S. Tangermann and M. Banse (eds), *Central and Eastern European Agriculture in an Expanding European Union*, CABI Publishing, Wallingford.

EBRD (2002), *Transition Report 2002. Agriculture and Rural Transition*, European Bank for Reconstruction and Development, London.

Fernandez, J. (2002), 'The Common Agricultural Policy and EU Enlargement. Implications for Agricultural Production in the Central and East European Countries,' *Eastern European Economics*, vol. 40, no. 3, pp. 28–50.

Girnius, S. (1994), 'Economies of the Baltic States in 1993,' *RFE/RL Research Report*, vol. 3, no. 20, pp. 1–14.

Grabher, G. and Stark, D. (1998), 'Organising Diversity: Evolutionary Theory, Network Analysis and Post-Socialism,' in J. Pickles and A. Smith (eds), *Theorising Transition. The Political Economy of Post-Communist Transformations*, Routledge, London, pp. 54–75.

Hejbowicz, S. (2001), 'The New Structure of Rural Economy in Lithuania,' in O. Ieda (ed.) *The New Structure of the Rural Economy of Post-Communist Countries*, http://src-h.slav.hokudai.ac.jp/kaken/ieda2001/pdf/hejbowicz.pdf

Hussain, A. and Tribe, K. (1981), *Marxism and the Agrarian Question, Volume 1: German Social Democracy and the Peasantry, 1880–1907*, Macmillan Press, London.

IFAD (2002), *Assessment of Rural Poverty. Central and Eastern Europe and the Newly Independent States*, International Fund for Agricultural Development, Rome.

Kim, H. (2001), 'World Bank Support for Agriculture and Rural Development in Latvia,' in C. Csáki and Z. Lerman (eds), *The Challenge of Rural Development in the EU Accession Countries*, Third World Bank/FAO EU Accession Workshop, Sofia, Bulgaria, June 17–20, 2000, The World Bank, Washington, DC, pp. 181–188.

Klimaðauskas, E. and Kasnauskiene, G. (1995), *Land Reform in Lithuania: Results and Problems of Privatization*, http://www.cbs.dk/centres/cees/publications/pfpb_project/book/litlan3.pdf

Lerman, Z. (2001a), 'Agriculture in Transition Economies: from Common Heritage to Divergence,' *Agricultural Economics*, vol. 26, no. 2, pp. 95–114.

Lerman, Z. (2001b), *Agricultural Output Index (GAO) for Transition Countries (1990=100)*, Based on Official Statistics,
http://departments.agri.huji.ac.il/economics/zvi-agriout.xls
Lerman, Z. (2001c), *GDP Index (1990=100)*, Based on Official Statistics,
http://departments.agri.huji.ac.il/economics/zvi-gdp.xls
Lerman, Z., Csáki, C. and Feder, G. (2001), *Agriculture in Transition, Land Policy and Changing Farm Structures in Central Eastern Europe and the Former Soviet Union*, http://www.agri.huji.ac.il/~lermanzv/book.html
Lerman, Z., Kislev, Y., Kriss, A. and Biton, D. (2003), 'Agricultural Output and Productivity in the Former Soviet Republics,' *Economic Development and Cultural Change*, vol. 51, no 4, pp. 999–1018.
Macours, K. and Swinnen, J. F. M. (2002), 'Patterns of Agrarian Transition,' *Economic Development and Culture Change*, vol. 50, no. 2, pp. 365–394.
Mathijs, E. and Swinnen, J. F. M. (1997), *The Economics of Agricultural Decollectivization in East Central Europe and the Former Soviet Union*, Policy Research Group, Working Paper, No. 9 (Revision of No. 1), Department of Agricultural Economics, Katholieke Universiteit Leuven.
Meyers, W. H. and Kazlauskiene, N. (1998), 'Land Reform in Estonia, Latvia, and Lithuania: A Comparative Analysis,' in S. K. Wegren (ed.), *Land Reform in the Former Soviet Union and Eastern Europe*, Routledge, London, pp. 87–110.
Ministry of Agriculture (2002), *Overviews and Reports, Agriculture and Rural Development, Overview*, http://www.agri.ee/eng/overviews/overview2002/
OECD (1996b), *Review of Agricultural Policies. Latvia*, OECD, Paris.
OECD (1996c), *Review of Agricultural Policies. Lithuania*, OECD, Paris.
OECD (2000), *Economic Surveys 1999–2000. The Baltic States. A Regional Assessment*, OECD, Paris.
OECD (2003), *Labour Market and Social Policies in the Baltic Countries*, OECD, Paris.
Pirksts, V. and Rozenberga, V. (1997), 'Tendencies of Changes in Agriculture and Food Production in Different Groups of Agricultural Enterprise's,' in *Structural Adjustment of National Agriculture and Food Industries within the Framework of Integration in the EU, The Seventh Finnish-Baltic Seminar of Agricultural Economists*, Vilnius, Lithuania 1997, Working Papers 13/97, Agricultural Economics Research Institute, Finland, Helsinki, pp. 57–62.
Poviliunas, A. and Batuleviciute, L. (1997), 'Structural Changes of Incomes in Lithuanian Individual Farms in 1990–1995,' in *Structural Adjustment of National Agriculture and Food Industries within the Framework of Integration in the EU, The Seventh Finnish-Baltic Seminar of Agricultural Economists*, Vilnius, Lithuania 1997, Working Papers 13/97, Finnish Agricultural Economics Research Institute, Helsinki, pp. 63–67.
Rudys, E. (1998), 'The State of Lithuanian Agricultural Companies and Perspectives of their Development,' in *Farm Economy in Baltic States and EU: Data Bases, Accounting and Planning, The Eighth Finnish-Baltic Seminar of Agricultural Economists*, Kuressaare, Estonia 1998, Working Papers 12/98, Finnish Agricultural Economics Research Institute, Helsinki, pp. 85–88.
Sepp, M. (1997), 'Agricultural Structures' Development in Estonia,' in *Structural Adjustment of National Agriculture and Food Industries within the Framework of Integration in the EU, The Seventh Finnish-Baltic Seminar of Agricultural Economists*, Vilnius, Lithuania 1997, Working Papers 13/97, Finnish Agricultural Economics Research Institute, Helsinki, pp. 36–43.
Spoor, M. and Visser, O. (2001), 'The State of Agrarian Reform in the Former Soviet Union,' *Europe-Asia Studies*, vol. 53, no. 6, pp. 885–901.

Stark, D. and Bruszt, L. (1998), *Postsocialist Pathways. Transforming Politics and Property in East Central Europe*, Cambridge University Press, Cambridge.

Swinnen F. M. J. and Mathijs, E. (1997), 'Agricultural Privatisation, Land Reform and Farm Restructuring in Central and Eastern Europe: A Comparative Analysis,' in J. F. M. Swinnen, A. Buckwell and E. Mathijs (eds), *Agricultural Privatisation, Land Reform and Farm Restructuring in Central and Eastern Europe*, Ashgate, Aldershot, pp. 333–373.

SYE (1994–2003), *Statistical Yearbooks of Estonia*, 1994–2003, Statistical Office of Estonia, Tallinn.

SYLa (1994–2000), *Statistical Yearbooks of Latvia, 1994–2000*, Central Statistical Bureau of Latvia, Riga.

SYLi (1995–2002), *Statistical Yearbooks of Lithuania*, 1994–2002, Lithuanian Department of Statistics/Statistics Lithuania, Vilnius.

Tamm, M. (2001), 'Agricultural Reform in Estonia,' in I. Alanen, J. Nikula, H. Põder and R. Ruutsoo (eds), *Decollectivisation, Destruction and Disillusionment – A Community Study in Southern Estonia*, Ashgate, Aldershot, pp. 407–438.

WB (1992), *Food and Agricultural Policy Reforms in the Former USSR. An Agenda for the Transition (Country Department III: Europe and Central Asia Region)*, Studies of Economies in Transformation, Paper Number 1, The World Bank, Washington, DC.

WB (1993b), *Estonia: The Transition to a Market Economy*, The World Bank, Washington, DC.

WB (1993c), *Latvia: The Transition to a Market Economy*, The World Bank, Washington, DC.

WB (1993d), *Lithuania: The Transition to a Market Economy*, The World Bank, Washington, DC.

WB (1996), *Staff Appraisal Report. Republic of Lithuania. Private Agriculture Development Project*, Report No. 14631-LT,
http://www-wds.worldbank.org/servlet/WDSContentServer/WDSP/IB/1996/03/07/000009265_3961019180755/Rendered/PDF/multi0page.pdf

WB (2000a), *Rural Development Strategy, Eastern Europe and Central Asia*, (Authors C. Csáki and L. Tuck), Official Documents, World Bank Technological Paper No. 484, The World Bank, Washington, DC.

WB (2000b), *Making Transition Work for Everyone: Poverty and Inequality in Europe and Central Asia*, The World Bank, Washington, DC.

WB/IBRD (2002), *Reaching the Rural Poor in the Eastern Europe & Central Asia, Rural Development Action Plan*, Environmentally and Socially Sustainable Development Unit, Europe and Central Asia Region, The International Bank for Reconstruction and Development, Washington, DC.

Wegren, S. K. (1991), 'Regional Differences in Private Plot Production and Marketing: Central Asia and the Baltics,' *Journal of Soviet Nationalities*, vol. 2, no. 1, pp. 118–138.

# Chapter 3

# Rural Communities in the Baltic States and Post-Communist Transition

### Rein Ruutsoo

## Introduction

This article examines the roots of the current strains of anomie in the societal structure of the Baltic States and seeks to analyse why the communities of the countryside, despite their critical role in the Baltic Revolutions of 1987–90 (sometimes called the Singing Revolution) appear to be the losers.

I have looked in detail at the development of Baltic civil societies in transition and have used the evidence of both quantitative and qualitative data in support of suppositions made. Qualitative data was largely collected by myself in interviews with country people in all three Baltic States.

What type of social order emerged from the 'revolutionized' societies of the Baltic States where there were profound breaks with the previous and radically different political and economic spheres? Traditionally two approaches have been used to address this issue: the economic and political.

The economic approach or 'the market solution?' This emphasizes reliance on the self-interest of all individuals to maintain public order. Order, in this approach, will arise naturally from a market economy.

The political approach. This emphasizes control and the ability of state power to enforce order.

Let us consider both approaches from the perspective of the Soviet societies of the former USSR. Soviet power in the Baltic States was repressive. The centralized command system supported an alien, colonising regime. Mechanisms to maintain order were not rational in the sense of modern western societies. In market economies individuals can satisfy self-interest by engaging in activities in the private domain. By contrast, in Soviet societies interests were largely pursued by cultivating ties with influential and powerful individuals rather that banding together and co-operating.

The fundamental illegitimacy of the Soviet system made any smooth transformation – property reforms, privatization, land reform a very complex task. It lacked the main foundation that could keep a society of disorganized people and collapsing systems together. This backbone was a civil sphere of shared basic norms and values (Lii, 1998, 115).

The rapid restructuring of the Baltic States, especially in agricultural production, is an interesting study in the transformation of rural communities and the shaping of modern societies. The emerging differences between the transformational trajectories of the Baltic States, which operated for fifty years on a common Soviet model is a very interesting object of study – a true laboratory of history.

The effect of the uniformity of the Soviet regime on the historically and culturally different Baltic States was deep. In terms of institutions the Baltic societies were, at the end of the 1980s, very much alike. But applying a totalitarian paradigm excludes the impact of 'society' and 'culture,' which are less accessible in any regime and much more resistant to transformation. The relationship between a regime and society cannot be viewed simply as political domination-subordination. This approach ignores regional and local differences and limits research of social organization in these societies.

The Baltic 'Revolutions' of 1987–1990 were crucial in shaping the organizational culture of transformation. They launched and sustained crucial economic and political processes that facilitated the transformation and consolidation of society. By transformation I mean the delegitimization of the Soviet regime. By consolidation I mean the creating of a nucleus of a new political community to facilitate the development of new social organizations and concepts of community.

There are a number of possible ways that societies may go when norms have broken down and they perceive a situation to be one of anomie: these are withdrawal, ritualism, rebellion, withdrawal and innovation (Merton, 1968, 230). All these types of reactions were observable in the Baltic countryside. There was a high crime rate, a high suicide rate, a high rate of alcoholism, and a high rate of rebellious and aggressive behaviour. Ritualism – an attempt to come to terms with change by means of symbolic practices – was also quite common during the Soviet period.

The tactics of transformation meant developing, initiating, organising and carrying out practical projects. This was not possible in the Communist state as people were prevented from developing organized interests in the public domain. Interest groups thus remained unofficial. The disintegration of the Soviet regime reduced control over people but is this enough for development of official organized interest?

The structure of the state, the nature and extent of its involvement in civic and corporate life and the organization of society are key factors in determining whether a country succeeds or fails in development. One possible conceptual base for this approach is the theory of social capital. Social capital is defined here as the stock of a society's productive assets, including those that allow the manufacture of marketable outputs that enable private profit and the stock of non-market outputs, including defence and education. The capacity of civil society for development is contained in four aspects of social capital (Woolcock, 1998), viz:

1. Integration at the micro level.
2. Linkage at the micro level.
3. Integrity at the macro level.
4. Synergy at the macro level.

All these four dimensions must be present for optimal outcome. By applying these performance indicators to transitional societies it is possible to measure progress. Successful interaction within and between bottom-up and top-down initiatives is the cumulative product of 'getting social relations right' and enabling sustainable growth.

Traditionally the qualitative aspect (hard indicators) – the number of associations or movements, the volume of membership of organizations, the frequency of participation, etc. is the accessible aspect of the social organization of society. For our study of exploring the different outcomes of transformation, – the observable differences in the performance of micro and macro factors in rural communities – are essential. Comparative data about this area, which includes the values, practices and public performance of Baltic societies at a community level, is quite rare. But, alongside, qualitative numerical data we have collected some quantitative observational information about important events, campaigns, the performance of institutions, the nature of attendance, etc. collected in interviews. They contribute to our task of mapping societal reactions and help us to flesh out trends reflected in general statistics. Public events, campaigns and rallies which flourished temporarily or became institutionalized and survived into the new social era – all tell about shared mentalities, collective memory, social bonds and reveal how people related to each other and their readiness to co-operate. All together they add up to important sociological resource (Morrison, 1998).

In all the Baltic States there is an observable contrast between political liberalization spearheaded by the 'Revolution' and the subsequent limited success of the nations to build balanced communities. There is a particularly large disparity between the contribution of the countryside to the Baltic revolutions and the current low influence of the voice of rural communities. Their empowerment, I believe, is crucial for general political and economic success.

**The Baltic Countryside on the Eve of the Collapse of the Soviet Regime**

To understand the difficulty of the social reconstitution of Soviet Baltic societies it is useful to do a recall their history and distinctive features.

The social reconstruction of Baltic societies after 1944 involved the subjection of individuals to a set of corporate norms as the base for social and cultural action and identity. This program was based on Stalinist mass terror. The forced collectivization of agriculture was just one aspect of the replacement of private property and an interactive society of individuals organized in groups (associations and collectives) according to interest. Collectivization was a direct attack on the social institutions of civic culture and the western cultural orientation of the Baltic nations.

Soviet structures introduced to replace banned 'bourgeois' associations, can only nominally be called 'civil society.' The organization of a society in which the state and society are related on the bases of effective, institutionalized citizenship differs from a society in which the mass of society is excluded from political recognition and participation, as was case with the Baltic States after occupation. Active and participative citizenship was undesirable in the Soviet regime. Bearing this in mind, the meaning of membership of associations in the Soviet Baltic States was quite different to those in the Nordic states (Siisäinen and Ruutsoo, 1996), even if membership levels were comparable. From the 1960s onwards, with de-fragmentation of the Soviet party-state, state-sponsored associations in the fields of culture, education and professionalism were able to take advantage of a relaxation of the system. They were able, despite repression, to develop a duplicitous nature and became a protest culture. During these years official associations started to operate as networks of opposition, and nurture a subculture of protest and dissent (Ruutsoo, 2002). In the years of the Baltic 'Revolutions' many of these duplicitous but formal organizations surfaced openly as leading centres of anti-Soviet and nationalist activity.

In 1988, on the eve of the 'Revolutions,' which brought to end Soviet regime in the Baltic States, there was a noticeable difference between the three Baltic States in terms of mobilization and participation of people in the associations of the voluntary sector, despite heavy standardization and state interference. Discussing these figures and data questionnaires in the light of empirical findings is one focus of this article.

**Table 3.1 Association membership in Estonia, Latvia and Lithuania in 1988[1]**

|           | N     | Capital City | | Other Towns | | Villages | |
|-----------|-------|-----|------|-----|------|-----|------|
|           |       | N   | %    | N   | %    | N   | %    |
| Estonia   | 1,493 | 454 | 41.0 | 629 | 45.4 | 414 | 50.7 |
| Latvia    | 1,636 | 567 | 25.9 | 639 | 31.0 | 430 | 35.3 |
| Lithuania | 1,485 | 250 | 33.6 | 760 | 34.5 | 475 | 30.3 |

*Source:* Baltic Report Data (1995).

As we can see, the biggest gap in association membership existed in the countryside. There were almost twice as many members of associations in the Estonian countryside than in Lithuania. It is hard to find a detailed explanation for these data. Accepting that traditional communities populated the countryside and collective farming was universal, country-specific and culture-specific explanations for these differences must be considered.

---

[1] All the figures in this section are based on a survey made in connection with the comparative research project 'Social Change in the Baltic and Nordic Countries' funded by NOS-S (1993–1994).

The level of general political mobilization in 1988 was considerably higher in Estonia than in Latvia and Lithuania. The usual reason given was the desire for restoration of historical tradition and pre-1940 style associations. Historically, the rural communities of Estonia, Latvia and Lithuania have very different political histories (Ruutsoo, 2002). Two of these Baltic nations – Estonia and Latvia – have a considerably longer modern civil tradition then Lithuania. As studies have indicated, despite the considerable impact of Sovietization on historical cultural (Raig, 1985, 101) the indigenous social practices of the Baltic nations remained strong. It is quite possible that in forty years of occupation each Baltic State developed its own type of civil culture, based on a fusion of traditional culture and Soviet standards (Siisäinen and Ruutsoo, 1996).

**Table 3.2  Memberships of associations in the Baltic States 1988 with ethnic breakdown**

| Memberships | Estonia | | | Latvia | | | Lithuania | | |
|---|---|---|---|---|---|---|---|---|---|
| | Estonian | Russian | Other | Latvian | Russsian | Other | Lithuanian. | Russian | Other |
| No | 51.5 | 61.5 | 53.3 | 65.2 | 76.0 | 72.4 | 66.9 | 64.7 | 70.5 |
| One | 26.8 | 29.3 | 10.0 | 23.1 | 19.4 | 19.7 | 25.1 | 25.0 | 21.2 |
| Two | 14.1 | 7.7 | 3.8 | 6.7 | 3.1 | 5.4 | 6.0 | 9.0 | 2.7 |
| Three | 5.0 | 1.0 | 1.9 | 2.6 | 0.6 | 0.5 | 1.4 | 1.3 | 5.5 |
| Four | 2.0 | – | 1.0 | 1.3 | 0.8 | 0.5 | 0.3 | – | – |
| More | 0.6 | 0.5 | – | 1.0 | 0.2 | 1.5 | 0.3 | – | – |

*Source:* Baltic Report Data (1995).

There are relatively small differences between levels of indigenous membership and non-indigenous minorities (mainly Russian speakers). This can be explained by the impact of Sovietization and the impact of the local (national) organizational culture.

In Latvia a large share of non-Latvians, whose association culture was limited largely to Soviet-modelled structures, very clearly shaped the civil culture. Compared to the other Baltic nations they had more contact with Russian speakers and more shared networks between the communities.

A research report from 1993 indicates that compared to the other Baltic nations and the bigger minorities in the Baltic States, the ethnic Estonians were considerably more 'socialized' in terms of the volume of association memberships. But in statistical terms the numbers of the Baltic nation active in associations was similar. An important finding is that in Estonia the bulk of people who had a membership of two to three associations were many times higher than in Latvia and Lithuania. For example: 14.1 per cent of Estonians 6.7 per cent of Latvians and 6.0 per cent of Lithuanians had membership of two associations. Examination of the figures for triple memberships shows 5.0 per cent Estonians 2.6 per cent of Latvians and 1.4 per cent of Lithuanians.

These differences do not yield enough evidence to talk about significant qualitative differences between the organized civil organizations of the Baltic States. But, the data indicate that in Estonia there had emerged a group of activists in the voluntary sector. In terms of social capital and resource mobilization the Estonia had, on the eve of the Singing Revolution a more developed network at its disposal. If we look for explanations on anti-regime mobilization, these differences must be considered.

## Protest and Self-Organization in the Baltic Countryside in 1989–1991, the Years of Anti-Soviet Nationalist Mobilization

The first phase of the Baltic transition, 1987–1990, was a 'mythological' period (Lauristin and Vihalemm, 1997). The first years of the awakening were characterized by a restoring of minimal trust between individuals, an increase of national self-confidence, and a commitment to societal issues.

The second phase of the Baltic revolutions was characterized by the emancipation of 'collective rationality.' This was the re-organising of people in terms of their every day and practical interests, both on a local and national level. Political acumen now became a central factor and took a key role in determining participation in politics. The next step towards the modernization of society was the organization of associations, which represented an organized interest. In this context 'stake holding' was a large problem. The critical issue was the creation of appropriate conditions for the emergence of modern political, social and cultural citizenship.

At the societal level citizenship surfaced as a problem in the context of restructuring old organizations and the building of a new collective consciousness. In the countryside it meant, first of all, redefinition by the people of their relations with kolkhoz/sovkhoz that were clearly outmoded institutions. In the sociological perspective kolkhozes/sovkhozes were not a legal entity but a social group with a social identity, and a social value, in relation to the actions of both individuals and the groups. The identity and status of the kolkhoz/sovkhoz was evoked through the use of symbolism – the name of the kolkhoz/sovkhoz, the name of the main building, addition to the roll of honour, state awards, monuments, the number of honorary workers, etc. The construction of the identity of the kolkhoz/sovkhoz had been one of the main tasks of the local CPSU cell.

Transition in the Baltic States was different to the neighbouring areas of Russia. In the States, the redefinition of relationships to the old kolkhoz/sovkhoz was impossible to separate from the restoration of property rights, and the declaration of Soviet 1940 citizenship as null and void. The old traditions of private farming used in re-defining kolkhoz/sovkhoz relationships in sociological terms also made an impact on the strategy of the reforms in the Baltic States.

The sociological element of the redefinition of social relations and 'kolkhoz/sovkhoz citizenship' became an essential part of reconstruction of the countryside. Transformation started with the identity, and the challenging of the 'substantive rationality' of the Soviet society. The basic Soviet mythologies (the successful building of communism, the consolidation of the brotherly Soviet nations, Russia as a bastion of socialism, etc.), forced onto the Baltic nations, were refuted. In a shift from one symbolic order and substantive rationality to another, nation-building projects became important step toward emancipation. Relatively well defined nation-building projects put the Baltic nations into a privileged position compared with the other former USSR nations, but, the Baltic nations had a longer way to go to being a western social democracy and procedural society than the old Soviet bloc Central European countries. Was it was possible to transform kolkhozes/sovkhozes, not only its methods of production with regard to a new forms of citizenship, new forms of political rationality (the nation state) and a new type of economic philosophy – a market economy. To my mind the conflict was not only legal, political or economic – a fundamental social displacement had to be tackled. A market economy is built on the different type of citizenship than a 'kolkhoz citizenship.'

What kind of identity was available in the countryside for social mobilization in late-communist society? Traditionally, class, profession, ethnicity and local connection were all considered likely bases for group identification. However, group identification in a class in its traditional sense was not possible. It was hard to differentiate the material interests of one group from other. The whole work-collective of a kolkhoz/sovkhoz, all groups that it was made up of, were subjected to the domination of the state. Thus the cleavages that were enforced were the Soviet state versus local society, and the workers versus local nomenklatura. While very obvious cleavages existed between the workers, specialists, administrators, party nomenklatura, they were not visibly reinforced by dominance of one group over another.

The period of the politics of values (articulated as a return to history, giving substantial meaning to the ethnic values, etc.) often produced in the camp of Western normative theorists a perception of a 'parochial turn' in the Baltic States. This was an unavoidable part of restoring some pre-Soviet identity and the USSR was perceived more and more as the main hindrance of development. This was a time when even limited economic reform projects such as economic autonomy or the internal market project were resented.

> The politics of remembering and forgetting different aspects of collective historical experience conducted by means of social symbolism is a form of social struggle in the field of culture (Halas, 2002, 116).

### 'Trouble with the Authorities' – An Indicator of What?

Discontent or trouble with the authorities, as it surfaced at village level had its target not so much the abstract system, Soviet power in general, but the performance of local government and the leadership of the paternalist kolkhoz system. Discontent and troubles with the authorities had in countryside a different pattern and configuration than they had in the towns.

Discontent is a negative side effect of any mobilization. It has not served the interest of scholars to see civil mobilization in the countryside as an important contribution to the Baltic revolutions. The countryside did not show up as an independent and original actor in the late 20th century revolutions as in 1905. But to what extent were rural communities activators or receivers of the revolution? To what extent were they able to become the authors of reforms? The experience of tensions and trouble does not necessarily turn a community into an insurgent movement. In order to determine what does necessitates a look at pre-conditions for self-examination and identity building.

Research reveals that popular mobilization in the different Baltic States gathered power at a different pace and civil activism surfaced at a different pitch. By 1988 the levels of discontent and the numbers of people who were in trouble with the authorities in the Baltic States reached a relatively high level. 12.4 per cent in Estonian 8.8 per cent of Latvians and 5.2 per cent of Lithuanians reported trouble with Soviet or party administration.

**Table 3.3 Reports of trouble with the authorities 1988**

| Reason | Capital | Other town | Village |
|---|---|---|---|
| ESTONIA | | | |
| Estonians | 16.8 | 13.6 | 8.6 |
| Russians | 3.8 | 5.6 | 17.9 |
| Others | 8.6 | 5.4 | 8.3 |
| LATVIA | | | |
| Latvians | 7.2 | 8.1 | 10.8 |
| Russian | 5.8 | 4.2 | 4.6 |
| Others | 9.1 | 8.5 | 7.1 |
| LITHUANIA | | | |
| Lithuanians | 11.3 | 5.2 | 3.3 |
| Russians | 1.7 | 5.0 | – |
| Others | 4.5 | – | 3.8 |

*Source:* Baltic Report Data (1995).

Trouble with the authorities in the first years of the unravelling of the communist system was reported as the most severe in Estonia. The amount of trouble with non-Estonians living in the countryside is an especially interesting factor – the number was rather small. The findings, in general, give a better understanding of the discontent which fuelled the civil mobilization that surfaced in Estonia a year or two earlier than in Latvia and Lithuania. A higher amount of troubles was reported in the Estonian and Latvian countryside.

Both in Lithuania and Estonia largest numbers of trouble with the authorities were reported in the capitals. The majority of anti-regime movements also surfaced in capitals. A lower level of conflict in Riga reveals a national oppression that has a lot to do with the ethnic composition of town. Riga was a heavily Russified town and the headquarters of the Soviet military power in the Baltic States. There were few towns in Latvia where the Latvians remained a majority. In Estonia ethnic Estonians were in the majority in the smaller towns. These towns were also regional rural centres and played an important role in civil mobilization. The large contribution of the smaller towns to social mobilization makes the map of discontent and the social configuration of the Estonian awakening different from that of Latvia and Lithuania. The Estonian towns of Võru and Pärnu had, by the autumn of 1987, been named 'frontline towns' (Ruutsoo, 2002). Our findings give grounds to speak about a geographically more balanced civil mobilization in Estonia and Latvia compared to Lithuania. These two Baltic States include a broad spectrum of society and both rural and urban communities. In respect of Lithuania – there are grounds to believe that the countryside there, with its relatively higher share of discontented people, became an important stronghold of political and national mobilization.

Areas of conflict with the authorities (as they were defined by the respondents) have, in the different Baltic countries, different configurations.

**Table 3.4 Reasons for trouble with the authorities**

|  | Capital | | Other towns | | Village | |
|---|---|---|---|---|---|---|
| ESTONIA, reason: | 1988 | 1993 | 1988 | 1993 | 1988 | 1993 |
| Political | 40.8 | 18.2 | 44.6 | 14.3 | 44.7 | 34.6 |
| Ethnic | 44.9 | 70.5 | 40.0 | 71.4 | 39.5 | 50.0 |
| Religious | – | – | 3.1 | 5.7 | 2.6 | – |
| Other | 14.3 | 11.4 | 12.3 | 8.6 | 13.2 | 15.4 |
| LATVIA, reason | | | | | | |
| Political | 59.5 | 30.0 | 59.6 | 26.3 | 52.4 | 31.6 |
| Ethnic | 40.5 | 67.5 | 36.2 | 60.5 | 38.1 | 68.4 |
| Religious | – | – | 2.1 | 5.3 | 9.5 | – |
| Other | – | 2.5 | 2.1 | 7.9 | – | – |
| LITHUANIA, reason | | | | | | |
| Political | 73.7 | 22.2 | 57.5 | 29.4 | 55.6 | 58.3 |
| Ethnic | 5.3 | 66.7 | 7.5 | 41.2 | 16.7 | 41.7 |
| Religious | 10.5 | 11.1 | 35.0 | 23.5 | 27.8 | – |
| Other | 2.1 | – | – | 1 | – | – |

*Source:* Baltic Report Data (1995).

Analysis of the 1988 data reveals that, in Lithuania, trouble was interpreted by the respondents in town and village predominantly as political conflict i.e. a problem related to the nature of the Soviet regime. In terms of the definition of trouble in the towns and villages of Estonia and Latvia there was relatively little difference. Despite the heated debates on language rights and legitimacy of 'immigration,' the main conflict area was not defined as ethno-cultural.

Unfortunately no question indicates an economic dimension of tensions. This possible source of trouble was not included into our 1993 questionnaire. Estonians also indicated quite a lot of other reasons. This signals that there were more reasons for troubles than the three listed specifically in the table above.

The debate on the contentious issue of the privatization of farming began in 22 March 1988, when the resolution on Individual Labour Activity in Agriculture was adopted in Estonia. The first owner-operated farms were established in 1988. By the end of 1989, 828 farms had been registered (Maide, 2000). In Lithuania the law that allowed the establishment of owner-operated farms was promulgated in September 1988. In Latvia a similar law was passed in 1989.

The number of Estonians who reported co-operational activities was twice as high in Estonia as in Latvia and Lithuania – 4.4 per cent in Estonia as opposed to 2.0 per cent and 2.1 per cent in the neighbouring countries. The more dynamic economic policy that was exercized in Estonia was one of the sources of tension. Interviews revealed that no local kolkhoz/sovkhoz administrations in any of the three Baltic States was in favour of the development of private farming not to mention privatization of land. But the conflict configuration was different. In Estonia and Latvia reports were dominated by tensions between the administration and new farmers who applied for land.

> I already in 1988 got land and when Sajūdis started, in 1989, I was the first member of local Sajūdis cell ... so; I became a national enemy for the local authorities, because I was the first farmer and the first Sajūdis member here. ... If we did not run into a conflict we would not started these activities. And I would not start private farming. When we started with Sajūdis activities and organized some meetings, local people supported us. In every meeting about 20 people used to come (A private farmer and former Sajūdis activist Saločiai, Pasvalys district, Lithuania).

There is, however, less evidence that private farming raised tensions between people inside the village community, as happened in Russia where tensions very possibly, were founded in the agrarian history of that country. They indicate a more profound integration of community, a deeper merging of the farm as a worker collective.

Only in Lithuania, compared to its neighbours, was the perception of tensions predominantly political. This bias had its source, probably, in the more *elitist* type of political identity of the Lithuanians. The Lithuanians perceived themselves first of all as citizens with rights and not so much as members of a cultural or linguistic community. This latter type of identity construction was more typical to the Estonians and Latvians.

An important role in the ethnic aspect of the conflicts was played by the size of the Russian speaking communities. Estonians and Latvian, with large Russian minorities living amongst them, perceived immigrants as a 'civil garrison,' which endangered their national existence. The most dramatic effect was in Latvia, where 52 per cent of the population was Russian, but, also, in Estonia a large share of immigrants gave prominence to ethnicity as a source of tension.

Religion played a more significant role in the Lithuanian transition than in Latvia or Estonia. As expected, the conservative smaller towns and countryside reported more numerous troubles related to religion. A lower level of political mobilization in the Lithuanian countryside illustrates, perhaps, the deeper alienation from society in the Lithuanian village. However, viewed in the light of a long and bloody guerrilla war of 1944–1953, the level of tension in Lithuania was low. One possible explanation for this is that a Soviet model of patriarchal domination (kolkhoz) fitted more with the historical Gemeinschaft type organization of Lithuanian society. The mentality of belonging to 'the same family,' as expressed by a Russian born chairman, was more deeply embedded in the civil culture of Lithuanians. But a relatively lower level of discontent may have also been another explanation. In more collectivist Lithuania troubles were shaped into more general ideologies, debates were delegated upwards and were not much personalized. But whatever explanation we apply it remains a fact that in 1988 the Lithuanian village and collective production were much more consolidated and the community more integrated than in Estonian and Latvia. In the same way we can say that in terms of conflict the patterns of the Lithuanian countryside were, in terms of societal mobilization, more stabile by the late 1980s.

Even more surprising than a relative Lithuanian silence in 1988, are the findings that in 1993 the average level of tension (in terms of reports of trouble with the authorities) was many times higher in Estonia than in Latvia and Lithuania. In Estonia 9.2 per cent of respondents reported trouble with the authorities, as opposed to only 5.3 and 2.3 per cent in Latvia and Lithuania. But, generally, a shift was taking place.

The main area of tension by 1993 was ethnic relations – the same was noted in all the Baltic States. The common historical heritage of large numbers of non-citizens played an important role here and tension was put down to political factors. The least tense atmosphere in 1993 was in Lithuania when the Lithuanian village looked to be the most idyllic place in the Baltic States. Only 1.9 per cent of people complained about any trouble with the authorities.

The Soviet collective farm model was universal in the economic-administrative structure of the Baltic States. But it is also a certainty that the application of the kolkhoz or sovkhoz model in Estonia, Latvia and Lithuania were not the same entities in socio-cultural terms. This was the situation all over the USSR – the standard system of collectivization was enacted in different historical frameworks. In Estonia and Latvia collective farms replaced market oriented and, in technological terms, quite developed private farming. In Estonia and Latvia the technological level of agricultural production in an Eastern European context was high. The number of lorries, tractors, harvesters, etc. was, in Estonia and Latvia, four or five times more than in Lithuania (Silberg, 2003, 53–54). Forced collectivization was much less relevant to modernization here.

Despite significant development in private farming in Lithuania during the period of the first independence the values of modern citizenship and individual production did not take a deep hold among the peasantry (Ambrazevičius, 1990). Therefore, it could be expected that peasants forced into kolkhozes would find much of its organization and values, to a large extent, 'natural.' The kolkhoz work collective was a Soviet substitute for the traditional rural familial structure – the fundamental base of organized agricultural production (Jowitt, 1992). This principle had more parallels in the agricultural traditions of Lithuania than in Latvia and Estonia. This emphasis refers not only to the long history of collectivism inherited from past centuries (Zvinkliene, 1995). Collective farming is also somewhat similar to the co-operative model of organization of agricultural production exercized in president Antanas Smetona's Lithuania, in the first period of independence 1918–1940.

Lithuanian economic policy of 1918–40 was developed along the lines of entrepreneurial capitalism with a strong element of political capitalism (Weber, 1958). President Smetona created an 'organic state' containing and an 'association of associations.' In this state the administrative hierarchy played an important role in business organization. Production was affected by the arbitrariness of political leadership, which largely sponsored agricultural state-sponsored co-operatives. It gave associations of producers' privileges in many respects comparable to individual entrepreneurs. Romantic historical perceptions of the relationship between the Lithuanians with their country and the land ('boden') facilitated a collectivist perception of reality. This romantic attitude also existed to some extent in Estonia and Latvia and politicians exploited it, but a more individualist approach made politics here more pragmatic.

## Collective Farms and the Politics of Identity

In very general terms it could be said that the relative success of the kolkhoz/sovkhoz system was why the growth of articulation of discontent in the countryside lagged behind the towns in all three Baltic States and why political development was slowest there on a national level. It should be remembered that after the Soviet annexation of 1944 the peasantry held out for decades and was a stronghold of armed and mental resistance.

In all the Baltic States there are reports (collected by us) of conflicts in the 1970s and 1980s between the local kolkhoz/sovkhoz administration and individual workers. There were various reasons for this conflict but very often the reason was related to second economy issues. The relaxing of the collective system in the late 1980s saw old resentments surfacing and conflicts fuelled. This led to criticism and instigated more general political debates. Our interviews show that all over the Baltic States the leading opposition figures in the village, the founders of new independent associations or activists in political movements had for different reasons, political, ideological, economical, personal, grievances with the administration or party-officials. But these tensions rarely gave a birth to a larger debate.

The collective farms had changed by the1980s. They were not the same work collective entities as originally intended. The conception of a collective farm as an ideological organization had lost its meaning. The main features of the collective were more structural than ideological in their nature. As a result of modernization accounting for individual contributions had became a central measure of pay. The emphasis on personal responsibility, the use of collective property (machinery, vehicles) as personal property conflicted with the very idea of the nationalization of property and the land. Individual households, especially in Lithuania, now produced a significant share of agricultural production in private.

In terms of opening up the internal conflict within the system, the years between 1988 and 1991 were crucial. As well as conflicts surfacing in the field of political interests, conflicts between supporters of collective vs. private production and changes in social identity were factors in mobilization. Mobilization developed as identity mobilization – group boundaries and consolidation of communities were drawn up along political and ethnic lines. National level identity politics was only partially relevant to local problems. The in-house constitution of the local community had little relevance to national elite politics. The countryside Soviet elite (chairs and top specialists) started to mobilize. There emerged in all the Baltic States groups of countryside people, which united into a rural administrative elite. In this sense a kolkhoz/sovkhoz model countryside became the equivalent of an anti-communist town elite [a *'Vendee'*]. The countryside was an important source of pro-independence mobilization in the nations. Identity politics as anti-Soviet, pro-independence mobilization became a relevant issue, defining and legitimising a radical change in land reforms and the restitution of private farming.

The years 1988–1990 were the period of 'movement societies.' The most essential role in the political re-structuring of society was played by the 'Popular Fronts.' The Estonian Popular Front (Eestimaa Rahvarinne), the Latvian Popular Front (Latvijas Tautas Fronte) and in Lithuania 'Lietuvos Persitvarkymo Sajudis' (Sajudis is the Lithuanian for 'Movement.' Generally the names of these movements are usually translated into English as 'Popular Fronts' but in Latvian the word 'tauta' has a very different implication than the Estonian 'rahvas.' The Latvian word is close to the German 'Volk' (Volksgemeinschaft) but the Estonian word is much closer to the Finnish 'rahvas' (and to Gesellschaft). A more accurate translation of Latvian 'Tautas Fronte' would 'National Front' rather than 'Popular Front.'

A lesser known, but very important role in popular mobilization movements was played by the likes of the Estonian Congress and the Latvian Congress. In Estonia the Heritage Protection Society played a unique role. In terms of legitimacy i.e. the ability to organize voluntarily private resources for official or public purposes, all these movements were highly legitimate at a national level. There is, however, very little information about their role in the transformation of the countryside. This last issue is another focus of this article.

## The Social History of the Collective Farms in the Baltic States

Modern history states categorically that collective farming ceased to exist in the Baltic States as a result of the official politics of nationalist rightist governments. Basically this is correct. A top-level political decision was made to stop collective framing and to disband kolkhozes/sovkhozes. But the situation becomes much more complicated if we want to answer the questions Why did it turn out this way?

Why was the kolkhoz/sovkhoz system destroyed when there were no alternatives available?

Research on the social and political history of the kolkhoz system in the Baltic States is in its infancy. In the Soviet years kolkhozes passed through so many stages that there sociological meanings became quite different. In terms of the Soviet system the kolkhoz had a very important institutional political-ideological meaning. In reality, the village Soviet (a substitute for local government) was an administrative unit without resources and this made the collective farm, the kolkhoz, the major institution in the countryside. It was the provider for local public space and educational, cultural and social institutions. Collective or state farms were the mainstay of society in the countryside. This kind of work organization broke down, according to official ideology, the gap between the world of work (private) and the public world found in the free market economies. Collective farms did more than just administer links between individual producers. The collective farm was a shared social world where private became public and public became private. To a certain extent this project was realized in the Baltic States especially in those areas with mixed population with leaders who had Slavonic cultural tendencies.

> There were Latvians and Russians, Ukrainians and Belo Russians in our kolkhoz. And we were like a big family – we were friendly, we helped each other and we also tolerated each other's faults ... as I said we lived like a big family. One of my neighbours had a birthday. There was a canteen in the kolkhoz. I went with a horse and carriage to his home and brought him here [into the kolkhoz centre]. All the kolkhoz people gathered here and we celebrated his birthday (Former kolkhoz chairman, Iecava, Latvia).

The kolkhoz naturally also paid the main costs of this birthday party. This story was told by the old chairman – the patriarch of the kolkhoz and a Latvian born in Russia. His family was oppressed, he himself was persecuted, but he remained loyal to the Soviet regime. He was in favour in principle of the kolkhoz system and he was positive about it in his interview. By the mid-1980s the picture was very different, but the disintegration of the collective farm had began much earlier. The old man chronicled some turning points in the history of the kolkhoz system, and the disintegration of the work collective.

> I started my work as carrier here in [Iecava] in 1946. I remember that at that time we were still held midsummer ceremonies. I remember that we were going around the fields, singing songs and checking if the fields are cultivated well and clean. Around the 60s the ceremony was banned [by CPSU] and I cannot understand why. Because it seemed to me that there was nothing wrong to have that fete. People came together just to enjoy their work, the summer and the result of their efforts. But later we began to come together again (Former kolkhoz chairman, Iecava, Latvia).

There are some grounds to state that in the 1960s and 1970s the collective farms reached a breaking point. They had become too autonomous in many senses in all the Baltic States. The process happened first in Estonia when the collective farms became comparatively rich and their chairs looked for the more important social role in 'their country.' They started to sponsor musicians and writers, hire architects and artists. Prizes on the best poem on the rural life, the best novel propagating ideas of protecting nature, the best song in Estonian, etc. were supported by the collective farms. Many of these cultural innovations, which challenged the monopoly of CPSU as a master of minds, were soon put on the ice. The giving of cultural prizes by kolkhozes was sometimes stopped by secret order, or their status was brought down to a regional level. The problem was, that these prizes whose winners were selected by special boards of the kolkhozes, with co-opted literary men and women, were prized more than Party sponsored titles and awards, which served friendship of the Soviet nations, achievements in building of communism, etc.

In the 1960s a practice emerged of 'collective farm days,' which actually served as a cover for traditional 'county or village days,' these boosted nationalist feelings, and encouraged local patriotism and ethnic solidarity. The Party frowned upon these popular movements that were sponsored by the kolkhozes in the 1960s and 1970s. It is even possible to say that the Soviet central administration started to sabotage revitalization of these important traditions of communal life because they became afraid of a new leadership at republic level and development of local solidarity.

The collective farm in the late Soviet period focused primarily on economic capacity. The social dimension of collectivization was subordinated to the economic. The doctrine was the bigger the better. The methods used were industrialization and the concentration of agricultural production. Small collective farms, usually centred on some villages, were swallowed into huge state-administered farms, turning the people from collective owners to just hired workers. This had a devastating effect on the original vision, or rather, illusion, of collective farming and the work collective. The impersonality of social relations characteristic of state farms, the increasing share of the workforce recruited from towns, the restructuring of patriarchal communities to 'paternalism' and the emergence of a technocratic elite destroyed the social fabric of collective farms. The old style collective farms had utilized deep historical bonds and appealed to collective memory and shared history. With the introduction of new working practices and social structure the kolkhozes started to fall apart.

As a result of forced mergers into huge industrial units instead of collective farms new leaders often emerged that did not care about social events and the tradition of sponsorship almost faded out in the 1970s (Ruutsoo, 2002). The supplementing of household economies from collective supplies and resources encouraged a policy of privacy in which the family and personal interests are emphasized at the expense of regime and societal interests.

In the second half of the 1980s the rise of such movements as the 'village movement,' which targeted restoration and reconstruction of local identity, were symptomatic of this disintegration. This kind of movement has not been the subject of much study but we have the an impression that they were more popular in Estonia where there are reports about 'village days' or 'county days' from very different places, than in Latvia and Lithuania.

Three main factors facilitated the revitalization of local traditions.

A romantic-nostalgic vision about a return to history and a pretext for public glorification of the period of first independence in contrast to the reality of everyday life in the kolkhoz.

The alienation and confusion resulting from the disintegration of the traditional home – people looked for more an integrated identity and social comfort – the lost paradise.

The support of the collective farm leadership for this kind of initiative. Kolkhozes financed these events. This shows the leadership using tradition to try to re-integrate the work-collective.

> The last chairman of the kolkhoz always participated in all the events launched in the years of the awakening period, when the national flag was taken out along with the others. We organized several meetings and singing events and the kolkhoz leaders always took part in them. I think that the last director wanted to show that he supported such activities (Agricultural specialist, Aizkalne, Latvia).

Our team studied some very impressive local festivals advertized as the 'parish days.' The Kanepi parish festival had a long historical tradition dating from the years of the first independence and beyond. During 1988, 1989 and 1990, the peak years of the singing revolution, the 'Days of Kanepi' attracted thousands of people – almost everybody living in the county and around. Choirs, orchestras, theatre plays, public discussion all contributed in different ways to the re-integration of the community. The nationwide political message of the 'Days of Kanepi' was unequivocal. It was a symbolic cultural demonstration supporting the restitution of national independence. Its local social function was to contribute to the restoration of the community. At the same time organizers demonstrated an excellent example of political mobilization through a valuation of culture and local history (Alanen et al., 2001).

There is less in-depth information about such initiatives in Latvia and Lithuania. Differences in national traditions also timelines had an effect. Political mobilization had its inner logic. The heyday of the village movement in Estonia was in 1988 and 1989 when there was still hope for change and the socialist economy had not yet started to fall apart. Funds were still available for cultural or political activism.

The content of the festivals began to change. In 1990 it became evident that the 'Days' lacked the earlier politically mobilizing role. Instead an economics conference was held, dedicated to the future of Kanepi. There were already clear signs of economic disintegration. The organization of the next festival was postponed and the political debates and identity consciousness raising, which had bees so much a part of the success of the 'Days' had been supplanted. The economic situation got worse and the singing revolution seemed likely to turn into a singing occupation. Instead of a common expression of general collective interests, people set about searching for a strategy of individual adjustment in conditions of increasing social-economic retraction.

The life cycle of the local festivals reproduced the more general dynamic of the national awakening. Step by step local 'identity politics' was replaced with a more pragmatic questioning about the future of communities and a search for practical solutions. The contribution of village and community movements was two-fold. To a certain extent it helped to stabilize society. The social sphere was injected with new impulses. The re-vitalizing of the social sphere created a shared living space in which the ideas and feelings of individuals were articulated. A sense of community and a cohesive societal fabric was developed. This social fabric in turn created an ethical dimension. This is essential for the maintenance of coherent public life.

In addition, the local movements had a huge impact on politics. Having censored national characteristics out of culture the Soviets had, in actual fact, created a counter-culture. This nationalist counter culture developed into a political anti-Soviet program. Thus the politicized cultural festivals organically fused into the political mobilization of the anti-Soviet revolution.

## Local Historical Memory – Constructed and Re-constructed

Effective civil activism and political mobilization has as its two main sources identity mobilization and mobilization of organizational resources. Identity-mobilization meant looking for shared ideas, collective memories, symbolic capital, which could serve for collective mobilization and the building of general future projects.

The first independence period of the Baltic nations was quite successful in managing history through symbolic objectivism of its achievements. The same aim was a goal of the Soviet regime. In the years of occupation Baltic history was totally falsified and too much larger extent than in the days of the Russian Empire. The history of the independent of Baltic nations was almost totally eradicated. A return to history, a return to 'roots' played a huge role in the Baltic revolutions – a much bigger role than in the East-Central European countries. A focusing on identity mobilization instead of resource mobilization was not a problem of parochialism of the Baltic nations but reflected a lack of organizational and practical resources to be mobilized.

There are reports that after the grip of the KGB relaxed during perestroika nationalist activists, who devoted themselves to 'spreading the truth' about history began to emerge in the countryside.

There was a 'Helsinki group.' Their leader was Vidinis. This was the leading [nationalist] association in the 1980s. They published some newspapers. I was studying in Jelgava when they were travelling around Latvia and giving lectures. They were not afraid to give lectures on history (Former agricultural specialist, Aizkalne, Latvia).

Specific to Estonia and without any parallel in other Baltic States was the surfacing of the heritage protection movement in 1986, which developed into a quite large network of local societies in the countryside. The total membership of the Estonian Heritage Protection Society (EHS) in 1990 was circa 10,000. Along with other movements it tried to mobilize people first of all around national cause and political identity. The large network of the society contributed to local politics at a parish and collective farm level. EHS focused on the collection of historical memories of the recent past, primarily the period of first independence and the crimes of the Soviets against humanity in the countryside.

The EHS ranks as one of the early risers, one of the first nationwide movements to inform the nation about the intentions of *perestroika* i.e. the conditions for political mobilization in Estonia the wider Soviet Union in 1986–1987.

> I learned, via the grapevine, that Mart Laar from the EHS intended to interview me, in order to record my GULAG experience. I was warned in the District Committee to hold my tongue. In a few days, there was Mart Laar in my study with his tape recorder ... I remember that Mart Laar spoke in 1986 or 1987 [it was 1987] in the secondary school of Kanepi, to introduce the activity of the Estonian Heritage Society, and also deliberate on the future of Estonia. It was a very impressive event. They were planning to come with a larger group to put the graveyard in order; however the meaning was to stir up the local population. It was not just a protection of past heritage, but also a political movement. The Society was the shield, the target being the repudiation of the authorities and the goading of the people in the direction of that line of thinking (Then public service officer of the village Soviet of Kanepi, Estonia. In 1992 Mart Laar became the first Prime Minister of newly independent Estonia.)

The organizational workings of society demonstrate the importance of linkages to neighbouring associations and nation wide cooperation (Ruutsoo, 2002). The contribution of activists of the movement – the 'missionaries from HQ'– was enormous for the effective launching of civil activism in Estonia. Country people who could not relay on the support and protection of academic organizations or trade unions were seriously in need of encouragement and support. The residents of Kanepi – among them veterans of the War of Independence, gathered after encouragement at the House of Culture (the Soviet civic centre), to set up the local branch of EHS. Houses of Culture were nationalized Civic Centres. These civic centres were, as was the case in Kanepi, built in the years of independence or even earlier, by the collective efforts of the people of the county. These historical civil society based community centres were typical in Estonian and Latvian rural areas but very rare in Lithuania. In the late 1980s not all in who came to the meetings dared write their signatures in an association charter as was typical at the time. The setting up of the local section of the Heritage Protection Society of Kanepi in June 1988 is typical of procedure adopted all over Estonia.

There is no information about an EHS style movement either in Latvia and Lithuania but the restoration of history played a similar role in the framing of national consciousness in these countries. The general assumption that the identity building of Lithuania was based on long-term history and not so much in the period of the first independence was also revealed in our study. But the difference was not only in a perception of a long term glorious past but also that the period of the first national independence had a different impact on national memory. Along many factors, a positive image of the period of the first national republic was much more effectively used in political framing in Estonia. The authoritarian regime of the first Estonian Republic was milder than in the other Baltic States and the economic prosperity of the countryside was relatively high. In the summer time thousands of Polish and Lithuanian field-workers had to Estonia and Latvia for fieldwork. In Estonia, where the independence period was heavily idealized, the project of restitution meant for many people literally a return to the past. It included the restoration of the privatized farm system, as it had existed before the war. The 'Restitution-fundamentalist' project – i.e. the restoration of the destroyed historical state and social system (civil societies, cultural networks, etc.) to exactly as it was, was much weaker in Latvia and in Lithuania. Sometimes the image of the past was even negative.

We did not want bourgeois Lithuania. I just wanted people to have more freedom (Private farmer and former Sajūdis activist in Saločiai, Pasvalys district, Lithuania).

All the people thought that it would be better than in Smetono's Lithuania. We had hope for an independent Lithuania, which would not belong to any military organizations. There was little talk about Smetono' s times. Some people said that it was good in Smetono's Lithuania; others were of a mind that this time was bad. It depended on their parent's standard of living. But my parents were dead and nobody told me about the past. Something always goes wrong after a coup, [at 1926 there was a coup in Lithuania], but now we hope for a better life (A geography teacher, Saločiai, Pasvalys district, Lithuania).

There were different moods. Those who had lived during the first independent Latvia remembered those times. But they had a twofold attitude. They remembered the difficulties as well as the good things of that time – about hard work and that there were poor and rich people. We were not ready for that because during the Soviet time as if we were all equal (Agricultural specialist, Aizkalne, Latvia).

So, as we can see – the kolkhoz system and the equalising of living conditions in the Soviet time had a more positive impact in Lithuania. This impression is supported by other studies (Rose, 1997). One of the nationalist-minded Lithuanian respondents even used the term 'fascist' when describing the regime of Smetona in Lithuania. Differences in the popular rhetoric in Lithuania and her Baltic neighbours when recounting the past are obvious. This was a kind of 'bourgeois Estonia,' that was expected to be restored by many people in Estonia. The word 'bourgeois' was almost dropped from public discourse, because Soviet propaganda had made the adjective 'bourgeois' a dirty word. In this context it must be stressed that both the quoted Lithuanian respondents represented conservative and nationalist-minded people. In Lithuania there was clearly a conflict – how to match the restoration of statehood and the restoration of property as it existed before the Soviet annexation. The idea of the restoration of national independence had important issues that could be addressed by international law. But to have a socially deeply divided nation was not an aim of a modern nation-building project. The same problem was, in principle, on the horizon in the other two Baltic States. People expected to have both economic freedom and social justice. In very general terms, it was a long time since people had been able to examine the past in order to use history as a tool to take a perspective of the future.

Recovering even basic facts in the historical memory was a big challenge for many people. There were many popular utopian elements that are typical in times of rapid social change. This was shown when the Baltic peoples started to mobilize their history in fight over independence.

On 23 August 1989, the 'Baltic Chain' united 1.5 million people – 25–30 per cent of the whole of the native population of the three republics – into a human chain from Vilnius to Tallinn. The Chain indicated in great depth the collective memory of all Baltic nations. The chain was intended to commemorate and remind the Baltic peoples and the world, about lost independence and about the Molotov–Ribbentrop pact which had been signed by Germany and Russia. The pact laid out the terms of the division of Estonia, Latvia and Estonia between the two powers and led to the annexation of the Baltic nations from Western Europe for fifty years.

With the organization of the Chain the countryside became, for the first time, an important player. The initiators of this historical event – the Lithuanian Sajūdis and the Estonian Popular Front – did the main organizational work, mapping out the route and arranging places on the road for groups who had declared their desire to participate. At the same it was revealed, that for the younger generation of the Baltic peoples it was difficult to realize the historical meaning this event. Across much of the Baltic States, especially in peripheral areas, where several generations had grown up as Soviet citizens, much had been forgotten. Many respondents from all the Baltic countries told us that they had not told their children about the family history, sometimes deeply scarred with deportations and atrocities, in order not to safeguard the future of the children, who might tell these stories in schools, where the staff was infiltrated by the KGB.

> Many of young people didn't understand what was going on. They just knew that they had to stand there [in the Chain]. Everybody went there and we had to go there too. That is why we went there. We didn't really know what was going on. We just knew that some wrong had been done to us in our history and something about a lie was related to this. Only after did we start to understand. My generation and those who were younger were not aware of it because there was no patriotism (The head of Regional History Museum, Kandava, Latvia).

This generalization reflects the difference between the state of mind of the Estonians and Latvians. In Lithuania the surfacing of the awakening was delayed but the extent of the nationalist anti-regime movements compensated rapidly for this. The organizers of the Baltic Chain were fully aware that heavily Sovietized Latvia was the 'weakest link' in the 'Baltic chain.' The success of Latvia, in bringing a heavily oppressed nation into the company of its Baltic neighbours was one of the greatest achievements of the whole event.

## CPSU and the Churches – A Clash of Identity

Interviews with Baltic country people suggest that membership of an association in the Soviet time had a very different social meaning in terms of loyalties, trade union/professional banding together (organizational interest) and network building. A membership in the loose networks of culture, sport or other leisure associations had much less social meaning than membership of CPSU or the Church. In the initial stage of the awakening (anti-Soviet mobilization) a change in the status and influence of these pillars of the community was critical. In the countryside it defined changes in civil identity and the configuration of political activism. In the towns, starting in the capitals of the Baltic States, the new political hegemony consolidated around intellectuals and their associations – the artists unions, academic association, journalist associations, etc.

The starting positions of the CPSU and the Church were, for the outside observer, heavily in favour of the CPSU. The CPSU was not a voluntary association – it was part of an international corporate organization. As such, it operated as the local infrastructure of central government, taking its orders from Moscow. It had an imperial ethos – closed, dogmatic and centralized. In the annexed Baltic States the 'national' church was more a kind of association than a public institution, as is the role of the Church in traditional Nordic countries. At the turn of the 1970s and 1980s, during the 'freeze,' the church was repressed and operated in an extremely limited capacity. It was handicapped by very high taxes and was not allowed any public visibility, let alone being allowed to perform the functions of a public institution. The attitude to the church among the common people was ambiguous. In Estonia and Latvia the Church was a publicly stigmatized institution with a small formal membership. It possessed little national-historical symbolic capital. Given its relatively low popularity the mobilization around the Church in Estonia and Latvia in the development Baltic Revolutions is amazing.

## National Branches of CPSU in the Baltic States – Losing a Grip

The regional organizations of the CPSU in Estonia, Latvia and Lithuania were the only nation wide institutions, which had branches and cells in the countryside. The latter included regular meetings and business discussions. According to data, an average 5–7 per cent of residents of the Baltic States between the ages of 16–60 were members of the CPSU. It was to be expected that Party preferences would be important factors in determining the winners and losers of economic reform and privatization. In this perspective not only political loyalty but also practical interests encouraged CPSU organizations to be active. In the early stages of the Baltic Revolutions there was little to indicate that the communist party would not, for the foreseeable future, be the only organized social power in the region.

Despite the same All-Union programme, a strict control on business, the same regulations and similar levels of party membership in all the Baltic counties, there was a variance in status of the CPSU organizations in the individual Baltic States.

The status of the Lithuanian CPSU was, for various reasons, higher than in Latvia and Estonia. Problems of the Soviet regime in the countryside included: how successful its administration was in establishing authority and a charismatic type of communist framework at all levels and sectors of community – in farms, departments and sections of farms. Effective domination of the CPSU required a type of societal organization, which can best be described as a 'neo-corporate' with the enthusiastic party elite integrated into the organization.

Interview with the people in the countryside who had been party-workers (apparachiks) and with CPSU rank and file members told us about the deep undermining of party ideas that has been called the 'routinization of charismatic organization' (Jowitt, 1992, 39). In all of the Baltic States the CPSU applied 'party–familial' concepts that profoundly eroded the authority of the organization. Typical to the countryside was a cadre-problem. On the periphery of the Baltic States party schools were not able to recruit people to train as party secretaries. CPSU has lost its charisma. Recruitment into the party rank and file members had long ceased to be a matter of convictions or commitment.

> My task was to work with people, organizing cultural events and working with public organizations. I belonged to the leadership of the kolkhoz. If there emerged problems with economy, we used to discuss them together. When I became secretary of the [rayon], party committee I was not subordinate any more to the kolkhoz chairman. We had different opinions but we always found consensus. As in every field, the local organization of the CPSU had a plan in respect of recruitment of new members to the CPSU. It was quite a big organization with 84 members. We had to affiliate 4 new candidates for CPSU membership every year. If I couldn't find appropriate people, and the local politburo found, there was punishment for bad work. To recruit somebody into the party was a hard task. I had to offer something practical to these people. The recruitment plan required a mechanic or agriculturist worker should join. I spoke with the chairman to find out who were the young, good workers. At that time apartment blocks were under construction and new technical equipment was being bought for the collective farm. There was a good mechanic and I had to offer him a new car or combine. A new apartment or a combine harvester for CPSU membership was a good offer. Another example, my neighbour got married and they and their two children were living a two-room flat. In the kolkhoz there were 18 combine harvesters and there was a need for a combine-driver. A neighbour replied that he would not apply for CPSU membership because he had rough living conditions, but finally he agreed and got a new combine and moved into a new apartment. You could, by way of punishment for failure to recruit be refused salary level 13 at the end of the year (Former CPSU secretary Saločai, Pavalio district, Lithuania).

The party of Lenin had lost its essential quality – organizational integrity: the ability to sustain a team spirit among political officeholders who act as disciplined, deployable agents (Jowitt, 1992, 121). The leadership of the CPSU had become subject to cynical manipulation, and the authority of party declined. Party apparachiks were, in the main, tolerated, according to word of an Estonia collective farmer if they 'let us work and kept themselves and their "advice" at a distance.'

It would appear also that party secretaries had lost much of their managerial role in the economic field.

The secretary of the local party cell had to do some administrative work as the Party leaders had to check the fulfilment of the [5 year] agricultural production plan. Their influence on community life was not very big. But they controlled and supported local cultural life. They used to organize celebrations in the local house of culture at kolkhoz level and sometimes they invited some guests from Latvia. Communists were the most active in cultural life (A director of the house of culture Saločai, Pavalio district, Lithuania).

As a consequence, it became hard to explain to workers why these people had to be paid! Along with bargaining power, which became the essential resource of success office, party officials enjoyed the privileges, which were attached to the local party elite – they got the best apartment, best cars, extra bonuses, etc. Their family members got the easier and higher paid job.

In interviews almost all party secretaries agreed that their job – the job of the political commissar at a kolkhoz – was not important. Because of the deepening alienation with the CPSU the being the chair and leader of a collective or state farm lost much of its status. A distinct quality of Leninist organization is the enmeshing of traditional status with modern elements (Jowitt, 1992, 16). The status of the kolkhoz chair and the professionals around him/her, however, was radically different from party *nomenklatura* status with regard to some of the modern elements required of its role. These included a sense of individual responsibility for the execution of tasks, ambition, an innovative approach, the development of a sense of personal-efficiency and the ability to manage projects as essential preconditions for a career. A nickname – the 'red barons' – that summed up this ideologized image of the patriarchal director or chair was not fair, according to all the chairs interviewed by us, who felt badly hurt by the sobriquet. By the time of the interviews the old men had devoted their lives to the building of kolkhozes, they had worked hard and their living conditions were only somewhat better than a good workers.

The low level of education of party secretaries and the low status of the local cell of the country CPSU constrained contributions to the discussion of reforms at a local level. In Estonia and Latvia the party also dictated that discussions were to be conducted in a spirit of 'reason, pride and consciousness of epoch' (the words on every party member's personal card).

A chair – 'a red baron' – was a key figure in the kolkhoz system with status and power but the humiliating nickname of 'red baron' was not known before *perestroika* because it had little grounding in practical experience. It did not gain any more ground in the years of *perestroika*. The popularity of the nickname has some parallels with the launching of the stigmatising term 'kulak' by the Bolsheviks prior to their liquidation of the independent peasantry. Both names symbolize modes of production – the private, market oriented mode and the state-administered mode.

The Baltic Revolutions did not produce any specifically communist *perestroika* style reform policies. There was no popular mobilization in favour of the extant system and the CPSU. Surfacing national revolutions became anti-communist revolutions. The tensions between the corporate structures of the top leaders and those of the work collectives became politicized. Communist party local branches split into the Moscow-minded and nationalist-minded sections. This happened first in Lithuania. The creating of independent communist parties had various reasons. It was seen by some as the only chance to maintain minimal authority for communists and nomenklatura among the titular Estonians, Latvians and Lithuanians. Lithuania was historically and culturally more sympathetic to charismatic leadership. This tactic did not work well in Latvia and failed totally in Estonia. Differences in the collapse of the communist party and its local cells in each Baltic country were symptomatic.

All proceeded simply. As far as I remember, the letter arriving from the [independent] CPE] bearing Väljas' signature. The letter said that those who wanted to were free to quit the Party. As a result nobody wanted to stay in the Party. The membership cards were delivered to the District Committee, and the school organization was also wound up (CPE activist, Kanepi, Estonia).

According to reports the party cells in Lithuanian schools in countryside stopped their existence almost in the way – no long debates, no questions. The most enthusiastic communists, including a teacher, moved to the town. Estonians and Latvians continued to participate in organizations in the towns somewhat longer – due to tactical reasons. It is reasonable to suppose that the total exit of Estonians and Latvians from the party would leave this 'instrument' in the hands of mainly Russophone reactionaries.

In Lithuania the disintegration of the local CPSU cells at regional level had, sometimes, a more complicated configuration than in Estonia and Latvia. Communist ideology had deeper roots and debates on its 'authentic' and not 'authentic' forms made sense. Quite a large number of people supported the 'national communist' Brazauskas and there was also a pro-imperial wing of the CPSU.

PS.: The Party [GPSU] split in the winter of 1990. When the GPSU and Lithuanian Communist Party split, the same happened here. On the side of Burokevičius [Moscow minded leader] went only the elderly people, mostly pensioners. It was not appropriate for the middle aged [people to support Burokevičius]. They [pensioners] did not like Brazauskas. The pensioners thought differently. And it is evidently the same now. You can talk with pensioners and all of them will say that in the Soviet Union it was better. And now, especially those, who are the bigger losers, at the election of Seima voted for Brazauskas.

Q: How many people left in the communist party local cell [totally 80 members] here, after the CP had split?

PS.: 60 members. What to do? If there is freedom, then they are free to go. 60 had joined LCP, the rest supported Burokevičius for about three years. Later on the basis of that [Moscow minded] party there was established a Democratic Labour Party. It was the same party just without name of Brazauskas. Then there were left around 36 members.

Q: And what then happened?

PS.: This party is still functioning [in our district]. There are 14 members. This is a big number, taking into account that in the district there are around 400 people. They are paying membership fees. They are coming together once per month and working out strategies with respect to elections and so on (Previous CPSU local cell secretary, Saločiai, Pavalio district, Lithuania).

It is also possible to see differences reflected in the retrospective interviews. In Lithuania people expressed more frequently positive feeling and even spoke sympathetically about the Soviet times or communists and felt bad in denying anything positive about past. The past was, relatively, a much more personal issue.

The collapse of the communist party in the Latvian countryside resembled very much the fate of the Estonian section of CPSU, but it also had some common features with developments in Lithuania. Communism seemed to have deeper roots in the Latvian village than it had in the Estonian countryside. It also seems that there were among the Latvians more pro-Moscow party members of CPSU then in Estonia, where their share was absolutely miserable. If according to reports, CPSU cells in Latvian schools just stopped, things were different in the collective farms.

The independent communist party [of Latvia] was established, which supported the idea of independent Latvia. It happened in the end of 1989 and in the beginning of 1990. When the CPSU collapsed, there was established at the same time the [pro Moscow] Front of National Salvation. Also here in [Aizkalne] such a group was founded. They were against the independence of Latvia. So, the previous CPSU became divided into two branches – one was the communist party for independence and the other was strictly against an independent Latvia ... The party secretary in our kolkhoz supported the idea of the independent communist party and she joined that. She also proposed to me that I join that party ... here in Aizkalne were communists who didn't support it [Latvian independence]. But there were also communists who registered themselves to the independent Communist Party of Latvia (Then agricultural specialist, Aizkalne, Latvia).

Communist parties in Estonia and Latvia were not forbidden as is sometimes perceived (Nørgaard et al., 1996). National Communist Parties emerged in 1990 and local organizations of CPSU continued to operate. After the failure of the attempted coup in August 1991 the CPSU was forbidden in all the now-independent Baltic States for good reasons. The reasons for the different story of the CPSU in Lithuania are complex. These include the more national character of the party, a different civil history, a historically more divided society and delayed modernization in the countryside. These peculiarities were reflected also in the support of communist or socialist ideas. In this context a successful transformation of Communist Party took place in Lithuania instead of collapse. As expected, disintegration of the Soviet agricultural system of production was also different. A question arises about the social forces and narratives: how was the Lithuanian countryside able to oppose its own socialist party of social reformers (Vardys and Sedaitis, 1997, 202–203) and delegitimize collective farming. After the collapse of CPSU the positions of national communist parties to effect reforms were very different. In Estonia they were heavily marginalized, in Latvia their chance to contribute to reforms looked much better and in Lithuania communists continued to operate in the political establishment. It could be expected that economic reforms in the Baltic countries would have gone in different directions.

Eradicating the communist legacy required action in two main directions – the separation of the social and economic domains and the limitation of the role of the state. In the different historical-political environments of the Baltic States the anti-communist and anti-Soviet narratives built up in different combinations. In Lithuania it was somewhat more complicated than in Estonia and Latvia. But in the Baltic States generally questions about the transformation of societies were radically different from other parts of the former USSR. The question was: is it possible for countries that had an exclusive political framework to invent ideological and institutional frameworks for integrating privately diverse and publicly shared identities (as it was in Russia)? There were issues of principle, of political restitution, economic reform, privatization, restitution of property justice and legality to be addressed. However, despite all national and historical differences the shared experience of occupation and the desire for restitution became the fundamental principles that shaped the political logic of reforms.

**The Church – 'The Living Dead'**

The Soviet regime made religious associations at best invisible. But the interviews reveal a quite different status and role of church in the awakening of the different Baltic countries.

Membership of the Church was very low in Latvia and Estonia where parish congregations had usually some a few dozen members – mainly representatives of the older generation who participated in Sunday services provided usually in cold and darkness (Ruutsoo, 2002). It was exactly the same in state in Latvia. 'I suppose that in the Soviet times in this small town of 5,000 people, the protestant congregation had 30 to 50 members (Priest of Protestant Church Kandava, Latvia). Peasant membership of the local church in Lithuania was much higher. A fundamentally different approach to the church is reflected in the fact that in Saločiai religion instruction became a part of the standard curriculum at school and classes were well attended by almost by all the students. In Estonia such attempts were an almost complete failure because of disinterest.

In 1987–1988 the grip of the atheist state over the church started to relax. Religion was perceived again as a 'private' matter and not so much an object of public policy. Religion was tolerated by *perestroika* as a bringer of comfort. Making religion a private matter was expected to decrease tension with regard to the church, which was perceived as a symbol of the innocent victim and unfair treatment. How does one account for the returning of new groups and associations into public politics at just the time when new spaces for political participation are being opened? The change in the status of the two pillars in the years 1987–1989 tells us about the opening of a space for deep changes in the power structure. Identification with the church was in the Estonian village traditionally, in the first years of *perestroika,* almost minimal. The church did not become a local or more general national symbol.

The position and status of the church in Latvia was quite the same as Estonia. However, in the second half of the 1980s a small group of very committed and enthusiastic young priests brought the church into the centre of the Latvian independence movement. It was the turn of these young priests to again make private public.

In the church emerged a movement initiated by the young priests – age about 30 – Juris Rubenis, Modris Plate and others. These people were very important for the whole society because of their contribution to the Atmoda ['Awakening'], Movement, Helsinki 86 Group. Many of the activists were pastors from the church (Priest of Protestant Church, Kandava, Latvia).

This group started to symbolize the spirit of national resistance.

In Estonia the clergy took a more hesitant position, however, the radical attitudes of a very small number of young priests came to be adopted by pastors working in the countryside. The Soviets had managed to penetrate the Lutheran church very heavily but despite this the role of the church in an almost socially bankrupt community became important. In the dramatic situation the church leaders made their choice.

'The old pastor, Peiker, regarded the Estonian struggle for independence with scepticism. Peiker had only low expectations for the Society for National Heritage and the Popular Front. He was convinced that it was unlikely that anything good would come of it all.' Notwithstanding, he participated in the restoration of independence, and infused boldness in the activists. 'Pastor Peiker had a crucial role in the creation in Kanepi of the Society for National Heritage and the support group of the Popular Front. He attended their meetings. Also, having lived here since 1945, and knowing as he did the local situation, he was often consulted,' says the young pastor of Kanepi county, Estonia.

The Catholic Church in Lithuania faced in practical terms the same kind of forced marginalization as the Protestant Church in Estonia and Latvia. But some of the Catholic clergy committed themselves actively to public dissent in the period of Soviet occupation from its very start and this supplied the church with relatively bigger symbolic capital. The much higher historical status of church in the countryside was reflected in the bigger role it had in everyday life. In the Lithuanian village the Catholic faith continued to be a part of personal and national identity and the self-esteem of the people. As the communist party lost its grip, the Catholic Church, which had traditionally been politically active got involved in politics and started to make use of opportunity structure.

On Sundays people started to crowd into churches, to come together and be confirmed. The church began to play an important public role. Believers, especially women, started to spread their opinion on who we should vote for about the village. There were some active believers, who suffered during the Soviet time and now they started to organize … I believe that the church was active not only here but also all over Lithuania, especially after gaining independence. Of course, there were people who didn't care, but some of the believers started vigorous active political agitation ... the church played a very important role [in winning support for Sajudis] (Teacher, Saločiai, Lithuania).

A deeper content of religion in everyday life made it harder to erect effective boundaries to put the church aside. The hidden debate between the public and private world was much more tense between the communist elite and the church in catholic Lithuania than it was in the two other, mainly protestant Baltic States. If the Estonian local administrative elite did not care about religious matters, in Lithuania the communist national elite faced problem fitting together private/religious and public/ communist.

They were afraid. They did not attend the church here, but maybe they went to church in other places ... there was such a case. The father of the previous chairman of the kolkhoz had died. The men brought the coffin into the church but somebody had reported it and they were punished because they had gone into the church. One of my acquaintances, he was the director, his father died and his children were hiding while their father was brought into the church. That man was also punished (Then CPSU secretary Saločai, Pavalio district, Lithuania).

The position of the Lithuanian elite was, because of historical reasons, much more ambiguous and it became a source of conflict in the ranks of the Soviet nomenklatura in the countryside.

Later the religious associations and church gave a birth to numerous right wing and liberal parties, which took or supported a restitutionist stand, especially in Lithuania. The right-wing nationalist narrative on national history and the building of future society had its source largely in religious identity.

The contribution of the church in building boundaries in the early arenas of revolution remained long hidden – both in respect of identity and resource mobilization. As the Soviet system started to unravel, the church had the potential to become a public forum. The attitude towards the church was not so much an issue of personal belief as a chance for re-alignment of individual political position and ideological loyalty. In the years following 1987 matters related to the church were systematically mobilized as symbolic capital in purpose of anti-communist campaigning.

As Juri Lotman has pointed out, during the periods of radical political change there is a noticeable increase of semiotics of behaviour (Lotman, 2000, 116). Once again a public institution, the church, and religion generally were used as tools for building demarcation lines, as symbolic boundaries, not easy for the communist elite to overcome without danger of loosing face. The church became, even very in the secular countries of Estonia and Latvia, a high profile public institution. The blessing of flags or ceremonies of anti-communist, nationalist associations and opening of their meetings by pastors became everyday practice; the church was demonstratively integrated into the fabric of restitution and anti-Soviet mobilization. The significance of these rituals was very clear – there was lot of 'magic' – this period is even called the mythological period of transformation. More important, was the setting up of discursive debates with excluded or included audiences and with the participants in these debates and decision-making processes enclosed within certain boundaries. A year later, when the church felt itself powerful enough, the radical wing of priests went into a more or less open political clash with the regime. National-radical religious leaders amongst the priesthood build became a bulwark of the radical movements – the EHS and later also the Estonian Congress.

The changes in status and the confronting in people's minds of different symbolic worlds gave a start to the restructuring of the public sphere and after that new hegemonies could shape political development. Re-definition of the symbolic boundaries of 'private' and 'public' supplied a device for constructing new exclusionary and new inclusionary narratives. A part of the private sphere as defined by Sovietism returned and became essential in the shaping of the new public sphere. Thus the 'legalization' of religion brought with it deep and unexpected changes in the power structure. The capacity of the church had it grounds in the fact that civil society is 'not merely an institutional realm. It is also a realm of structures socially established consciousness, a network of understandings that operates beneath and above explicit institutions and self-conscious interests of elites' (Alexander, 1992, 290). The legalized church started to function on the basis of a 'civil symbolic sphere'. It supplied the structural categories of 'good' and 'bad' into which every member of society is made to fit. The narratives around the church (persecution, repression, stigmatization, etc.) constructed a code that started to distinguish 'ours' from 'theirs,' 'good' nationals from betrayers, the Soviets from the non-Soviets, patriotic behaviour from the not patriotic. The people in Baltic villages became the subject of a deep re-configuration.

The differences in the remote history of the Baltic States revealed that there is an intimate relationship between political participation and the manner in which the boundaries between private and public are historically articulated. In Estonia and Latvia religion was historically more private and less public, national business. This limited the effective re-definition of private and public. The Lithuanian countryside was, in the Soviet period ideologically much more encamped. Political cleavages were defined more in ideological terms and the two big narratives – Christianity and communism – both had stronger positions than in Estonia and Latvia.

## 'Movement Society' and Social Capital in the Countryside, 1988–1991

The push that was needed for the surfacing of the popular movements in the countryside demonstrates the importance of the links and linkages outside of the community for encouraging the people to start movements. The original coming together of local people needed support to take the shape of an effective group. The Popular Front of Estonia (PFE), launched in April 1988, started as the first national large popular movement in the Baltic States. The movements did not knowingly start to build a network. The bottom up grass root mobilization across the country met later a top down support from the central organization, which took hold. The importance of the Popular Front in the initial period supplied a real public political alternative to CPSU and a comprehensive program of political change namely the abolition of the power monopoly of the CPSU and the restoration the of independence of the Baltic States. But political opposition is not the same as a societal project.

The tactic of the Popular Front can be called 'movement surrogacy' (Dawson, 1996, 19). The political intention was hidden behind apparently non-threatening causes in order to take the first steps towards mobilising people to support more radical platforms. If both the explicit and implicit goals of the movement targeted the same audience, then this surrogacy strategy might be viewed as the most rational strategy given the conditions.

Despite the self-limiting approach and cooperative stand towards official frameworks, open mobilization was in the beginning too revolutionary for the village people. Interviews given to us in all Baltic States reflected deep frustration and anger of collective framers. The awakening of the countryside people who were the most conservative and anti-Soviet minded segment of the population was a hard task because people needed encouragement. This was why the creating of rural cells of PFE was delayed.

> Pastor Raimond Peiker was of the opinion that the creation of a support group of PFE would be successful if the meeting could be associated with a visit and a speech by some public figure. Peiker reached an agreement with Jaan Kaplinski [a well-known dissident, poet and essayist], and Kaplinski delivered an impressive speech ... Deputy chairman of the village Soviet Olev Pärna was elected chairman of the support group (A farmer, Kanepi, Estonia).

Only with the bringing together of almost all the practical and intellectual resources which were at the disposal of small communities, could movements start and make the people optimistic. The Popular Front really was popular – it united people from very different institutions – from the outcast church to the top of the village Soviet. There were no limits for the participation of CPSU members in the PFE, but in practice, people belonging to the CPSU nomenklatura (party secretaries) were expected (later forbidden) not to take the top-positions in the organizational structures of the Front. Local urban cells like the cell of the PFE in Kanepi were small. A dozen people made up the nucleus. Local cells in villages were quite similar all over the Baltic States. A common complaint of the respondents (more intensive in Latvia and Lithuania,) was about the small number of people who were able and interested enough to contribute to the movements. This finding is supported by data on social mobilization displayed in other studies. If, in the towns, political clashes between different elite groups organized into PFE and EC was fierce, in the village both movements cooperated. It is interesting that confrontations did not developed in areas where real material interest and options for restructuring strategies were open.

Cells of the Lithuanian Sajūdis emerged in the countryside more smoothly. Sajūdis started six months later, when the Soviet regime was more disintegrated. It seems as if the visible role of Lithuanian communists in the founding of the Sajūdis initiative centre in Vilnius encouraged the CPSU elite to pushed ahead also in the countryside. Sajūdis, like PFE, was formally founded for support of *perestroika* but the element of 'surrogacy' came in because of the different setting of Lithuanian polity.

> There was an organization of 20 people. They came together and held meetings in the house. Of culture 25 per cent of founders were party members. The leader of Sajūdis was my deputy ... I was also a member of Sajūdis. Later I left it, when I saw what kind of people had joined to it. It started with good ideas. But when I saw how they were thinking, I did not like it. As it usually happens [with the popular movements], they [local Sajūdis members] also went to extremes. The activists of Sajūdis were ex-members of communist party, most of them. When they were party members they said one thing, but when they joined Sajūdis they started to say the opposite. Thus, I initially thought that Sajūdis would support the party line. But in the end they were against the party (Then CPSU local cell secretary, Saločai, Pavalio district, Lithuania).

> Some of communist just changed their minds 180 degrees and started to propagate the ideas of Sajūdis. It made people laugh. The Party secretary of our kolkhoz became the chair and also changed his mind. But nobody approached him. In some cases in the meetings they were reminded about their communist past but later nobody cared (Teacher, Saločai, Pavalio district, Lithuania).

Confusion was expressed by a then CPSU secretary, who, as he himself described it, had tricked people into the party, and it is very informative about the state of mind of the nomenklatura. It was exactly due to his contribution that in the countryside it was very difficult to find and mobilize anti-regime elite, active people, who had not been recruited into the party corrupting them. This was one of the main purposes of CPSU recruitment – to make people mental serfs of the system. Joining CPSU (tricked or not) members pledged an oath in writing to give up their citizens' right to think and operate independently. The main innovative idea of the new Popular Fronts was that it became possible to recruit back some, usually the most talented, of the tricked people. The rapid transferral of human resources lead to an uneven distribution of these resources among the competing parties and movements and favoured the Popular Fronts.

Also, in Latvia, the PFL cells emerged in almost every county and village we visited. In the areas where the kolkhoz was composed of mainly of Russian speaking immigrants, (replacing Latvians in the areas that were emptied by deportation), Interfront, the KBG sponsored anti-independence movement emerged. But according to reports it was not very active.

The importance of close contacts with the national top elite and at least some minimal concentration of independently thinking people in the area – to maintain autonomy against the pressure of the regime – was revealed also in Latvian interviews.

> In the kolkhoz not all people shared the same ideas. Employees of kolkhoz had different opinions. The current leader of our municipality had been engaged in the nationalist movement in Preili. He was the agronomist in the kolkhoz. He talked to me and also

with other people. And the director of the Rainis' museum Baiba Ducmane also brought these [Popular Front] ideas to Aizkalne. She was from Riga. She brought and delivered newspapers of the awakening movement [newspaper of PFL – Atmoda]. Several people became interested. There were also some farmers involved. We didn't know what would happen. I had grown up in the Soviet time and I didn't know much about the independent Latvia. But the head of our municipality and his family was repressed in the Soviet period. They were deported. Maybe that was the reason why he became one of the initiators of the independence movement here … The head of the municipality gathered us together. My colleague and I went simply to hear what was going on. This event was reported in the local newspaper – we heard about the establishment of the Popular Front. And Peteris [Rosenthals] said that in Preili also there would be a branch of the Popular Front and those who are interested in joining should go and take part in this meeting. And as we were old enough we decided that it wasn't too big a risk for us to participate. We were attracted by the prospect that the Soviet system would be changed … Our school collective of teachers comprised of very different people, among them also members of CPSU. But I was not afraid: I asked myself: what could they to do me? They couldn't sack me. I was already quite old. I had left my position of director because I run into conflict with the head of collective farm who was the real boss in those times (Then teacher, Aizkalne, Latvia).

As the fight for political leadership in Lithuania heated up, so also relations between the CPSU and Sajūdis, which had overlapping memberships, became intense. Soon after their founding both Sajūdis in Lithuania and the PFL of Latvia started to radicalize and became more and more anti-communist. The Popular Front activists in Aizkalne were eager to deny members of the CPSU the right to participate in the workshops of the Popular Front. The nationalist stand of the PFL and the tough anti-communist rhetoric of the renewed Sajūdis (the radicalization of Sajūdis took place four times according to some scholars) made the life cycle of the movements different compared with the more moderate PFE. The moderate and calculated politics of the PFE had its background in a more diverse political domain in Estonia, where a nationalist-fundamentalist movement – the Estonian Citizens Committee – was launched in the autumn of 1989. The movement denied the legality of the Soviet regime and called for the restoration of the Estonian Republic. The Estonian [National] Congress was a body elected by the citizens of the annexed independent Estonian Republic and their children. The sister movement to the Estonian Congress – the Latvian Congress – was also founded but for various reasons it never became as popular and effective as the Estonian Congress.

Interview respondents, among them national-radical religious leaders, who were, largely, the backbone of the radical movements in Latvia, denied hearing about the operations of Latvian Congress representatives in the countryside. In Lithuania the ethnic composition favoured native Lithuanians, but also, the historical arguments about legal continuity were less important. Sajūdis dominated and no other influential movement emerged.

The Citizens' Committees – the local cells of the Estonian Congress movement emerged in almost every Estonian county. A radical activist described to us how the practical and organizational support of the local official, Soviet apparatus – the village Soviet – was decisive for the Congress in Kanepi. 'The people would not by themselves have registered at the election of the Congress were it not for "old Leib," the then chairman of the village Soviet, who led them.' Besides starting the local committee of the Congress where one could have one's name registered, the so-called 'passport office' of the village Soviet of Kanepi also registered 'historical' citizens of the country. If in the cities and on a national level relations between the PFE and the EC were tense they cooperated quite smoothly in the countryside.

The idea for an Estonian Congress reached Kanepi via the same Olev Pärn [deputy head of the village council and member of CPSU] who had laid the foundation of the Popular Front. He was an excellent organizer. Also actively involved in the movement was the teacher Vahur Tohver [he was also the deputy to the Congress of the Popular Front] the forest warden Rait Hirv and one more teacher, whose name eludes me. They organized the registration and held the election (Entrepreneur, then activist of EC and PFE).

The PFE and EC, like their sister-organizations in Latvia and Lithuania, elected their representative bodies, had regular sessions, issued bylaws and made regular political statements originating from effective contacts with local cells. Lithuanian Sajūdis was even able to create departments in regional centres all over the country, which were staffed by full time, paid activists. The similarities in organization of all the Baltic popular movements had its source in their close cooperation. In all the Baltic counties groups, boards, etc. of popular movements operated on county, district and national level.

The members of the Popular Front [of Latvia] met at sessions at district level ... There was a Preili town branch and here in Aizkalne there was a support group. The local initiative groups went to Preili to participate in demonstrations and to celebrate national days. We helped build the barricades in Riga in January and we went to Riga for Independence Day ... In Aizkalne there was a kind of nucleus of LPF. There was a committee of the Popular Front in Preili, it called together the support groups and in that way people in pagastas [parishes] were kept informed. Every pagasta had its own local leader (Then agricultural specialist, Aizkalne, Latvia).

But not everything went smoothly. The more hierarchical and corporate organizational model of Lithuanian society was reflected in the top down organizational culture of the popular movement. There were repeated complaints by interview respondents related to the big distances between the city and periphery – the peasants and the elite –, which complicated cooperation in many respects. Our questionnaire results showing that the Lithuanian village was the most peaceful place in the Baltic States, was supported in interviews.

Representatives from the Sajūdis cells in some villages of Pasvalys district, Salociai, Vaskai, Joniskelis, had joint meetings. We [a small number of activists] discussed problems and we surmised that people were afraid about the future. If nothing changed, nobody knew what would happen to the Sajūdis activists. Maybe Lithuanians are more cowardly than other Europeans. I think that there was very weak management from the upper echelons in Vilnius. They thought that they could do everything without any help from the lower level. But I think that they needed our support from the villagers (Former general economist of kolkhoz, Sajūdis activist Saločia, Pasvalio District, Lithuania).

Yet, despite differences in levels of participation, divergence of interests and clashes with each other, all these nationwide movements played an important role in bringing the countryside into politics. At the same time the network of associations, there were tens and tens of small strongholds across the country, made it very difficult for the Soviets to suppress movements. But none of the urban elite initiated and supported movements had in their programs any specific vision of agricultural reform or the future of the kolkhozes. A vague rhetoric about legality and restitution left the programmes of the Popular Fronts open to all modes of agricultural production. The disruption of the kolkhoz economy started to produce disappointment among the people. The lack of a positive programme confused the villagers. It was a big danger to the national movement. In the Latvian countryside people sometimes did not believe the Popular Front was a constructive player.

Part of the people didn't want the representatives of the Popular Front to take power. Because they thought that they [the Popular Front activists] were only able to destroy the previous system and not create a new one. It was a general mood throughout the countryside (Then agricultural specialist, Aizkalne, Latvia).

The Popular Front of Latvia was politically close to the Estonian Congress. The radical, restitution-oriented Estonian Congress also generated a widespread feeling of disappointment and despondency, because it failed to achieve anything substantial. According to the reminiscences of the deputy:

One of the leaders of the Congress visited Põlva [a district in the South Estonia], giving speeches: some interested people gathered in the county centre. They were expecting direct, practical help to cope with their daily problems. Rumessen's talk turned out to be far-fetched and much too general. These tillers of the soil expected a solution or alleviation of their problems. They were still waiting for the 'white ship' [a mythological bringer of good fortune equivalent to a magic wand] that would come and help. Most of all they hoped for protection from the Russian Army and the Russians (Agricultural specialist, Kanepi county, Estonia).

In the opinion of a central organizer of the elections to the Congress in Kanepi, the latter had no action plan and the 'role of Congress to put life into Kanepi and get it going was next to nil.' Herein lay the causes for the erosion of the influence of the Congress – the inflated hopes and the practical incompetence of the Congress. The position of the people of Kanepi, and more widely, the position of the people of Estonia itself, was controversial. The activity plan of the Congress was nothing but unsophisticated restitution of property. They agitated for an automatic restoration of property of land to what it used to be before 1940. The motives and arguments of Congress were exclusively political. The Kolkhozes were perceived as the symbols of Soviet annexation, the strongholds of socialist mentality and even the power base of the 'red barons' – the chairs who supported collective farming. The confusion of villagers in respect of the Estonian Congress meant that, at the turn of the 1990s. It was possible to say that the restitution of private farming was not the favoured programme of people in the countryside.

It is not possible to get a just and practical solution to property reform when it is not supported by other economic measures. The ironic reference to the arrival of the 'white ship' clearly implies the utopian nature of the EC vision for the future of the villages. The solution of the Estonian Congress for future rural development was uncompromising restitution of property reform. After this Sajūdis and LPF also started to call for rural political solutions based on the Estonian Congress.

## Identity Building, Symbolic Politics and the Restoration of Historical Monuments

Almost all captured and Sovietized nations turned back to their roots but overall identity of the Baltic nations was very fragmented. During the last century Baltic States shared a close history – the Wars of Independence against Russia and Germany in 1918–1920, the periods of national independence during 1918–1940, the Soviet annexations from 1940, the years of anti-Soviet guerrilla war in 1944–1953 and the forced collectivization and deportations of 1949–1951. This shared history in many respects gave them a similar 20th Century identity.

Whole campaigns contributed to the reshaping of local historical and identity that posed a challenge to local authorities and collective farm leaders. The restoration of historical monuments destroyed by the Soviets became especially important in the countryside. The restorations were symbolic actions on a large list of inclusive-exclusive practices whose agenda was dictated at a national political level, which helped develop anti-Soviet and pro-national identity mobilization. The process of restoration highlights the power structure in the countryside, the role of the chair and management practices.

The restoration and renovation of old memorials and the taking down of Soviet monuments marked a specific period in the Baltic history of transition. Usually it was a spontaneous act defined by local characteristics. Specific to Estonia was a specific national-level campaign for the restoration of plaques and monuments whose aim was the protection of historical heritage. This campaign became an association in itself. Supported by the church, the heritage protection movement started renovating graveyards and cultural monuments. This later became the focus of an openly symbolic fight. But, as everywhere in the countryside, the different movements – the 'fronts' and 'congresses' – cooperated and contributed to fund raising or helping with paperwork or building materials. In Estonia every county had put a monument to the fallen in the War of Independence. During the Soviet time more than 250 plaques, monuments and crosses were destroyed! Parish monuments in Estonia provided a strong community connection with the locality and also had personal meaning for individuals whose relatives' names were engraved on the memorial stones. The destruction of these monuments by the Soviets sent out a powerful symbolic message and their restoration now sent out an equally powerful message back.

The destruction of the monuments usually happened in the darkest period of the Stalinist terror, at the turn of the 1950s. In Lithuania it carried on to as late as1972 and many bad memories remained with local people.

We were 3 members of Sajūdis cell, who did [restored] it. I was one of them. Also a former communist and a woman took part in it ... We decided to restore it because I had remembered this monument from my childhood. I was 7 years old, when this monument was pulled down. It was pulled down late, in 1972. The monument was built to commemorate Lithuanian independence. And somebody did not like the words 'Lithuanian independence.' We started to unearth the old monument on the 4th of April [1989]. They were laughing at us and thought that we would unearth it and then abandon it. But we continued to work every Saturday and we restored it. They could not believe in it in the beginning (Private farmer and former Sajūdis activist in Saločiai, Pasvalys district).

According to our sources the number of plaques and monuments restored in Lithuania was not large. This was because of Lithuania's different administrative structure. Monuments were mainly built in the middle of the districts. Symbolic significances in the different countries was different: but the restoring and re-designing of crosses was, in a Catholic country, of major symbolic value. 'Why did they [the Soviets] destroy the monument to our fighters for independence – it was not a cross!' – cried out one of our Lithuanian respondents.

Among the first restored in Estonia were more than a dozen ton-weight plaques in Kanepi parish. The process of symbolic restoration was already under way in 1987 when people secretly started to bring flowers to the place where the broken parts of the memorial lay. When the EHS was established in Kanepi in the summer of 1988, people were still very cautious and 'nobody dared say outright that the monument must be restored.' The role of historical memory became important. The veterans of the War for Independence had the matter closest to their hearts. They brought the pictures of the plaques to campaign meetings. The honoring of the shattered monuments became an essential part of the 'Days of Kanepi' festival. Flowers were laid at the plaque while a women's choir sang. There was also the practical problem of locating powerful equipment and machines needed to do the job.

In Estonia as early as summer 1988 the official treatment of the War for Independence underwent a drastic change. For example, in Kanepi, in August, someone started digging a hole, without an official permit.

We got the machines and started to dig. When it started raining, nobody left. ... When we were digging, everyone else stayed indoors. Nothing moved in the streets. We decided that, come what may, we must have the hole dug by daybreak. ... The largest problem was how to get six cubic meters of mortar quickly. Only a couple of men were available; however, when the invitation was circulated, quite a few people arrived, everybody equipped with tools. There were construction workers carrying boards; everybody did their best, and everything was a success (Then agricultural specialist at Kanepi kolkhoz, Estonia).

Along with the restoration of historical monuments by nationalists, other campaigns were aimed at identity mobilization. Soviet monuments were taken down and historical names restored to streets, settlements and bridges. New monuments were built to the memory of people deported or murdered in the beginning of the Soviet annexation.

In the second half of 1990 the grip of the Soviet army over the Baltic countries, including the countryside become tighter. In Estonia effective professional explosions were done at night and destroyed some of already restored and newly established monuments. They were restored again. A kind 'war of monuments' began. To provoke trouble the Soviets destroyed some of their own monuments! There is plenty of information that the USSR was prepared to occupy Estonia once again. The stand of countryside now became important. The officers of the occupying army, usually from local garrison units, started to inspect parish centres, to make visits to officials of the village Soviets and 'warning.' Intimidation was clearly the aim.

The ground was still covered with snow when someone knocked at the door. There was a Russian officer standing in the corridor. We offered him some coffee. He told us his version of the situation. The officer alleged that there had been street riots in Võru [a small town in Southern Estonia] and that the Army had to prevent conflict and bloodshed. He explained the reason for his visit as rumours that in Kanepi and the whole of Estonia there was evidence of an anti-Soviet state of mind, as a result of which in Kanepi, a monument to Soviet soldiers had reportedly been damaged. A moment later it turned out that he had photographed that same monument and found everything in perfect order ... I am sure that he knew very well what we were up to, and that the Soviet authorities had been tipped off about us having gone too far. He was evasive, but he hinted that in Kanepi some people were engaged in anti-Soviet activities ... The officer also gave his view of the Molotov–Ribbentrop Pact, which he claimed to be enemy propaganda and he also talked about the voluntary entry of Estonia into the Soviet Union ... He wondered how a Soviet official dared speak like I did. I said that I used to be a member of the CPSU, but there was no Party left now. The empire was collapsing and the party had admitted defeat and called it quits. He was very surprised ... When leaving, the officer reportedly said to the chair: It is highly possible that a Soviet Army unit will be stationed in Kanepi (Then Mayor of Kanepi, Estonia).

It was at that time that the restored paramilitary associations – the Home Guards of Oma Kaitse in Estonia and the Zemessardse in Latvia started. A much more ambivalent relationship of Lithuania with its first period of independence obviously complicated the utilization of the historical experience and the *Tautininkas* movement was rehabilitated only after 1991. In tune with the old guerrilla movements experience the detachments started to prepare to hide in the forests, to built bunkers. They also started to hide documents such as lists of residents in order to make conscription and deportation more difficult. The Soviet military offensive became the most visible and aggressive in the Latvian and Lithuanian countryside where they could rely on and legitimize their activities supported by the Moscow-minded part of a divided CPSU. It was clear that Moscow had not given up. The disciplining of the countryside, where action was not so visible to countries abroad impressed by the phenomenon of Gorbachev and Glasnost was a starting point.

At that time when the CPSU was forbidden and the independent Latvian Communist Party started to function, the nearest company of the Soviet army took a very active part in events. They sent their solders to guard the buildings and the institutions of CPSU in order to keep their confidential documents safe. They also guarded Lenin's monument in Preili. They guarded them for several months. They knew that there would be a town council decision to demolish Lenin's statue. Therefore the Army sent solders to guard it. But the decision to demolish was taken anyway ... I think that those Soviet Army actions couldn't affect the nucleus of the national movement. But I felt that among the people there was a quite different mood. Maybe there was still disbelief that it was possible to overthrow so solid a system. There was a period of some doubts [about success]. Especially when the people saw that Soviet solders had been sent to Preili. But however, we continued to gather in front of the communist party building and organize demonstrations demanding that army should quit our territory (Then agricultural specialist at kolkhoz, Aizkalne, Latvia).

It is quite revealing that the restoration of monuments and the organization of demonstrations and pickets remained, in the main, the only achievement of many local cells. Baltic movements remained in these terms protest movements. An explanation for the fact that they did not develop into larger and more permanent structures or interest oriented organizations is very much explained by the identity-mobilization nature of the movements. Looking for more detailed explanation prompts a sociological look at Soviet societies.

## Collective Farming, Popular Mobilization and Civil Organization

Despite differences in the agendas of the movements and progress made in the periods of awakening of 1986–1991 there are striking similarities in the Baltic States in this first period of anti-communist revolution. Namely:

Popular mobilization undertook, structurally, almost identical stages in terms of identity mobilization and the formation of alternative symbolic capital.

The annexed status of the Baltic States made the main aim of mobilization a state vs. society conflict. The dominance of one or another form of identity politics – political, ethno-cultural or religious identity – depended on the national-cultural nature of the particular country.

The effective shift from the 'identity politics' to shared or organized 'interest politics' required much a more profound restructuring of the state-society relations than afforded by *perestoika* politics. It is hard, however, to construct any clear class interest base for an explanation of the logic of the Baltic revolutions. It should be expected that previous owners, deported people, perceiving themselves to be the main winners of restitution should be the main 'motors' propelling social movements. They were active in the movements and had their own associations but these *associations* remained marginal. More detailed analysis is needed but it appears that the main bulk of popular movement activists were agricultural specialists, who, quite often, were people with higher education. The class position of the agricultural specialist class was ambiguous. As well paid professionals they enjoyed good salaries and privileges. It was in their interest to keep a Soviet production system (though not the political system) running. The efficacy of the movements is shown in the fact that they too were ready for reform.

The disinterest or low capability of administration at a local level to deal with a bottom-up mobilization (with good top-down policies) was the main deficit of the Soviet system. The stagnation of the system meant it had lost all dynamism. But in the other hand the kolhozes/sovkhozes, even if they had been able to react more or less adequately, were doomed because they had two main deficits.

Almost all economic, social and cultural resources were concentrated into the collective farms and the head of the farm and *not* the head of the local Soviet operated as the local 'sovereign' (there were different descriptions of this role by interviewees).

The disintegration of the party-state put collective farming for many years into a 'twilight zone.' The emergence of the popular movements, which quite soon became openly anti-regime, made the position of the kolkhoz leadership complicated. It is very interesting that, despite relatively large intellectual and organizational resources, there is no indication that popular or other movements and their leaders went into constructive co-operation with the leadership of kolkhozes/sovkhozes. There is little or no information from the years 1987–1990 about discussions on local reforms, about the restructuring of kolkhoz management, about increasing democracy or any policy of inclusion, which would integrate opposition into the top echelons of kolkhozes/sovkhozes. On the whole, cases of open debate at local level about the prospects of de-regulation were rare.

Popular movements, which did not have any specific program in respect of future of the kolkhoz system, had a very obvious anti-system identity. Some factions of the kolkhozes/sovkhozes, however, identified themselves with the Soviet economic system. The Estonian Maa Liit (Rural Union) was founded in 1989 during the maturing years of *perestroika* and was the only one of the new independent associations which publicly sided with CPSU and shared it's political line. It was composed of chairs of Estonian collective farms and directors of sovkhozes. The Rural Union challenged the nationalist radicals of the Estonian Congress movement claiming its activity was provocative and destructive. The question was not about the tactics of revolution (which was the source of clashes within the Popular Fronts) but about the conflict of substantive interests and the maintaining of collective farming. On the contrary, the Association of the Directors of Factories that managed the majority of Estonian workers supported the nationalist radicals! Two different logics of reform were in place. The directors of factories were potential winners of privatization, potential new owners. The chairs of the collective farms were clearly in danger of becoming losers. The main issue for reform in the countryside became *perestroika politics* i.e. CPSU politics and conflict arose here about the issue of legalising private farming.

A typical situation in all the Baltic States was that the *leaders* or *activists* of radical movements, sometimes even moderate new movements such as the Popular Fronts, were people who were repressed by the regime. They were the victims of deportations or individuals, who, because of personal qualities or bad luck had got into conflict with the leadership of the collective farms i.e. their activity was shaped into anti-regime or anti-kolkhoz identity.

The future prospect of kolkhoz collectives as legal and community institutions coloured the position taken by the top leadership of the collective farms in respect of very clear anti-regime symbolic-political actions. The form and content agenda of action was pitched at national level with the clash of the popular movements with CPSU. It must be remembered that the heads of the kolkhoz system were members of CPSU and had to follow the party line. Two the most important actions in this respect were the call for the restoration of historical monuments destroyed by the Soviets and The Baltic Chain. During the dramatic winter of 1990–91 there was a very big difference in Latvia and Lithuania commitments for the defence of public and governmental buildings and district centres. The repressive actions of Moscow (symbolic and real) revitalized memories from the 1940s and 1950s and opened old wounds and radicalized nationalist movements in Latvia and Lithuania. Estonia was lucky to avoid terror not because of clever politics but by the fact that there were not enough internationally minded communists with Estonian ethnic origin, who could make up a pro-Moscow communist party and administration.

The position of the kolkhoz leaders was the most complicated in Estonia, because anti-Soviet popular movements there started as early as in 1986–1987, when the old regime was still in relatively good shape. There is also evidence that the middle district level administration, not so sensitive to public opinion, conspired against popular movements in all the Baltic States.

I was one of the people who restored a monument [in Salociai]. I had good relations with the Razukas [Latvia's nationalist movement activist]. He helped us to get materials needed for reconstruction work, because it was not allowed for Lithuanian enterprises to do such kind of work (Former general economist of kolkhoz, Sajūdis activist in Saločiai, Pasvalys district, Lithuania).

Dissidents or the national counter-elite defined the agenda of symbolic politics at a national level. The way that the local counter-elite used this agenda for challenging the official local elite, the chairs or head of village Soviets is revealing. There is not one single report that kolkhoz/sovkhoz chairmen run openly into public conflict with the local counter-elite leaders and national movements by forbidding an action or discouraging people, for example. The leaders of the kolkhozes were, against expectations quite cooperative officially. The then headmaster of the school at Kanepi remembered that the chair of the local collective farm was not against the restoration of a monument.

'We visited the chair and discussed where it should be set up, e.g. in front of the school, maybe … I also heard that the collective farm gave money to the project …' The then headmaster of the school, Kanepi county, Estonia.

The leader of the HPS recalled that the head of construction at the collective farm charged the costs of the work (mortar, stone, the use of cranes, etc.) to the account of the Inter-Collective Farms Building Office of Põlva, the Road

Construction team of Võru and the collective farms. The stone was put into its proper place in the centre of Kanepi, with a powerful tractor.

There are similar reports about a cooperative stand from Latvia and Lithuania, where an analogous campaign gathered power about a year later.

> I had a disagreement with the kolkhoz director and we decided to establish a Sajudis cell. Our first task was to restore this monument in front of their office. When we started it, the local church was the first, to support us. When we had almost finished our work, the kolkhoz gave us 3 thousand Roubles. Later they even supported me with some machinery (Private farmer and then Sajūdis activist in Saločiai, Pasvalys district, Lithuania).

The interviews also indicate, however, that that kolkhoz' chairs, even if they were good nationalists, were quite hesitant to give open support to anti-communist minded actions. Kolkhoz leaderships had much at stake. There is not a lot of information about what happened behind the scene, about what kind of instructions was received from Moscow or what kind of blackmail was used by the CPSU, for instance. The CPSU as the main manager of economic resources in a deficit economy had a strong armoury for disciplining local leaders. They never took public positions and usually did not give public speeches. They followed the Soviet tradition – they were not public persons but managers. In the 1960s and 1970s there had been an important change in the role of chairs. They evolved from *apparatchiks* into managers 'It is leadership competence that distinguishes the political manager from both the command-oriented *apparatchik* and the rule-oriented technocrat' (Jowitt, 1992, 98). They stood in the background and took a conservative stand from the margins. There is no information that any of them ever took an initiative or were the initiators of events. They were worried about saving their face in a rapidly changing and heavily unpredictable society.

Alongside the campaign of the restoration of monuments and names, the other essential event which revealed the role of collective farms, was the 'Baltic Chain.' Organized by the Popular Fronts, the human chain passed through many villages and outlying areas and had, to some extent, a larger meaning for country people than for the towns. It became a sort of test of the loyalties and resources of countryside residents – activists, opposition and administration. Without the contribution of the countryside the Chain would have been less impressive and successful.

> The creators of the Baltic Chain were local people – members of different organizations that operated at a district level. There were many organizations, among them Sajudis. At that time there was still a kolkhoz and the administration gave a bus. People went with their own cars too (Teacher, Salociai, Lithuania).

> People here were quite active. During the Baltic Chain I was in Jekabpils with my family. But I know that people from Aizkalne also took part. A special bus was organized for those who wanted it (Agricultural specialist, Aizkalne, Latvia).

It must be remembered that it was 200 km from Aizkalne to the route of the Baltic Chain. A very similar picture in terms of the cooperation in organizing people was seen in Estonia. For example, three busloads of people arrived from Kanepi, Estonia, to stand in the Chain in a place reserved for them by the organizers, alongside those who had come in their own cars. The buses were supplied by the kolkhoz.

There is some evidence to support the idea that the attitude of the chairs of the smaller kolkhozes and sovkhozes were different. In some cases, for example, in Estonia, the chairs of the largest, most industrialized, modern agricultural complexes did not respond sympathetically to their workers expressions of interest to take part in the Baltic Chain. Participation was defined as private interest and was no excuse to take a day off work. Two factors were at work here. Firstly, chairs of these big complexes were heavily dependent on Moscow as their industries were directly integrated into the agricultural production system. Secondly, in the sovkhozes, the personal and ethical relation between work collective and the senior leadership of the production machine was lost. The professional qualities of the leaders of the sovkhozes were closer to technocrats than to managers.

### The Military Offensives in Latvia and Lithuania –
### A Challenge to the Countryside

The Baltic Chain became a kind of rehearsal for the much more serious challenges which people feared were to come – the bloody clashes, which they knew would be the response of a tyrannical regime. Good communications and a new form of national pan-Baltic integrity, which resulted form the Chain and contributed very much to the civil integration that sustained the Baltic peoples during the 'bloody days' of Riga and Vilnius. The 'bloody days' bore a considerable weight in the countryside. Facing catastrophe, the nationalist communists and nationalist radicals found themselves on the same of the barricade. The managerial-communist administration and the radical popular movements started to work together.

The contribution of kolkhozes and village people to the independence movement was at its most dramatic and decisive in January 1991, when the 'spetnaz' occupied public buildings, the TV and Radio centres and printing houses in Vilnius and Riga. The using of force by the Soviet military was a signal that Moscow had made a decision to suppress the Baltic revolutions. Our interviews show that almost everybody understood what was at stake.

> There was a rush to Riga. Students from schools also went there – hundreds of people. But the main organizers were the Greens. Everybody organized people to go to Riga – the secondary school, The Technical College, the city (Head of Museum, Kandava, Latvia).

> I know that our people went to Zakuzala [in Riga] to defend our TV centre and building. Those who stayed at home gave them food and money. I remember that two parties went

[to Riga], but I don't know just how many people participated. These people were mostly ordinary people, later Zemessardze [Regional Defence] was organized ... The organizer of these popular detachments was the mayor. The last kolkhoz director helped too. He became the director of the kolkhoz during the very last years of its existence (Agricultural specialist, Aizkalne, Latvia).

In Lithuania too the village people rushed to Vilnius. The progress made by the popular movements during six months of civic nation building from scratch was impressive. Zemessardze (Regional Defence) was a voluntary armed detachment restored by the Latvian independence movements for defence of the Latvian Republic. Zemessardze in Latvia played the same role as the Popular Front Home Defence and the Congress Defence League (Kaitseliit) in Estonia. The Soviets had destroyed both organizations – Zemessardze and Kaitseliit – in 1940 along with other national defence l institutions. According to an activist of a local detachment of the Defence League (a veterinary surgeon, Russian by nationality) 'the Defence League found support in Kanepi first and foremost among older men who had been members before the 1940 occupation.' Later its membership grew to several dozen. The kolkhoz in Kanepi contributed to the launching of both popular civil defence initiatives – the Home defence and the Defence League. The founding of the Defence League meeting in Kanepi was held in the repair shop of the collective farm. The equipment for mobilization issued by the Soviet administration to the collective farm were now assigned to the Defence League. One activist reminisced that 'part of the outfits, the high boots and uniforms, were bought with money from the collective farm.' In Lithuania the same kind popular defence association operated mainly in towns.

The position of the collective farm elite remained ambivalent. The sincerity of the commitment of kolkhoz chairs was questionable. An activist told us: 'I think that the last director wanted to show that he gave support to such kind [nationalist] of activities,' Agricultural specialist, Aizkalne, Latvia.

One of the retired kolkhoz chairs in Latvia expressed his confusion about the decision of his successor to the post of the chairman who had sent kolkhoz workers to Riga, to defend the Latvian revolution.

I know that people had taken big lorries and tractors from the kolkhoz when they went to Riga. I wonder if people now would do the same. At that time chairs gave a command and the workers implemented it ... Of course it was painful to bear [that people were killed in Riga], but on the other hand it was not anything special because every day somebody is killed in some road accident or catastrophe. The question was what will happen next? If I knew then that the future would be empty buildings and uncultivated fields, I would have taken another attitude (Then kolkhoz chairman, Iecava, Latvia).

This comment is very interesting. The retired chairman is someone with the divided identity (Russian born Latvian) revealed schizophrenic position of many of his class and many of Sovietized people. Supporting the national revolution the nationalist kolkhoz leaders could expect that if nationalist succeed there would not be kolkhozes after some years. But they had little alternative. Failing of Baltic revolutions had meant disaster. Interesting new fact is that the command to send kolkhoz workers to Riga 'was given from the Iecava district.' Interviews from Latvia made a strong impression about the very deep commitment of countryside people from all classes to events in Riga. It appears that the Latvians gathered their final resources from the countryside in order to muster more ethnic Latvians in their capital, where they were a minority, to bolster the pressure of the uprising.

Kolkhoz leaders, then, faced a conflict of loyalty and personal identity. In this socially schizophrenic position, the majority made a choice for their country. The patriots in the countryside mustered enough resources and legitimacy to get people on the move – 'to march on Riga' (and on Vilnius). In the days of the struggle for liberation the mettle of social capital, the integration and communication at micro level, the synergy at macro level and the shared hopes for the future of the communities of the Baltic nations were severely tested. It is possible to say that the destiny of the Baltic revolutions at that time was very much decided in Riga and Vilnius. Latvia was the weakest link, which made it a prime target. The patriotic stand of the country people was a critical resource that enabled the Baltic States to withstand enormous pressure.

In this context the growing popularity of the stigmatising discourse on the 'red barons' is surprising. Historically 'barons' composed a class of exploiters in the Baltic nations with an alien origin (mainly German). Their estates were nationalized in the twenties and given over to the peasantry. The content of this discourse was defined by the mainstream discourse of restoration of property rights. Notwithstanding its ideologically motivated rhetorical origin, the popularity of the nickname reflects the alienation between the people and their 'masters.' When the sovkhozes withered away and with them the patriarchal-style collective farms the fundamental flaw of kolkhoz system was exposed. They had been installed by force and represented illegitimate, imperialist power. The illegitimacy of the Soviet regime in the Baltic States was exposed fatally in 1990–1991 when Gorbachev sent the Soviet military to discipline the national Supreme Soviets who were operating on an absolutely *legal* basis – even in terms of the Law of the Soviet Union.

The transformation of a society based on a 'collective is good, private is bad' ethos to one based on procedural rationality and a substitution of charismatic-managerial leadership with interest-oriented impersonalized management was not, in principle, an impossible task. But international mainstream influences favoured something very different. East Central European countries in transition faced a powerfully constructed discourse that put the return to the west, the gaining of independence, ultra liberal economic policy and liberal philosophy into the same basket.

## Conclusion

In the Baltic revival the role of the countryside was significant in all of the States, but especially in Estonia and Latvia due to ethnic composition.

The tasks of dismantling and transforming the Soviet system involved issues of both de-regulation and social mobilization. In the emancipation of civil society two main stages – identity mobilization and interest-based mobilization – have been observed.

The countryside never took a leading role in the awakening because mobilization was delayed there. Identity mobilization was complicated because a counter-culture – the main framework that was used to develop an anti-Soviet mentality in the cities – was weak. Scarce social capital in the villages meant that it was practically impossible to challenge a CPSU cell there. The Soviets maintained control over the official associations for longer preventing them operating as networks of opposition, and the nurturing of a subculture of protest. That is why the essential role in the re-awakening of the villages was due to networks and personal contacts with the intellectual circles in towns.

In the village power conflicts, long hidden, surfaced as an open contest between two main hegemonic structures in the village – the church and local cell of CPSU. In the countryside the church created an important alternative public sphere, it was an organizational and spiritual-cultural resource for the surfacing social movements of political opposition. The greatest role of the church in identity mobilization was in Lithuania because of its deeper entrenchment into national culture and history.

The crisis of the Soviet regime in the countryside was deeper because of the fundamental nature of the conflict created by forced collectivization, which, when conflict became public, surfaced as a vital issue. An interest based class conflict of was at work in the countryside in some places but the final outcome of the struggle was decided by the political solution. Other observable phenomena were that the life of interest-oriented protest was limited to the period of the awakening and interest-based cooperation did not survive the first periods of transformation.

The situation of the country people, the identity built into their normative world and their positioning in the Soviet regime was more compromized and ambivalent than that of urban people. Instead of a 'peasant citizenship' a 'kolkhoznik citizenship' had taken hold. The protest potential of the countryside was lower not only because of a missing effective counter-elite with social capital but from the perspective of the integration of the community in ethnic terms and structure of work, viz:

- The collective farms remained relatively isolated from the harmful disintegration of the nation that irritated urban residents in the 1970s and 1980s day after day. Russification, ideological indoctrination and mass immigration from Russia did not affect their worldview to the same extent and it remained more on the horizon.
- Restricting of personal freedom, censorship and political control were experienced less dramatically by ordinary people living in the village.

The Baltic rural communities, isolated from modern Western trends and affected by the historical-ethnographic shaping of socialist culture suited the conservative tendencies of country people. Song and folk dance festivals, traditional music days and the like were, despite being a part of Soviet political mass culture, much in tune with a popular model of entertainment, which made repression more tolerable.

Many kolkhozes remained islands of relative stability in a disintegrating system and sometimes maintained a capacity to operate as autonomous surrounded territories. The general standard of living in the countryside grew in the 1960s and 1970s and was, in many aspects, higher than in the towns (Raig, 1985). It created a class of agricultural specialists (agronomic workers, veterinarians, technicians, cultural officers) with an interest in sustaining the system. The Soviet industrial farming oriented system was quite effective in shaping a generation trained in Soviet schools and universities whose identity favoured modern technologies and industrialized agriculture. These peculiarities help to explain why social mobilization in the countryside was delayed.

But collectivization also had fundamental flaws. These were:

* The Soviet-introduced pseudo-collectivist forms of cooperation (kolkhozes), which mobilized different forms of familial structures such as work brigades were not able in the time allotted to them to re-shape the identity of the older generation of private farmers and to outlive the withering away of social memory. The experience of mass terror made people silent but they did not forget.
* The legacy of forced collectivization that delegitimized the kolkhoz/sovkhoz system in Estonia, Latvia and Lithuania from the very beginning. The Soviet-constructed identity conflict between 'old-primitive-outdated-private' versus 'modern-industrial-collective', did not work in the Baltic States. It was not a concept that could prevent eventual de-collectivization and the restoration of legality.

In the 1970s and 1980s, along with the withering away of the organizational integrity of system, the way of life and norms in the countryside became segmented. Gradually household production (legal and illegal forms of a second economy) became incompatible with the substance and spirit of collective farming. As a non-systemic element it operated independently from the official forms of public cultural and social activities. The world of the new forms of the non-systemic 'household' belonged to the 'private' and thus broke the continuum. It could not inject new integrity into the old system. The norms and values that were effective in the running of the new 'household' conflicted with the official cultural model and standards of interpersonal relations. The kolkhoz community was divided into groups defined by individual achievement norms and the institutionalized sector that had adapted to a 'learned handicap' kolkhoz collectivist mentality. New achievement-oriented norms were framed by 'amoral familial' links and the free market. Under 'amoral familialism' no universal shared social ethics can exist. 'Amoral familialism' was limited but it was the only accessible bottom-up strategy for achievement-oriented people in a closed Soviet society. New 'household' building strategies started both in symbolic and mental terms to marginalize kolkhozes and sovkhozes as models of community infrastructure.

Thus the kolkhoz/sovkhoz system as a community model lost its substance – it stagnated in the official language – as the Soviet system lost its organizational integrity and viability. New generations of administration (chairs, party secretaries, leading specialists) were socialized into a new organizational framework, one characterized by a shift from ideologically maintained domination to pragmatic manipulation (managerialism). But the collective farm system remained the base of their 'class interests.'

The surfacing of the new familialism based on household economy damaged the substance of the kolkhozes very badly. The kolkhozes, however, continued to operate as collective organisms because they had maintained their privileged access to social synergy – to resources, which defined the ties that connect citizens, public officials and resources across the public-private divide. At the same time patriarchal dominance revealed its flaws. The community of a legitimate regime has been defined as 'able to assume, in some basic respects, the voluntary provision of private resources for official or public purposes' (Jowitt, 1992, 41). This capacity – the 'voluntary provision of private resources for official or public purposes' – is one of the main preconditions of modern civic culture and society. The promotion of procedural rationality and the maintaining of public order dominate over any other supportive framework of success in modern civic-ness. The kolkhoz system did not have this.

As sham patriarchal collectivism disintegrated, new, modern social civic-ness could not get a foothold. The surfacing cultural manifestation of individuality (not to mention the political one) remained private with regard to ideas, values and directions. But it was powerful enough to fragment the shared social world of the collective farm. The collective farm system did not have a mechanism to allow private interests to absorb and legitimize themselves or one that could facilitate developing interest-oriented cooperation towards a 'procedural' society. In terms of social capital, the linkage of social organization was missing. To make bottom-up development successful, linkages to political organization and broader extra-community institutions have to be constructed.

The post-communist history of the countryside of the Baltic States is usually described by scholars writing on modernization as one where individuals have obtained freedom and the opportunity to participate in the building up of their community, but, they lack the stable community base to provide guidance, support and identity i.e. they have linkage but they do not have societal integration. Durkheim identifies this as one of the hallmarks of modernization. A lack of moral or social standards of conduct and belief in community or the individual results in anomie. The result is not only heightened cognitive dissonance for individuals but also increased rates of disaffection, suicide and violent crime across the society which became observable in our earlier studies (Alanen et al., 2001). In the case of the Baltic States the disintegration of late Soviet society into 'amoral familialism' started the development towards 'amoral individualism' as described by anthropologists (Woodcock, 1998, 172). In both cases there was neither familial nor generalized trust, narrow self-interest permeated all social and economic activity and members of society were isolated from all forms of cohesive social networks. A large percentage of a village community was thus characterized by the absence of both integration and linkage.

The collapse of the Soviet regime and the restoring of national independent states, which legalized markets and individuality in formal terms created the opportunity and space for the developing of linkages. The nature of state-society relations in the post-Soviet Baltic States is crucial for understanding the views of economic groups and their efficacy in shaping the willingness and ability of the state (and the other actors) to act in a developmental manner. But the internal dynamic and development of these initiatives occurred in the context of a historical framework that undermined the capacity of independent groups in civil society to organize themselves in their own collective interests. The fundamental issue of the restitution of property dominated all other arguments. There was a powerful surfacing of historical memory. The most abused and suppressed aspects of Baltic history – deportations, nationalization, forced collectivization, the guerrilla war in the forests of 1944–51 – which harmed almost every single family, cried out for redress.

The community stock of social capital, which had emerged spontaneously in the years of *perestroika*, was weak but it was enough for a launching of different nation wide actions. A Baltic political elite with relatively small differences emerged and adopted not only a policy of restitution but also a neo-utilitarian 'minimal state' concept of government. This meant that, despite the rhetoric of restitution, there was no attempt to restore the sophisticated network of cooperatives and bank mortgage systems, which were the backbone of successful private farming until the end of the 1930s. The Estonian government of 1924 had introduced a 'New Economic Policy' of tight fiscal control and made managerial decisions at macro level. For example, it cut state expenditure, made limited loans and favoured the development of agriculture over industry.

Effective interest-oriented mobilization requires a supply of new forms of social capital and cooperative top-down society building strategy, which involve all strata of a society. In the years that came after the countryside in the Baltic States became a victim of free market doctrine and a simplistic restitutionist approach. There was little interest in exploring the infrastructure that supports effective cooperation between state and civil society. The flimsy illusion of re-creating a strong civil society based on traditions from the past became fatal – it justified ignoring a need for effective support for restructuring. The rhetoric of 'a minimal state' legitimized the disintegration of the countryside into numerous losers and very few winners.

## References

Alanen, I., Nikula, J., Põder, H. and Ruutsoo R. (eds) (2001), *Decollectivisation, Destruction and Disillusionment. A Community Study in Southern Estonia*, Ashgate, Aldeshot.

Alexander, J. (1992), 'Citizen and the Enemy as Symbolic Classification: On the Polarizing Discourse of Civil Society,' in M. Lamont and M. Fournier (eds), *Cultivating Difference: Symbolic Boundaries and the Making of Inequality*, The University of Chicago Press, Chicago, pp. 201–231.

Ambrazevičius, J. (1990), 'Leedulase vaimne pale iseseisvas Leedus,' *Looming*, no. 8, pp. 1117–1127.

Baltic Report Data (1995), Collections of Processed Material, Manuscript.

Dawson I. Jane (1996), *Econationalism. Anti-nuclear Activism and National Identity in Russia, Lithuania and Ukraine*, Duke University Press, Durham.

Halas, E. (2002), 'Symbolic Politics of Public Time and Collective Memory. The Polish Case,' *European Review*, vol. 1, pp. 115–129.

Jowitt, K. (1992), *New World Disorder. The Leninist Extinction*, University of California Press, Berkeley.

Lauristin, M. and Vihalemm, P. (1997), 'Recent Historical Developments in Estonia: Three Stages of Transition (1987–1997),' in M. M. Lauristin, P. Vihalemm, K. E. Rosengren and L. Weinbull (eds), *Return to the Western World. Cultural and Political Perspectives on the Estonian Post-Communist Transition*, Tartu University Press, Tartu, pp. 73–125.

Lii, D.-T. (1998), 'Social Spheres and Public Life. A Structural Origin,' *Theory, Culture and Society*, vol. 15, no. 2, pp. 115–135.

Lotman, J. (2000), *Semiosfäärist*, Vagabund, Tallinn.

Maide, H. (2000), *Transformation of Agriculture*. Http://www.ibs.ee/ibs/economics/tee/maide1.html

Merton, R. (1968), *Social Theory and Social Structure*, Macmillan Press, New York.

Morrison, L. (1998), 'Protest Potential of European Republics of the Former Soviet Union,' *International Journal of Sociology*, vol. 28, no. 2, pp. 12–35.

Nørgaard, O., Hindsgaul, D., Johannsen, L. and Willumsen, H. (1996), *The Baltic States after Independence*, Edward Elgar, Cheltenham.

Raig, I. (1985), *Maaelu sotsiaalseid probleeme*, Valgus, Tallinn.

Rose, R. (1997), *Baltic Trends: Studies in Co-operation, Conflict, Rights and Obligations*, Studies in Public Policy, No. 288, University of Strathclyde, Glasgow.

Ruutsoo, R. (2002), *Civil Society and Nation Building in Estonia and the Baltic States*, Acta Universitatis Lapponiesis 49, Rovaniemi.

Ruutsoo, R. and Siisiäinen, M. (1996), 'Restoring Civil Society in the Baltic States 1988–1994,' in M. Szabo (ed.), *European Studies: The Challenge of Europeanization in the Region: East Central Europe. Vol. 2*, Hungarian Political Science Association and Institute for Political Sciences of Hungarian Academy of Sciences, Budapest.

Silberg, U. (2001), *Agraarpoliitika*, Tartu Ülikooli Kirjastus, Tartu.

Vardys V. S. and Sedaitis, J. B. (1997), *Lithuania: The Rebel Nation*, Westview Press, Boulder.

Weber, M. (1958), *The Protestant Ethic and Spirit of Capitalism*, Scribner, New York.

Woolcock, M. (1998), 'Social Capital and Economic Development: Toward a Theoretical Synthesis and Political Framework,' *Theory and Society*, vol. 27, no. 2, pp. 151–208.

Zwinkliene, A. (1995), 'Réforms Agraires En Lithuanie: Fidélité a la Tradition,' *Études Rurales*, pp. 103–115 and 136–140.

## Chapter 4

# Constructing Capitalist Firms: Former Socialist Industrial Complexes and their Struggle for Survival

Jouko Nikula

## Introduction

The restructuring of former socialist industrial complexes into competitive capitalist firms has proved to be a much more difficult task than many western advisers anticipated. Privatization in industry has not necessarily resulted in the rebirth of the capitalist class or the middle class, nor in improvements in work culture or work organization. Over the past couple of years there has been much debate on the causes for the failure in the conversion of industrial complexes into modern market-oriented firms. According to Dunford (1998, 90–95), the ultimate reason has been the failure of the reformers to understand the nature of the Soviet type of economy as well as their 'failure to appreciate the nature of the institutions on which capitalism depends and of the mechanisms involved in the creation of capitalism.' The first failure (to understand the nature of the Soviet type of economy) refers to the fact that during socialism, constant shortages and waste meant that firms were dependent on informal relations, such as barter and negotiation. This was why socialist enterprises operated on completely different principles than capitalist firms, or even firms in less developed countries in Latin America or Africa.

The second failure refers to the highly simplistic understanding of the social preconditions for privatization and restructuring, which require an institutional infrastructure that facilitates the transition and that acts as a shock absorber for enterprises that are set to lose their markets and financial stability. The third failure refers to the failure of western advisers and probably domestic politicians in many post-socialist countries to grasp the nature and mechanisms of so-called original accumulation: what kinds of strategies and resources, both financial and social, did different social groups have at their disposal when the privatization programmes were launched. The fourth failure is related to the conversion of so-called original accumulation from a struggle over money and other material resources into a battle which destroys the mental foundation of society (its values, norms, etc.) and which has adverse consequences for the development of society at large.

The result of industrial privatization was not the rebirth of the capitalist class or the middle classes, but rather a politically guided process, which saw the national wealth, appropriated by the former political elite. For the purposes of this article, the key issue upon which we focus is the second point, i.e. the nature and type of institutional infrastructure that is needed for the successful transition from socialism to capitalism.

Specifically, this article is concerned to analyse the strategies of adaptation of former Soviet industrial complexes. Representing different branches of production, these complexes are located in areas outside the capital regions of the three Baltic countries.[1] Since these countries have adopted different economic policies, their operating environments are also different. The general theoretical frame for the analysis is provided by regulation theory, in which the emphasis is on modes of industrial regulation as well as the preconditions, which give rise to certain strategies of adaptation.

The first part of the article provides a brief description of the nature of socialist economy, the main mechanisms and modes of coordination and control of economic action and exchange. The second part proceeds to introduce the locations, focusing on their strategies of adaptation.

## Embeddedness and Transition

One of the most important novelties in the theoretical discussion on social transformation has been the notion of embeddedness, which emphasizes the role of individuals and social relations in economic exchange. The notion of embeddedness broadens and deepens our understanding of the concept of market beyond a simplistic and economistic view, and brings to the fore such issues as interests, informal social networks and informal institutions. Kosonen and Salmi (1999, 139) emphasize the role of inter-enterprise coordination in transitional societies where central planning institutions have been removed and no formal institutions have as yet emerged to take their place. 'Underdevelopment' of formal institutions reflects a situation where the formal institutions of the 'clan state' or 'captured state' (Wedel, 1998) have become corrupted and lost their ability to act as a third, impartial organ. The analysis of different levels of the market is thus focused on the exogenous environment, the governance structure and the production system and their interrelationships. From this point of view the analytical principle of regulation theory is too narrow and too system-centred, neglecting the role of social networks and the informal institutions as rules, which guide and restrict the decisions and actions of different actors. As Kosonen and

---

[1]    The material for this article was collected in Estonia and in Latvia during 1999–2001 in the framework of the research projects '*The Social and Psychological Consequences of the Decollectivisation of Agriculture and Rebirth of Substituting Structures in the Baltic Countries*,' and '*Socially Embedded Small Enterprises in Rural Estonia – the Role of Social Networks in their Formation and Development*' (1999–2001). Both projects were financed by the Academy of Finland.

Salmi (1999, 144) note, the 'institutional formal and informal rules are to a large extent internalised by economic actors.'

For our analysis of the strategies adopted by former socialist enterprises to adapt to the market economy we need then to look, first of all, at the ways in which their relationships to their external surroundings – other firms, state organizations, clients, etc. – was coordinated during the socialist period and what institutions and what kinds of practices have replaced the previous practices and institutions.

**Enterprises and Economic Regulation in Socialism**

One of the pillars of socialist economy was the system of large industrial complexes, which in many localities were the only sources of employment. Based on the availability of raw materials and labour, these complexes were each assigned a specific role in the nationwide division of labour. Production was not based directly upon market demand but rather on a set plan determined by the central planning authorities. The requirements of that plan depended largely on the complex's earlier performance as well as its management's negotiation skills. The responsible ministry bought their products and their workforce was either appointed by officials or obtained from the local pool of labour reserves. In a sense then, these complexes operated in autarchy and had no competition, nor did they have to worry about profitability.

According to Smith (1995), the pattern of accumulation in socialism represented an extensive regime of accumulation – growth was guaranteed by expanding investments in forces of production, especially in previously predominantly agricultural areas. The industrial economy that resulted from this policy 'produced distinct regional formations of enterprise regulation. Vertically integrated, large enterprises dominated many regional economies and created local "cathedrals in desert" characterised by high levels of local autarchy' (Smith, 1995, 763). As Smith notes, there were two kinds of linkages of integration, one local linkage and another to industrial association and to the respective ministry.

Alongside these monofunctional localities there were also diversified regional economies in which different branches of production had almost equal economic significance. Smith describes these economies as relatively self-sufficient, locally integrated and metropolitan economies. A third group of industrial economies in socialism was represented by localities that were more marginal industrially and dominated by agriculture. Here branch plants did not dominate industrial economies and industrial employment was low. Production linkages were localized, centred on the processing of local products.

**Interpretations of Transition Inspired by Regulation Theory**

The key concept and main object of analysis in regulation theory is the regime of accumulation, the historical conditions of accumulation. The mode of accumulation can either be extensive or intensive. Another closely related concept is the regime of regulation, which refers to the set of institutional forms required by each specific regime of regulation. Together, the regime of accumulation and the regime of regulation give rise to a mode of development. Each mode of development has its typical crises, which are non-threatening and self-regulated. In contrast to the purely economic understanding of regulation regime, the spatially oriented school has it that the regulation of the accumulation of capital takes place through many sites of regulation and that these sites are influenced both by state and non-state institutions and interactions between agents that constitute an economic system (Smith, 1995; Dunford, 1998). One of the regulatory sites is the industrial economy within which it is possible to identify sets of regulatory relations and institutions that produce mode of enterprise regulation. The constituent parts in each mode of regulation are: (1) suppliers of input; (2) buyers of output; (3) the enterprise itself; (4) the finance/banking system; (5) other producers and (6) the state (see Figure 4.1).

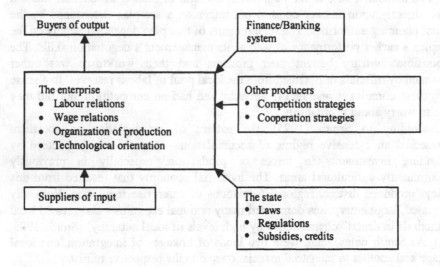

**Figure 4.1  Conceptual model of modes of enterprise regulation (adapted from Smith, 1995, 762)**

Together, these relations have a major impact on the mode of development since enterprises located in particular regions with particular forms of production, skills requirements, decision-making capacities, etc. have a significant role in creating spatial divisions of labour. The embeddedness of regional economies means that enterprises are integrated into socio-spatial networks of cooperation and competition and structures of regulation, which affect the ability of industries and regions to transform themselves into emergent forms of local capitalism (Smith, 1995, 769).

**The Localities**

The three localities selected for closer examination in this study share some similarities in common: all of them are comparatively small and all used to have a local collective farm that had an important role in the local economy. Two of the localities are situated in northern Estonia, some 40 kilometres from Tallinn. Both had small paper mills that were important employers alongside the collective farms. In K1, the paper mill was the only industrial employer during socialism, while in K2 there was also another industrial plant. In other respects, too, K2 had amore diversified industrial economy.

The third locality is situated in the northwestern part of Latvia, traditionally a prosperous area on account of its agricultural production. In K3 there was just one branch plant during socialism, but that plant played a major part in the Latvian electrical engineering complex.

**The Companies**

*K1*

The paper mill in K1 was originally founded in 1938 as a pulp mill. During the war the mill was controlled in turn by both Germany and the Soviet Union. In its early years the mill used the most modern technology available at the time, imported from Germany. In 1940, when Estonia was annexed to the Soviet Union, the mill was nationalized. The shift from pulp to paper took place after the war when new production machinery was installed.

In the late 1980s the management of the paper mill tried to get exports to western markets under way, but 'we had little success because of quality problems.' In 1992 part of the machinery at the plant was damaged in an accident. The manager of the company noted that 'by that time we didn't have an owner, the state was not at all interested in finding the resources we needed to carry out necessary repairs so that we could have restarted production; so we just closed down. Without that boiler we could not continue production.' The manager suspects that the reluctance of the state to finance the necessary investments was due in part to ethno-political reasons: 'we had Russians working here . . . and it was more a political than an economic decision.' Shortly after, the Estonian government decided to declare the paper mill bankrupt. This was one of the first big state enterprises in the country to go under: 'it was really a test to see how the legislation works.' The state of bankruptcy lasted almost three years, during which time the mill was visited by several foreign companies, but none were interested in taking over. During the bankruptcy the mill had a staff of no more than 150 workers, compared to 750 before the closure. The factory kept going by producing doors and windows and selling paper products. It was not until 1995 that the commission managed to persuade a foreign investor to take control of the plant.

## K2

The paper mill in K2 is much smaller than that in K1. Established in 1907, it started out with 20 workers, producing mainly low-quality paper (mainly for newspapers). In the 1970s production lines were opened for better quality paper grades and a new department for consumer goods (writing paper, exercise books, wallpaper, etc.) was set up. Staff numbers peaked in the 1960s and 1970s at about 400. Russian workers were mainly recruited in the late 1970s and the first half of the 1980s: the plant was now doing three shifts, which obviously increased the demand for workers. However the local state farm was thriving, and many workers who were reluctant to do shifts decided to leave the factory to work at the farm. The new vacancies were filled by people recruited with the help of Russians already working in the factory. Most of the chief operators of the paper machines were Russians: this was one of the professions for which there was no training in Estonia. The newcomers came mainly from Russia and the Ukraine, persuaded by the better wages and living conditions in Estonia. At the end of the 1980s Estonians accounted for 80 per cent of company management, on the shop floor for half of all workers.

As well as a paper mill, K2 has had a leather goods factory that was established in 1964 as a department of a major manufacturing combine. Staff numbers at the factory have been around 25–30, peaking in 1985 at 102. The manufacturing complex employed a total of over 1,000 workers. Its headquarters and one factory were located in Tallinn, while its other units were spread out across Estonia. The design department was based at the central factory in Tallinn. The organization of production was not always as rational as it could have been: 'in those days, whether you had ordered [spare parts] or not, something was coming in all the time.' On the other hand, the plant did not have to concern itself with marketing or customer relations, because it produced women's handbags for the All-Union market, mainly in Russia: 'We produced 30,000 bags a month. A lorry came; the container was loaded and taken to the railway station. That was all we would see of our products.' According to the last director of the factory their products were in such demand that local people could not buy them even if they wanted to, except when production exceeded the plan level.

All work at the factory was done on contract: each worker was assigned a specific amount of work to do; nothing more or less was accepted. There was no incentive to work harder because of these limitations and because:

> During the Soviet era we did originally have a wage scale, but then later on all wages had to be the same. The dressmakers kept a close eye on one another to make sure no one earned more than anyone else. And the trade union made sure the materials were so distributed that everyone got equal amounts to work with.

This wage system led to a situation where some workers worked no more than three or four hours a day, while the slower workers would spend more than 10 hours to get their share done: 'there was this one big department with all the lights on and no one else at work except one worker.'

In early 1990s the factory was transformed into a cooperative, which rented the premises from the state. At this time staff numbers declined to 60.

**K3**

The factory in K3 was established in 1963 to help alleviate the problem of women's low employment in the countryside. Initially the factory was one of the workshops of a bigger complex producing blocks and pulleys, but no ready-made wares. In 1975 the factory was formed into an independent plant producing record players. A production association was set up that comprised four enterprises, although each enjoyed relative independence. The development team in the production association would work on creating new models for production, which were then evaluated within the association. Each enterprise would make its own decision on whether it would be able to produce them. The final decision on the producer was made by the association.

Production plans were endorsed both among the enterprises themselves and by the 4th central board in Moscow. The plans more or less based on the actual production capability of the plant. The factory produced around half a million transistor radios a year, with two models usually in production at the same time. In addition to ready-made radios it also made around 15,000 sets for export to Cuba where they were fine-tuned and fitted. In addition the whole of the Soviet Union, the factory's main market areas included Cuba, Bulgaria, Romania and Czechoslovakia. The factory also produced main condensers for all the radio sets produced in the Soviet Union – several million of them a year. These components were also exported to capitalist countries in small quantities: this provided an important source of foreign currency with which the factory could buy the equipment it needed. All trade with foreign countries was organized through *Sovexport* in Moscow, for the enterprise was not allowed to sign contracts direct with foreign customers. Once a deal had been closed, factory management was informed that there was now money on their bank account and they could buy what they needed.

During the stable decade of the 1980s the enterprise had a staff of 1,000. However not all of them worked in the main factory, but the company had several branches in nearby districts. In addition some of the work was done outside the plant as the local kolkhoz helped with winding the spools.

The factory provided workers with a number of social services, including holidays at the trade union's villas in Jurmala as well as at the Sea of Azov and the Black Sea. The plant also supported amateur art, choirs, etc. The factory provided accommodation, building most of the apartments for its workers itself. In addition the factory contributed to infrastructure development by renovating the local cultural centre, the post office building, and various plants.

The work organization suffered from the typical problems of the Soviet type of work organization:

> We did have some problems with work discipline at the time. Every brigadier and worker knew how many sets had to be produced a month. Some work was also done in

shifts. But there were also people who drank and who didn't turn up for work every day. In some specialized professions at our plant, such as that of tuners, there was always a lack of staff because there were no people with the right training. If such a specialist did not turn up for work, administration would go out to his home to find out why. If someone failed to show at work that meant others would have to do overtime – not 8, but 9 or 10 hours. They were of course paid. There were even cases where some employees had to work for two days on the trot.

According to one manager the reasons for the problems with work discipline had to do with the economic system:

Perhaps it was all due to the planned economy. If you have to produce 50,000 sets a month, then at the end of the month your workers will be tired and perhaps they were not too concerned about the quality. Therefore they would even monitor when the sets were produced – at the beginning or at the end of the month.

At this time problems were also caused by the scarcity of specialists. Knowing that the attainment of production targets was primarily down to their skills, they would begin to demand more money.

Furthermore, the enterprise was dependent on different providers of spare parts from around the Soviet Union, including Samarkand or Irkutsk. While the plant in Latvia was in business all year round, Samarkand would close down when the harvesting of cotton began. Consequently the Latvian plant also had to cease production. Nevertheless they were required to meet the targets of the plan. The solution of the Latvian management was to transport workers from Latvia to Samarkand to produce the parts they needed. Workers were also sent to Kirovograd in Ukraine to produce other vital radio components.

The first 'shake-up,' as the last director describes the process that led to the complete privatization of the factory, started in 1989: it was at this time that the previous management decided to transform the company into a cooperative. This decision was mainly motivated by the desire for greater economic freedom: 'spending in state enterprises was very strictly controlled – that is for salaries, that goes towards culture, that towards social costs, etc. – and there was also price restrictions.'

The cooperative was formed by 100 employees, both engineers and ordinary workers, each investing 1,000 roubles. During the two years that the cooperative was in business, it rented all its premises from the state. Productivity increased after this changeover and continued to grow until 1992, when the Russian markets were closed.

K4 is a food-processing factory in northern Estonia, supplying a state-owned bakery and run as part of that bakery. In the late 1980s the bakery was taken over by a state-owned distillery. Towards the end of the decade there was an accident in the factory when the main boiler exploded. This resulted in the director being sacked and a young engineer being hired in his place – who happened to be a friend of the son of the director of the biggest bakery.

**The Processes of Privatization and Restructuring**

Smith (1995, 762) says that there are principally three different strategies of adaptation to market economy: the first he calls globalized enterprise regulation; the second, mercantilist enterprise regulation; and the third, de-industrialized enterprise regulation.

*The Globalized Model of Enterprise Regulation*

In the globalized mode of enterprise regulation enterprises secure access to western markets through strategic partnerships with western companies, or receive direct foreign investments that allow them to acquire new technology. This in turn makes it easier to lay off employees and adopt new forms of work organization and managerial structures and practices. The role of the state in this mode of regulation is to control the process of restructuring through privatization, regulating foreign investments and securing the development of small and medium-sized enterprises.

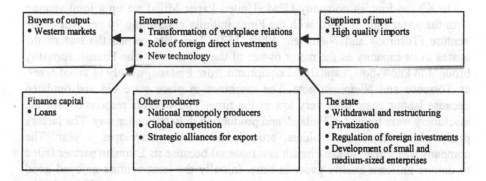

| Buyers of output <br> • Western markets | Enterprise <br> • Transformation of workplace relations <br> • Role of foreign direct investments <br> • New technology | Suppliers of input <br> • High quality imports |
| Finance capital <br> • Loans | Other producers <br> • National monopoly producers <br> • Global competition <br> • Strategic alliances for export | The state <br> • Withdrawal and restructuring <br> • Privatization <br> • Regulation of foreign investments <br> • Development of small and medium-sized enterprises |

**Figure 4.2 Conceptual model of globalized enterprise regulation (adapted from Smith, 1995)**

Only two of our cases fit into this mode of regulation. In K1, direct foreign investment helped the factory survive when in 1995 its director managed to persuade a foreign investor to take over. The investor had no previous experience in papermaking. Ownership of the factory was transferred to a group of enterprises that comprises almost 30 factories around the world. Following the privatization of the factory, the new owners began to streamline the organization of production and to shed some parts of the operation. For example, most social services were transferred to the local municipality. In addition, the rigid hierarchies among higher clerical personnel were removed: '... some of the departments have been done away with, there is no planning department any more and also the vacancy of head engineer has gone. We now share all the duties amongst ourselves, me, the production manager and the chief technician'. In spite of the changes in the work organization and the heavy emphasis on marketing and quality, there still remain motivational problems on the shop floor: 'people still do not understand that the quality of production is all-important, there is still that Soviet mentality.' The organizational changes are reflected in everyday work in many different ways: managerial staff no longer get Saturdays off and working days can be up to 12–15 hours long:

> In Soviet times I started work at 8 am, and I left home at 6 pm. Now, I come in at 8 am, but often I don't get to leave home until around 2–3 am and still come back to work at 8 am. And Saturdays are normal workdays now; I cannot remember the last time I had a Saturday off. And in Soviet times you could go to the Black Sea for a month without any problem. Now we can have a week off at the most.

The factory has major investment plans for the future, intending to gradually replace all its machinery. A new production department has already been opened with a new paper machine. The factory has also acquired an influential position in local government, the 'factory party' (the name is the same as the factory's) is presently the biggest single party in the local council and the factory manager is also chairman of the council.

In K2 the Finnish company UPM (United Paper Mills) set up a joint venture with the paper mill in 1992, with the Finns initially controlling 70 per cent of the venture (Törnroos and Nieminen, 1999, 132). The state owned the rest of the shares in its capacity as the major owner of the paper mill. The Finnish company brought in know-how, capital and equipment from Finland, mainly to avoid taxes, as Törnroos and Nieminen note. The machines in place were old and outdated because labour costs were very low at the time. The workers responsible for the machinery were trained in Finland, and production duly got under way. The factory produced envelopes and folders, around 150 million envelopes a year. The company was forced to use Finnish raw material because its Estonian partner failed to deliver uniform quality paper in time. Initially the joint venture enjoyed good success, and the number of employees and production increased steadily in 1993–1994. Production was to be marketed mainly to Sweden, Holland and Finland; only one-third was to be sold in the eastern markets. However, as it turned out the paper mill acted as a stepping-stone to all markets in the east (Törnroos and Nieminen, 1999, 135). The mill proceed to gain almost monopoly position in the Estonian markets, controlling around 75 per cent while the rest consists of either Russian 'grey imports' or small local producers.

The paper mill remained in Finnish ownership until 1998, when UPM retreated and the company was sold to a Norwegian firm, which is now continuing the production of envelopes in K2. The major problems faced by UPM, according to Nieminen and Törnroos, were bureaucracy and individual-level problems between the Finnish owners and local decision-makers. In the initial phases of the joint venture the company had to negotiate both with local interest groups and its Estonian partner. A second problem, according to a company representative, was the 'Soviet mentality', the reluctance to take action and to bear any responsibility. The third problem, according to Nieminen and Törnroos, had to do with the same phenomenon: the quality of the labour force was reflected in a low work morale and lack of respect for the rules and regulations concerning the terms of employment. Additional problems were presented by high crime rates as well as difficulties with infrastructure and related logistics services. However the difficulties that were identified by the Finnish party in their joint venture in Estonia and that they saw as the reasons for its failure were not necessarily at the heart of the problems. Adam Swain (2001) has observed that 'the rules and conventions that organise and structure the economy, then, are not reducible to market rules and price signals. Rather these rules and conventions are broader in scope and include ideas about quality, technology, and organisation that are influenced by the social, political and cultural context' (Swain, 2001, 13). This means that the problems in transferring 'best practices' or western models of management are related to the fact that "they are filtered through the lens of local cultural formations and historical experiences" (Ibid).

In K4, the food-processing factory in northern – Estonia came under severe pressure in the early 1990s as a group of local people demanded that the factory be closed down on account of the environmental risks it presented. In response the factory began to modernize its outdated purification systems, but since the project started during the rouble era and was funded by the Soviet state, Estonian independence and the introduction of the crown brought the whole project to an abrupt end. As a result production volumes dropped from around 230 tons to 40 tons a year. Morale and motivation in the workplace was also adversely affected, and there was a growing drink problem. The present factory director recalled:

> People lost their motivation since they often felt that everything they did was in vain. In some sense it was the dictatorship of the proletariat: one day when the workers were changing a pump, they suddenly said hey, we're out of fuel! What they meant was that they wanted more vodka before they would complete the job!

At some point the threat of closure was imminent, but in 1992–1993 the factory gained an independent status and shortly afterwards it was privatized. Bids were received from one Estonian and two foreign companies. The Estonian and one of the foreign bidders joined forces against the third company, concerned that if its bid were to win, the factory would in fact close down and all production would be moved outside Estonia.

Following privatization the new owners reinstated the present director of the company and began to reorganize production. However, 'when a new owner comes into the country, it brings along its own rules and regulations that are hard for us to understand, and initially they no doubt made a number of wrong decisions – and had to pay the price for learning the rules of the game in Estonia.' The company director is here referring to fact that the new owners invested heavily in areas that were not the cause or the solution to the problems; the solution would have been something that did not require money but more practical intervention: 'order in the company was restored as soon as the rules were made clear to everyone.'

Under foreign ownership, the factory's organization of production has been streamlined, the number of auxiliary labourers sharply cut, and the number of both clerical staff and production workers reduced, albeit only marginally. In its heyday the factory employed a staff of 70, today there remain no more than some 30 workers. The cuts in labour force have been made possible by investments in new automated production lines.

Production volumes rose back to their pre-crisis level within three to four years. Most of the production goes to the Baltic markets and to some extent to Russia, although sales to the Russian markets did suffer some setbacks, mainly on account of credit problems, the collapse of the Russian rouble and the heavy duties imposed on Estonian goods. In the Baltic markets, too, exports in the fiercely competitive marketplace got off to a poor start. During the crisis the company lost 70 per cent of its market share, but now they control some 70–80 per cent of the Estonian markets and 15 per cent of the Latvian markets. Their efforts to break into the Finnish markets failed against strong local competition. However, as the director pointed out: 'We have our gun ready in the cupboard waiting, and we can probably make a nuisance of ourselves to our Finnish competition.'

Terms of employment in the company are defined on the basis of its own internal regulations. According to the director, 'the workers know their rights, working conditions have to be normal, and wages are regulated according to the factory budget, which is like a bible for us.' For this reason there is no need for a trade union, and monetary rewards are so high that workers are well motivated.

Informal social networks have played an important part in the development of the firm. Initially all repairs and maintenance jobs were farmed out to a friend's company, 'that's how things are done all over the country.' Social networks are also used in recruiting new staff:

> Earlier we tried to recruit workers through newspapers ads but that was extremely difficult and inflexible, we now ask people if they could recommend friends who might be interested in working in the factory. We ask around so we know what kind of people we are taking on.

The price of labour is not a consideration in company employment policy, all that matters is productivity and the quality of work: 'good labour does not come cheap, the use of cheap labour is an expensive strategy.' However, there is an acute shortage of good and skilled workers in the municipality. The main problem, according to the factory director, is that most young and qualified people prefer to go to nearby Tallinn where they can earn more money. The people who remain are old and low skilled: 'these people used to work on the collective farm, which used to be number one here in every respect, we came only fourth or fifth. These people have a Soviet mentality and they will never fit into our company.'

## De-Industrialized Mode of Enterprise Regulation

Since the collapse of socialism large parts of post-socialist countries have seen industrial output and employment decline and consequently an accelerating process of de-industrialization. Industrial R&D and marketing have come to a virtual standstill, managerial networks and skills have disappeared. Former associations or unions of industrial branches (textiles, leather, electronics) have vanished and individual plants have had to carry the full burden of their miscalculations in developing production, investments and global competition. Throughout Eastern and Central Europe, there has been a major reshuffling of markets. All this has resulted in extensive layoffs, cutbacks in production, shorter working hours, etc. As the process continues, capitals are being devalorized en masse.

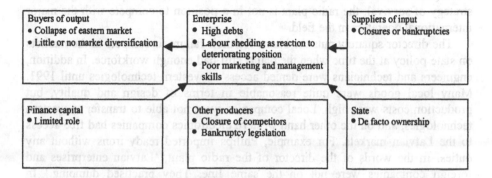

**Figure 4.3 Conceptual model of de-industrialized enterprise regulation (adapted from Smith, 1995)**

For the leather goods factory in K2 and the electronics factory in K3 in Latvia, the mode of regulation has been that of de-industrialization, a slow process of degradation. The factories suddenly found that they were no longer parts of the united Soviet industrial space, but minor actors in newly independent countries. Initially both factories tried to change their product range so that they could compete in the western markets. For example, the manager of the electronics factory in Latvia recalls:

> There have been several projects around the development of production for western countries. From 1993 through to 1996 the product range was completely changed, but it was and still is impossible for these kinds of goods to break into the western markets. During this period the company also changed Russian components to Japanese ones, including Philips. At that time there was still enough workforce here. People came to work at the plant and they were interested because there were competent people here who could do the job.

According to this manager, the company's failure was due to the reluctance of the Latvian government to support their efforts to restructure. Since 1993, production volumes have decreased tenfold, and since 1997 there has been no change. The factory lost all the money it has deposited in Bank Baltia, and the next major setback came with the crisis in Russia in 1998–1999. The company's main markets remain in Latvia and Lithuania. In addition, they still have small barter deals with Russia, exchanging their products with materials they need, such as metal or speakers. However business with Russia only accounts for 2 or 3 per cent of total sales.

With profitability declining, the company has not been able to make any new investments. In the words of a company director, 'we haven't bought any basic machinery, even though we should have.'

The company currently employs around 40 workers in radio production. As well as manufacturing radios, the company also offers transportation services. Furthermore, a fish-canning firm operates in the radio plant, representing another strategy of survival: the radio plant is not in a position to compete with the major international operators in the field.

The director squarely puts the blame for their failure to compete in the markets on state policy at the time when the plants still had enough workforce. In addition, engineers and technicians were denied access to western technologies until 1991. Many local goods were quite reasonable in terms of design and quality, but production costs were high. Local companies were not able to transfer to western technologies, and on the other hand western electronics companies had free access to the Latvian markets. For example, Philips imported ready irons without any duties. In the words of the director of the radio plant, 'Latvian enterprises and foreign companies were not on the same line. They practised dumping.' In summary, the company did not only lose markets during the 1990s, but it also lost a large part of its tacit skills as the most skilled members of the labour force left the company and some of the existing skills became useless. In the end, the processes of de-industrialization and de-professionalization became mutually reinforcing mechanisms.

In K2, the leather goods factory tried to hold on to its market position by producing the same models at the same quality standards as they had done for years. But then 'the Finns came and said your factory will be closed down because your chief designer has been trained in Moscow and cannot design a single product that would have been competitive in markets.'

The factory also produced briefcases, which at the end of the Soviet period were very popular in Estonia: they had a wooden frame, were covered with artificial leather and were very inexpensive. Later on the subcontracting factory, which produced the frames, was destroyed in a fire. The factory also subcontracted in the production of helmets for motorcyclists by upholstering them with textiles. When the helmet factory went bankrupt, 'the firm was left with four tons of plastic helmets, together with 150 kilometres of band and 11,000 handles for the briefcases.'

In the director's assessment one of the key reasons for their failure to restructure was the labour intensive technology: all their western competitors had new, more advanced technologies and were able to produce at lower costs. As a consequence the factory lost some big contracts with department stores in Estonia. Their own stores in Pärnu and Tallinn also had to be closed down due to dwindling sales. The company's director explains: 'Our products were not in demand, people wanted to have foreign goods and then came the goods from Hong Kong and then they started coming to us and ask, please could you repair this and that.'

Another serious obstacle, the director continues, was 'the Soviet person, they didn't want to work but demanded to be paid immediately.' This created a serious cash problem and at least temporary problems with wage arrears. With dwindling orders and cash flows, the company started to go into the red during 1995–1996.

Before the introduction of the Estonian crown the company bought their own premises and they were relatively profitable during the first couple of years. However as soon as the economic upturn bottomed out and imports started to flow in, the company rapidly lost its markets, especially after their main markets in Russia began to falter.

To try and keep afloat, the company started to produce anything that was in demand, for example sheets for export to Belgium. That experiment come to nothing, though, because the production of sheets required a different kind of technology and skills their workers did not have. As a result the company lost part of its workforce and had to make further cutbacks in production. The best workers moved to a Finnish company in the same branch, which offered better wages. The company made a last effort to survive by job-order manufacturing of various products for different customers. One of the new openings was the production of bags for mobile phones, but again that came to nothing as mobile phone models kept changing so rapidly. The company designed the products themselves, but in the absence of any market research they soon found their stocks were piling up. After six years of declining prospects, the company was forced to close down. Nowadays the company's only source of income is from renting out their former premises to two smaller companies, one Finnish company which produces bags and an Estonian textile firm. The company also sells heating energy to the surrounding community. The company still has 40 shareholders, mainly former company staff who are now either working in the Finnish firm or retired. The director of the company is the only paid employee. The premises are occupied mainly by the Finnish company, which moved in 1997. Initially the firm was a middleman for an Estonian subcontractor. At that time the Finnish company occupied just one room and then rapidly expanded to two stores. Today, the Finnish company owns the majority of shares in the Estonian company. The success of the Finnish company, according to the last director, is explained by the fact that it has a known brand and therefore has large markets in the Nordic countries. Their own trademark was and will not be known by anyone, even if their quality was superior.

## Conclusion

The cases reported above clearly highlight the difficulties that former socialist industries face in seeking to restructure under market economy conditions. Baltic industries have received no support at all in these efforts from government policies. Especially in Estonia, government policy has been geared to destroying the remnants of the socialist command economy, of which the former socialist industrial complexes are seen as prime examples. This policy of destruction has at once knocked the bottom out of the social networks and systems that were so crucial to the survival of the industrial complexes. Furthermore, nothing has been brought in to try and replace the institutions destroyed, but firms have been left to their own devices. The former integration of these firms into socialist markets meant that they had no skills in either marketing or design, both of which are essential for successful competition in capitalist markets. And when they lost their foothold in the eastern markets, the companies had nowhere to go: unable to renew their product range, they had no chance of breaking into western markets. With their dwindling profits, the companies also had no chance to make necessary investments in the means of production or to pay their workers enough to keep them on payroll. For municipalities like K1, the bankruptcy of their sole employer was a complete disaster. During socialism the company effectively maintained these kinds of municipalities: the company built houses, schools and provided all the necessary social services and benefits for its employees. With the process of transition, these services have become the municipality's responsibility. With the bankruptcy of the main employer in the region, unemployment has soared, municipal services have had to be cut back and tax revenues have dwindled. At the same time other companies in the areas have suffered as the purchasing power of the local population has been severely curtailed.

The companies that have lost out most in the transition are those that have been unable to update their production mix, that have no or only minimal skills in marketing and that have been producing consumer goods in areas where competition is most intense, such as radio sets, accessories, etc. The main reason for management failure is perhaps that during the socialist mode of regulation, the manager's main tasks and the main measure of his or her skills was the ability to acquire raw materials, spare parts and labour force under conditions of permanent shortage. Another priority was to meet the production targets set for the unit. Since the collapse of socialism, the priority has shifted to finding new markets and to learning how to sell and how to guarantee quality standards.

Successful transition has been possible for companies that either have a national monopoly in their branch or that are producing goods in less intensely competitive markets (e.g. paper sacks). These companies have managed to attract foreign investments and renew their machinery and establish export relations with foreign companies. Many of these companies are subsidiaries of major foreign enterprises. The necessary management and marketing skills, and often a ready-made market as well, have been imported from the parent company. The managerial strategies are also 'imported goods,' even though that has created some tensions and conflicts within the company, requiring workers to accommodate another type of management, instead of a mixture of egalitarianism and Soviet-type paternalism.

# References

Aglietta, M. (1979), *A Theory of Capitalist Regulation*, NLB, London.

Dunford, M. (1998), 'Differential Development, Institutions, Modes of Regulation and Comparative Transitions to Capitalism. Russia, the Commonwealth of Independent States and the former German Democratic Republic,' in J. Pickles and A. Smith (eds), *Theorising Transition: The Political Economy of Post-Communist Transformation*, Routledge, London.

Granovetter, M. (1992), 'Economic Institutions as Social Constructions: A Framework for Analysis,' *Acta Sociologica*, vol. 35, no. 1, pp. 3–11.

Granovetter, M. S. (1993), 'The Strength of Weak Ties,'. American Journal of Sociology, vol. 78, no 6, pp. 1360–1380.

Kosonen, R. and Salmi, A. (eds) (1999), *Institutions and Post-Socialist Transition*, Helsinki Business School, Helsinki.

Smith, A. (1995), 'Regulation Theory, Strategies of Enterprise Integration and the Political Economy of Regional Economic Restructuring in Central and Eastern Europe: The Case of Slovakia,' *Regional Studies*, vol. 29, no. 8., pp. 761–772.

Smith, A. (1998), *Reconstructing the Regional Economy: Industrial Transformation and Regional Development in Slovakia*, Edward Elgar, Cheltenham.

Swain, A. (2001), *Restructuring the Ukrainian Coal Mining Industry: "Networks of Restructuring" and the Geographical Portability of Organisational Theory*. www.sunderland.ac.uk/~os0hva/swain.htm.

Törnroos, J.-Å. and Nieminen, J. (eds) (1999), *Business Entry in Eastern Europe. A Network and Learning Approach with Case Studies*, Kikimora Publications, Series b4, Helsinki.

Wedel, J. R. (1998), *Collision and Collusion: The Strange Case of Western Aid to Eastern Europe 1989–1998*, St. Martin's Press, New York.

# Chapter 5

# Networks, Skills and Trust:
# The Necessary Ingredients
# of Rural Entrepreneurship
# in the Baltic Countries?

Jouko Nikula

## Introduction

Since regaining their independence (1990–1991) all three Baltic countries have pursued radical reforms of agriculture as part of their overall goal to create a free market economy. As Ilkka Alanen, with many others have noted, these reforms have predominantly been disastrous – resulting in a steep decline in production and employment, along with the destruction of productive and human capital (Alanen et al., 2001; Unwin, 1997).

Indeed, the decollectivization of agriculture was a truly radical social change in a deep sense for all of the Baltic countries. Aside from the fact that it was an economic revolution that changed the relations of ownership in one full swoop, it also was symbolized the destruction of previously existing forms of community, as well as the cultural and moral fabric that had previously been based on a sense of the community. Nevertheless, in the process of destruction, something new was born – a group of private, small, non-agricultural enterprises. Many of these were simply spin-offs from kolkhozes or sovkhozes, such as construction firms, sawmills and machine-stations – but a significant quantity of new private entrepreneurship also saw light at that time.

The roots of first private firms date back to the late 1980s, when Gorbachev issued legislation concerning small state enterprises and cooperatives. From these beginnings, a number of small, privately owned firms were established. The material basis for future successful firms was laid in that moment, as these spin-off firms could legally appropriate various assets from state-owned large enterprises (Clarke, 1994; OECD, 2000). In addition to this, managers and workers from all of the Baltic countries were favoured with supportive legislation in the early stages of privatization. The level of required start-up capital was very low, which made it easy for enterprising people to establish their own firm.

Ultimately the prevailing political line supported the dismantling of large state-owned companies, and the transfer of assets to privatized or new small and medium-sized firms. Indeed, such firms, which were created during and after the decollectivization of agriculture, might have acted as the basis for a new industrialization of rural areas, since they included forestry and construction firms, as well as trade and services. The initial preconditions concerning the technical level of agriculture in the Baltic countries during socialism and in particular the markedly different results of adopted reform policies in the respective countries (see Alanen in this volume), strongly influenced the way in which enterprises have adapted to an overall market economy and the consequent successes or failures involved in achieving this goal.

The simplified idea of rural modernization in the Rostowian sense, according to which land reform (meaning the creation of family farms) provides a decisive boost to modernization in rural areas, resulting in the growth of food production, urbanization and industrialization[1] (Rostow, 1960).

## Debate on Transition and Entrepreneurs

One of the key issues in the study of transition economies has been the role of the small- and medium-sized enterprises within the transition to both a market economy and a modern democracy. A common element in much of these studies is the rather simplified belief in the existence of nearly perfect markets as well as objective, calculating entrepreneur: 'the ideal-typical firm/entrepreneur at the centre of the neo-classical market model is a rootless, un-embedded actor, making each market exchange in an impersonal, neutral way' (McIntyre, 2001, 3). This kind of theory, derived from neo-classical economics, in fact propagates 'some sort of a market automaticity in the process, which leads to mass emergence of successful small- and medium-scale family enterprises, to privatization of large enterprises in a transparent, competitive process and the result will be a system, which emerges and functions under nearly perfect competition conditions (which exist nowhere in Western Europe)' (Ibid.).

Scase (2000) takes a much more critical stance towards the socio-economic role of all forms of small- and medium-sized enterprises within the post-socialist countries. According to him, for the lion's share of the SMEs the motives for entrepreneurship do not represent true entrepreneurship in the Weberian sense, but just motives of proprietorship. In the former, personal interests and well-being are sacrificed for the success of the firm, while in the latter case the relation is diametrically opposite. The positive functions of the SME's within the post-socialist countries are devoted to sustaining free-market ideology and free-market ideals for customers (Scase, 2000, 11), Scase argues quite bluntly that 'Indeed, in many economic sectors entrepreneurship is undoubtedly the exception. It is for this reason that the small business owners do not constitute a viable force for rejuvenating the growth of these economies' (Scase, 2000, 6).

---

[1]   The Rostowian theory is analogous to the idea of crushing the power of owners of latifundias in Latin America or the "red barons" in the Baltic countries as a precondition for development of democracy and economy.

McIntyre shares Scase's critical opinion concerning the majority of the SMEs in post-socialist countries, but sees the structure as a continuum, rather than opposing alternatives. At the lower end of the continuum is the survival/small trading types of business, while at the top there is entrepreneurship.

**Table 5.1 The Continuum of small- and medium-sized entrepreneurships**

|                        | Income                    | Types of business                            | Growth                      | Personal orientation                            |
|------------------------|---------------------------|----------------------------------------------|-----------------------------|-------------------------------------------------|
| Entrepreneurship       | Systemic entrepreneurship | Economic entrepreneurship                    | Long-term goals             | Capital accumulation; personal austerity        |
| Proprietorship         | Maintenance orientation   | Niche entrepreneurship, transient            | –                           | Surplus mainly for personal consumption purposes |
| Survival / small trading | Low income              | Trading only                                 | No cumulative growth        | Damage to health                                |

*Source:* Adapted from McIntyre, 2001, 17.

I would argue that the classifications presented above are parts of a simplified ideal-typical picture of the state of entrepreneurship that exists in most post-socialist countries. In reality, these types are mixtures of the traits that McIntyre treats as separate entities. In the early days of private entrepreneurship, a large part of enterprises belonged to the survival category of entrepreneurship and only a minority could even be classified even as proprietorship types of entrepreneurship. Nowadays, the minority of entrepreneurs in rural areas is still close to the definition of real entrepreneurship in the Weberian sense.

*Why it is Difficult to Promote Entrepreneurship?*

Kolodko (2000, 6) argues that 'One may claim that the main purpose of the whole transition exercise is the creation of entrepreneurship.' The difficulties in creating a viable and strong SME sector in most post-socialist countries is related mostly to the lack of coherent policy: liberalization and stabilization are not enough, but require two additional elements; institution building as the core of the transition and microeconomic restructuring of the existing capacity (Ibid.). This state of institution building entails the creation of organizations that both create and maintain the economic rules. As North and many others following him have noted, this process is necessarily a long one. Such microeconomic restructuring necessitates new investments, capital, and re-training and rearranging skills structures (Kolodko, 2000, 6). McIntyre (2001) also adds that small enterprises require the support of large enterprises as a source of input and market for the output and most importantly, as sources of individual entrepreneurial leadership (McIntyre, 2001, 6).

The European Bank for Reconstruction and Development (EBRD), notes that there are specific obstacles regarding the development of rural non-agricultural businesses:

> The rural investment climate is also less business-friendly in terms of access of finance and quality of infrastructure... it is often similar to the disadvantage experienced by small firms... rural enterprises recording less growth, investing less and restructuring more slowly than urban firms' (EBRD Transition Report, 2002).

In conclusion, one can say that the recent discussions cited above point to very important issues regarding the situation and development of small- and medium-sized enterprises within the post-socialist countries, through describing the pathologies of entrepreneurship and indicating the possible reasons for these pathologies. In this chapter I will analyze the difficulties that rural enterprises face, and the strategies that they have adopted as regards conducting their businesses. This analysis is based on data concerning non-agricultural enterprises in the Baltic countries. The data was collected from 1999 to 2001 within the framework of a research project entitled 'The Sociological and Psychological Consequences of Decollectivization of Agriculture in the Baltic countries' (see Alanen in this volume; Nikula, 2001). In our analysis the classification presented above is loosely used to point out the core characteristics of the group of the entrepreneurs under analysis, but the aim is to analyze the decisive factors, which promote or inhibit the success of the enterprise.

Before elaborating on the analysis of the specific enterprises, it will be necessary to provide a short description concerning the general trends of economic transition and production structures, based on national and international statistical material.

## The Transition and the Change of Economic Structures within the Baltic Countries

The Baltic countries have historically pursued different types of transition politics since the beginning of the 1990s, when the Estonian governments adopted the most ardent neo-liberal shock-therapy line while in Lithuania governments followed a more stepwise strategy. The Latvian strategy was, on the other hand, more modest than the Estonian one, yet more radical than the Lithuanian version. During the latter part of the decade, there has been a clear convergence of policies between these countries.

Nevertheless, in all of these three countries the large-scale industrial production, such as metal industry, chemical industry and mining are branches of industry, where the largest job destruction has taken place, negative growth rates have been recorded. Some relevant examples might be the production of office machines and computers (-74.2 per cent in Estonia), radio- and television production (-69.7 per cent in Latvia) and the production of motor vehicles (-42 per cent in Lithuania). Branches where increased job creation has taken place are wood production, the production of furniture, and the manufacturing of clothing and apparel. In these branches, most of the growth has taken place among the small- and medium-sized enterprises (OECD, 2000; Statistical Yearbooks of Estonia, Latvia and Lithuania, 2002). Figure 1 depicts key trends, in terms of changes in the employment structure according to the economic sectors.

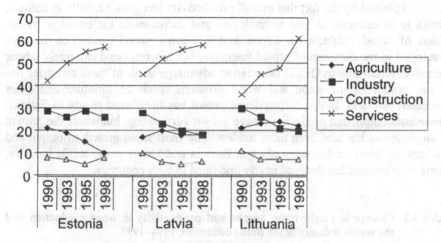

**Figure 5.1 Employment by sector in the Baltic countries (%) 1990–1998**

*Source:* OECD Economic Surveys; The Baltic States – A Regional Assessment, 2000.

The employment trends in the three Baltic countries have been quite different. In Estonia the agricultural employment has plummeted rapidly since the early 1990s, while in Latvia (and especially in Lithuania) such employment has remained more or less stable until the end of the decade. The picture is the opposite when it comes to industry. In Estonia the industrial employment has remained stable, while in the other two Baltic countries such employment has declined significantly. In Latvia, for example it has declined from one third to less than one fifth. In all of the Baltic countries the employment in services has increased rapidly, and this sector has grown particularly rapidly in Lithuania.

Reasons for such intensely divergent paths of development might be found in the way in which national governments have realized their agricultural reform policies (see Alanen in this volume). Another reason could be related to the differences in 'modernization strategies,' which were applied throughout different Baltic countries. During the Soviet period, all of the Baltic countries were rapidly industrialized, especially Latvia. Lithuania remained partially agricultural, as well as highly industrialized (electronics and energy) country during this time. In Estonia light industry developed more strongly than it did in other Baltic countries, which gave Estonia an advantage during the beginning of the transition, since it 'left the economy in a better position to adjust to market than other Baltic States' (OECD, 2000, 28).

The internal restructuring of employment within industry shows us that in every Baltic country employment has remained stable, and has even grown somewhat within consumer goods production, while regarding the production of industrial inputs and the engineering industry, the drop in employment has been very steep. Within consumer goods production, manufacturing of wood products has been almost the only sector, which has shown growth in both Latvia and Lithuania. In Estonia the industrial employment has been stable throughout the last decade. This can be explained by the fact that overall productivity has grown rapidly in Estonia, thanks to investments in new technologies and increases in the extent of value-added of wood products. In Latvia and Lithuania wood production has not developed in the direction of capital intensive, but has remained extremely labour intensive. According to OECD the relative advantage in all of these countries lies in the production of wood and wood products, such as furniture and other processed products. In this sector, development has been most steady In Estonia, where both output and productivity have grown very strong, but where the growth of employment has also been more modest. The most rapid growth of output and productivity however has taken place in Estonia and Latvia in the textile industry, where employment has declined nearly one-third in both countries.

Table 5.2   Change in employment, output and productivity in wood production and the textile industry in the Baltic countries, 1994–1997

| Wood products | Estonia | Latvia | Lithuania |
|---|---|---|---|
| Employment | 37.0 | 50.2 | 7.0 |
| Output | 131.9 | 104.6 | 18.9 |
| Productivity | 123.9 | 36.2 | 11.5 |
| **Textile industry** | | | |
| Employment | –31.1 | –27.0 | –19.0 |
| Output | 106.6 | 64.7 | 1.8 |
| Productivity | 202.3 | 125.8 | 25.6 |

*Source:* OECD Economic Surveys; The Baltic States – A Regional Assessment, 2000.

The conclusion that the OECD report drew from this data is that

important strides in the production and trade of new output ... have been achieved only in Estonia ... In both Latvia and Lithuania, important changes in RCA (revealed comparative advantage) are found in commodities related to their natural and human resource endowments, such as forestry, in the case of wood working, and female skilled labour in the case of garments. Therefore ... the pre-war patterns of trade-specialization have re-emerged (OECD, 2000, 179).

The OECD report also notes that technologically advanced output is still rare, and that even though restructuring is still taking place, all of the Baltic economies are vulnerable in relation to business cycles. In addition, with rising wage levels, they are most likely losing their competitive edge to Asian countries.

## Occasional Entrepreneurs

As shown by the data concerning employment trends in different branches of the economy, an extremely rapid growth in employment has taken place in the services sector of every country. The OECD report, along with numerous previous reports, has noted as well that the area of services, which has experienced the most radical growth, has been trading, while in other sectors such growth has been relatively modest, particularly in the public services. In rural areas the preconditions for a successful firm in the services are generally very thin. The purchasing power of the local people is typically very low, due to poor wages and wide spread unemployment. The ubiquitous nature of social problems, such as alcoholism and criminality, make enterprising mentally challenging, and for many entrepreneurs starting up a firm has not taken place merely out of hope, or the belief in rapid success and a lush life, but rather out of necessity. Many of these entrepreneurs have a higher education, as well as a significant work history as professionals in the public sector. Therefore, for some part of them there has been a question of decline in social position and at least the partial loss of previous skills and prestige. Rural entrepreneurs in the service sector have, on the basis of their own savings or financial aid from relatives, initiated their own firms, whether they are shops or bars. They have developed their business gradually by avoiding too many risks. Social networks have been of crucial importance to the majority of businessmen in the initial stages of setting up a firm; such networks have provided them with knowledge, labour and financial resources.

## Ladies in the Shadow of Giants – Bar Owners from Central and Northern Estonia

The village in question is a small, agriculturally dominated community. Nearby is a small town. The village is a typical 'one-firm community,' where in the past almost all were employed by a collective farm, which changed the form of ownership into a closed agricultural company during the era of decollectivization. The company is still the biggest employer in the village, and there are only a couple of private firms.

Three siblings, all of whom inherited money from their grandfather, own the local bar. A precondition of the inheritance stated that the money should be used for establishing a local business. Initially, the family members planned to start a forestry business, but then opted for opening a grocery store. In the beginning, their company merely consisted of a food kiosk. Their competitor was (and still is) a cooperative shop, which had a smaller range of choice and higher prices. Later, the siblings renovated the building where the kiosk had operated, and constructed a second floor, upon which they opened their new bar.

Soon after, they gave up the mobile shop and concentrated more on the catering business. The firm currently employs seven people, of whom four are relatives; the owner, her brother, the brothers' wife and her mother. Occasionally also the owner's husband also helps in the bar. The business is not very profitable and therefore the owner has kept her original occupation as an accountant at a school that serves a neighbouring town. The reasons for the bar's meagre profitability are obvious; the inhabitants of the village are poor, and the agricultural firm also operates a similar kind of business; a shop and a canteen in the village. It also gives a discount to their own workers and a couple of years ago the company paid a part of their employee's wages in foodstuffs, pork, milk, flour, etc. The bar owner thinks that the competing firm is trying to strangle her family business by dictating the amount of goods that are necessary to buy in order to get a proper discount. The competition has also brought along some positive effects; the competing cooperative was compelled to enlarge their supply of goods, and to lengthen the opening hours of their shop.

Local people primarily visit the family bar only when there is some party going on; otherwise the regular clientele consists of workers from the neighbouring town. Local people visit very seldom, perhaps on payday. It has even happened, says the entrepreneur, that 'the bar is empty even if there has been a band and music. Locals do not come here at lunchtime just to eat. That would be a real miracle, if such thing would happen.' She admits: 'Well, you cannot get rich with a bar. But I cannot go bankrupt either, because we have so many employees – probably we will eke out a meagre living for a year or two, we are so dependent on this locality.' Another big obstacle is the lack of skilled employees: 'If you need a worker, there is nobody to hire, there are only poorly skilled workers, and many have problems with alcohol, those old tractor drivers. The fact that there is unemployment is not enough, because there is the also lack of competent and diligent workers.'

Even though this entrepreneur has not faced the threat of racketeering directly, her enterprise was indeed once visited by a group of thugs. They appeared during the beginning, and according to entrepreneur, were estimating what the economic possibilities of the firm were regarding payment for 'security services.' Due to repeated incidences of criminality, the firm hired a guard. Another manifestation of social problems that is causing inconvenience for the firm is alcoholism. According to the entrepreneur: 'A couple of times a month, on pay-days, if a decent person would like to visit the bar, there is such an evening of drunkards that he or she decides that he or she will never come here again.'

Another bar owner tells a similar story from a small town in northern Estonia. She used to work as a teacher at a local school, even after she had started her own business. For her the teaching work was (like her colleague in central Estonia), a necessary safety net in case the business might fail. She admits that:

> The wage of a teacher was low and I imagined that with this [bar] I certainly could earn more than I would from the income from the school. Now the wages in schools have risen, and if I had the same wage [as now] at the school, I would have never started this [venture].

The bar in northern Estonia was established in the basement of a block of flats, which was on sale following the bankruptcy of a local paper mill, which currently owned the building. The transaction was financed by a loan from friends and relatives. During the first year, the proprietress was almost the only employee in the firm. She cooked, cleaned, washed dishes, and took care of the bookkeeping and orders. During the next year, she employed one person and repaid her debts. Now she has six persons on her payroll, and she can devote herself to managerial issues.

This Estonian entrepreneur has developed her business very carefully; step-by-step and only after serious analysis of the contingent possibilities and risks. She has taken out some loans, but only for herself; to buy a car, but not for the bar: 'Otherwise, I have not taken out much credit; it is too complicated and becomes very expensive. If you have a big company and something valuable as collateral, then it is easy to maintain credit.'

She does not have competitors in the town, since the bar that existed in the neighbourhood of the factory went bankrupt, *because* '... there were some young men who drank a lot and used their money in that way [for drinking] and therefore they had no money to make investments, and there was also a lot of disorder.'

The lack of skilled labour is for this entrepreneur the biggest obstacle in developing her business, in combination with the low purchasing power of the local residents, many of who are unemployed. Most of those, who still have work, are employed by the paper mill, which pays rather low wages. The main incomes of the bar, the same as the bar in central Estonia, come from the sale of alcohol (i.e. beer and spirits, especially during the week ends) and the food acts only as an enticement for people to drink. This fact causes a lot of trouble for her:

> In the evenings, especially at the turn of the month, when people have money, Fridays and Saturdays are very tough for me; there are a lot of difficult customers. Young people come and drink, and then there are fights and all sorts of disturbance. People smash dishes. I cannot stop wondering why it is so necessary to break and steal, toilet paper, hand dryers, paper baskets and countless number of glasses. Everything is being stolen. And tables are demolished.

Through the years, however, she has succeeded in establishing herself as a serious entrepreneur. She has studied bookkeeping, ordering goods from wholesalers and local producers. Her plan for the future is to gradually expand her business to other activities in the catering field. One could convincingly argue that she has, thanks to her cultural capital (and especially her social networks), learned the rules of business, as well as what it takes to become a successful entrepreneur.

Her strongest supporter and adviser has been her relative, a successful local shopkeeper. He has acted as a consultant for her in many matters regarding running the business. This is related to the fact that attitudes towards male and female entrepreneurs differ significantly.

> To be alone as a female entrepreneur, it is rather complicated. Let's say some maintenance [needs to be done to] the building, if there is a need to repair windows. As soon as the workers see that the person who orders the repairs is a woman, immediately things are done differently ... When my relative had some repairs done in his shop, he was controlling what was done and everything was done perfectly.

Another difficulty is the attitudes of people towards entrepreneurs in general.

The bar owner from northern Estonia says: 'there is the attitude like what do you have to complain about? You can afford to live well.' The owner of a textile firm has also noted the same attitude: 'Here you are not accepted and therefore I do not have that many friends in this town.' The mayor of a different town observes that this attitude has its roots in the past: 'and of course the Soviet-period has its role until the new generation steps in. Here [in Estonia] people think of entrepreneurs as if they were a kind of kulak, that they have it good, they have money, but they do not think that these people offer employment to other people.'

According to entrepreneur from northern Estonia the local community and its leadership is not encouraging entrepreneurship: 'I don't like this town, the people act somehow ... like this stealing ... people have low wages and they do not go to the theatre or to parties, because life is so expensive that all the money goes to food. Almost everybody has a plot of land, and it takes a lot of time and energy.' The city administration is too passive, and is not interested in promoting entrepreneurship, she says. She wishes that 'there would be somebody active a man or a woman who would take leadership in the city. Then things might start to change. But wise and diligent people do not take the job of municipal boss, because the wages are low and the responsibility is big, and there are a lot of duties.'

These two stories illuminate the patently disadvantageous position of rural entrepreneurs in the service sector, who are dependent on local demand. There is practically no market for their services, due to the large state of poverty in rural communities. There is also no labour market, and there are very few possibilities of getting support and help, besides relatives and friends. Entrepreneurs are forced to hold two jobs in order to cope with the low income that their firm provides. One extreme example of this is a hairdresser from southwest Lithuania, who formerly worked as a masseuse at a local sanatorium. Following the disintegration of the collective farm, she lost her job and was compelled to set up her own firm. This business provides hairdresser services, massage and health consultation services. Yet, none of these practices provides an income good enough to make ends meet, and she must work as a part-time teacher in a nearby school as well as visit Spain regularly as a guest worker, picking grapes.

With that in mind, one can argue that even though these entrepreneurs in many respects fulfill the criteria by which Scase characterizes them representing proprietor type or even survival type of entrepreneurship, it is not the entire picture. Far too often the entrepreneurship is just a personal endeavour, taken on by coincidence, or compelled by sub-standard levels of income. These kinds of 'accidental entrepreneurs' have very narrow. perspectives regarding business, which for them acts only as a safety net, or second job. Their possibilities for growth are marginal. Their incomes are low, since most of the profits go into the upkeep of the firm, as opposed to personal consumption. Still, they are definitely hard working, and in many respects fit the definition of the 'Weberian type' of entrepreneur. Yet this is not proprietorship in the sense that Scase describes as dominant in most post-socialist countries. The fact is, this is not proprietorship at all nor is it true entrepreneurship. Rather, it is merely a way of survival. In a market economy, private entrepreneurship does not succeed by relying on the help of relatives or friends. In stead, it requires *institutionalized support* in the form of loans with low interest rates, either from banks or from the State.

## The Competitive Edges and Disadvantages of Rural Enterprise

In rural areas, much of the local entrepreneurship is based on spin-offs from former collective farms where the general level of technology and infrastructure is underdeveloped, and production is by necessity labour intensive, based on local material resources and traditional skills. The regional enterprises have limited resources to train their own workers to use new technologies and materials.

In many cases, the only competitive edge of such enterprises is the wage level, not the quality of the product. Together with primitive technology and work conditions, this scenario creates serious motivational problems among employees, as the stories below will readily testify.

### A Toy Factory in Northern Estonia

The toy factory in question was established as a family firm in 1991. It started out as a reforestation firm, but changed its line of operation within a year. The premises of this factory are in an old cowshed that was purchased in an auction during the era of decollectivization. There are three brothers who are both owners and employees, along with five paid workers. The best times of the company were during the mid 1990s, when the company employed twenty workers. Originally the company planned to engage in furniture production, but since there is a larger, and more developed furniture factory in the same municipality, the company decided to begin producing wooden toys. An additional initiative for the brothers was the interest of a Swedish corporation, which wanted to buy their products. Nowadays the toy factory exports mainly to European Union countries, and only 5 per cent of their products remain on the local market. The brother who is a director of the firm, admits that there have been some problems in finding new clients:

> All of the potential clients we have offered our products have refused to buy. When we made our offers, immediately a frantic bargaining begins and they think that we can produce our goods for nothing. Only those customers who have found us themselves have remained our clients.

The main problem for the toy factory is that there are other vendors, with more sophisticated technology who have decades of tradition and brand recognition, especially in foreign countries. The technology of brother's factory is simple, labour intensive, and according to the entrepreneurs, this is practically the only competitive edge they have: 'The worse things get with the Estonian state, the better chances we have of staying in the market. It is a question of wage levels, because if the wage level gets very high, we shall be kicked out of the market.' The entrepreneur estimates they will survive over the next ten years, because the fact is that:

the number of firms is going to decline. We can talk about saving the small enterprises, but I cannot see any real help anywhere. If real help was coming, it would require firm decisions from the Parliament, but it is dependent on big companies to such an extent that one could easily call it political demagogy.

One of these entrepreneur brothers thinks that not even new investments would help, because of their limited incapacity to compete with foreign producers, and 'to get investments to such a small firm, even if I had money, I would not invest here. The chances of an Estonian firm gaining a foothold in such a branch are only possible for a short period of time. It shall not last.'

So, even if he, as an entrepreneur acknowledges their defeat in the market, and the fact that firm provides only a living for them and their employees, he has not become bitter, but he personally has learned the basic rule of the entrepreneurship.

I know, after being actor in the market economy, that what I have gained yesterday does not really have any value tomorrow ... and I do not feel offended if I see tomorrow that I do not have anything left. Every morning one must start everything all over again, 365 days every year.

## A Sawmill Owner from Eastern Latvia

The municipality where this particular sawmill operates is one of the poorest in Latvia: agricultural enterprises and farms are extremely small, and the unemployment rate and levels of poverty are high. The sawmill owner is in his thirties. He was born in a small town, graduated from a local elementary school, and studied to become a carpenter in a near-by city at a vocational school. After a stint in the Army the first job he got (with the help of acquaintances) was in the police. He worked there for less than a year. This was during the time of the collapse of socialism (1991), and like many others, he had the idea of becoming a private entrepreneur.

Initially, he engaged in trade, and had a kiosk. Later, he began also to rent rooms for the shop. He bought the kiosk with small rouble loan, but this business ended abruptly when his competitors set his kiosk on fire. His first business lasted four years, until 1995. Regarding the reasons for the failure of the kiosk business, he notes:

I am not the son of any communist party leader or a kolkhoz director who can support my own business. Thus, I was alone in my business and all the money I could invest in my business was my own. I also regret that I did not have good education. I have not studied in a high school or university. Without any acquaintance or friend in the state institutions, it was very difficult [to be an entrepreneur], especially with a small firm.

After giving up his businesses, he returned to being a wageworker at a Swedish-Latvian joint venture in forestry, where he worked as the manager of a department. During his two-year career in the company, he bought some basic equipment from them, partly with his wages and partly on credit. He established his new sawmill in a small neighbouring rural village. The sawmill was located on abandoned pig farm, which had been privatized during the decollectivization of the kolkhoz. But before making the decision to start business, he visited a number of sawmills in neighbouring municipalities and regions, as a form of 'market research.'

Nowadays, he has one workshop with 20 to 25 workers, most of who have earlier worked in the kolkhoz. He says that he worked with more than 50 people before he was able to find people actually capable of doing the demanding work.

> Those people are the best workers who do not have their own farms and land, because they do not really need extra money. It is not good to pay anything in advance, because it is for certain that the next day there will not be any worker. Maybe on the third day somebody will arrive.

He thinks that in cities, however, the labour market is better, because there is more competition within the workplaces, and the workers must be responsible. In rural villages 'I have to employ those who are living here, and I have no choice. Some of them I have hired already three times – but I cannot fire them because there is nobody I could hire to replace them.'

The sawmill was more profitable during its initial phase of operations, when there was less competition – even though the entrepreneur notes that 'there are around 100 sawmills in the neighbourhood, and maybe five of them are competitors in woodcutting. Most of these 100 sawmills have only one very simple saw, and they sell planks to local wholesalers.' The most serious sawmill competitors are foreign – predominantly Swedish companies who can sell at much lower prices than a Latvian firm can. He also says: 'I am worried about those bigger sawmills. Those bigger sawmills can saw around 500–600 m3 per month, but I can saw only 150 m$^3$.' After five years he thinks that there will be only perhaps five sawmills left out of one hundred, because the smaller sawmills are just too small and technically backward.

At the moment the sawmill owner dreams about starting up furniture production:

> I think that I could invest that money in processing, perhaps carpentry workshop to produce something for export. The market is overcrowded with planks for floors, but there is a deficit of pre-glued plates for producing furniture. But then I would need a dry-house, a plane and a special fixing for gluing and so on. I am interested in the possibilities of developing such business, but the price even for second-hand machinery is huge.

## A Furniture Factory in Northern Estonia

The furniture factory that belongs to this particular company operates in the cowshed of the former collective farm, which was purchased at auction. The sawmill is less than one kilometre further, across the river. The company itself is situated in a peripheral area of the community, in a rather rickety and dilapidated infrastructure. The cowshed is also fairly dilapidated, since one of the sheet-metal walls has been stolen. There are, altogether, twelve men who work for the company. The furniture factory employs four of them: one man works with a drill, one operates the machine saw, one uses the planning machine and the fourth is the shop director. All of the equipment is old and worn-out, dating from the woodworks department of the collective farm. The factory produces legs for bedsteads and the gratings for beds. They are currently the sub-contractor for a larger Tallinn-based furniture factory. Their sawmill produces material for a nearby summer-cottage construction company. According to the director the key problem of the company is the quality and motivation of the labour force. It is extremely difficult for him to find skilled and reliable workers. The larger, more modern and successful furniture factory in the municipality has already employed the best workers. Those who are left over seem to always have problems with alcohol and proper work discipline. The director complains:

> If you need a worker, there is nobody to hire. There are only poorly skilled workers, and many have problems with alcohol, those old tractor drivers. The fact that there is unemployment is not enough, because there is a lack of competent and diligent workers. I cannot even find a reliable supervisor and therefore I have to do everything by myself: organizing the work, doing the marketing, selling, transporting and bookkeeping. My workers can only do the simple work.

## A Brewery in Southern Estonia

The brewery we will examine currently operates on the outskirts of a small rural town near the Latvian border. This brewery was established during the 1960s for the production of sweet fruit wines, which is still a part of the production profile. The other, larger part is dedicated to production of beer. The brewery was formerly owned by the local kolkhoz, and employed approximately 130 employees during late 1980s, which was historically the best time for the brewery. But the glory days of the brewery ended during the early 1990s, when the privatization of the kolkhoz took place.

The brewery was then sold to a group of businessmen from Tallinn, who had a wholesale business, but no experience in running a brewery. The new company is currently under the supervision of two hired managers. One is responsible for the production of goods, and the other is the director of finance.

The nature of restructuring of the company has been cautious. The production lines have remained almost the same, as they were when the company was purchased. Some investments have been made in the quality of their beer production. The company received more modern production technology from Germany. This technology was already second-hand, and was donated to the company as humanitarian aid. The number of company employees has declined rapidly, and the company currently employs only less than one-half of the peak year's labour force. The biggest reductions have taken place in auxiliary work and transportation. The basic structure of the labour force has remained the same; most of the employees have been in the company more than ten years.

This is both good and bad. The good side of the situation is that basic knowledge capital has been preserved, which is crucial for a brewery business. The bad side is that the employees are not accustomed to new times.

The prevailing mentality and attitude towards work is totally different here from that which exists in Tallinn. Here, the kolkhoz mentality prevails and is very deep – there is a lack of initiative and thinking. People need very comprehensive guidance even in the simplest things, which should be obvious to anybody.

Yet, the problem of work motivation is not a generational issue, the director of the company argues. It is more of a general attitudinal problem, which could be solved by a raise in wages. The only real solution, according to him would be stricter administrative measures, such as more surveillance at the gates when people come and go. 'Total control,' as he says.

All the firms described above share similar features. Their technological level of production is low, working conditions are poor, and the products are not very competitive. Therefore, the firms are not able to pay decent wages or offer benefits other than work. Their employees are predominantly old, and do not possess any particular skills that could be sold in the labour market. Their age, low wages[2] and lack of skills prevent them from moving to cities, where the level of living costs – rents, payments and prices – exceeds their financial means. In this situation the workers lack the motivation to work as diligently as an employer might prefer, and the employer lacks actual means to improve employee motivation or working conditions. What both parties probably share is a common aversion towards trade unions: for workers, the trade union represents an organization that cannot (and is not really interested in) help them, while for the employer the unions represent needless and potentially costly 'middlemen' in their relations with the employees. The rate of unionization is in all Baltic countries very low, and despite the fact that there are indeed unions, which are trying to organize employees; the development of the trade union movement has been slow. Unions mainly operate within large companies and the state sector, but in the private sector (and especially small and medium sized firms) unions are practically non-existent or actually banned. According to the results of the Working Life Barometer in the Baltic countries (1999) only 8 per cent of employees in the Estonian private sector currently belong to trade unions, and the percentages in Latvia and Lithuania are not much higher; 10 per cent and 11 per cent respectively. In the public sector the rate of unionization is significantly higher. In Estonia it is 20 per cent, in Lithuania it is 21 per cent and in Latvia an impressive 42 per cent of all employees are unionized (Antila and Ylöstalo 1999, quoted from OECD 2000, 165). The OECD report notes:

Less than one-fifth of all workers are aware of being concerned by collective agreements...More generally the surveys suggest that most workers regard wages and

---

[2]   The rural wage level in all Baltic countries is less than three quarters of the wages in the capital areas (OECD, 2000).

other employment conditions as being determined essentially by 'individual' rather than 'collective' procedures [with or without trade unions] (Ibid.).

The prevailing attitude among entrepreneurs is surmised succinctly in the opinion of a textile firm owner: 'No, there is no trade union here. It is one of those things that I do not need. It is such, how should I put it, those unions, they take us back to socialism, again.'

The organizational weakness of the workers in terms of interest representation and defence to some extent provided employers with almost unlimited powers to hire and fire, especially during the early years of transition. The textile factory owner from northern Estonia provided a most striking example of this. According to her during the beginning of operations her Finnish partners had completely different ideas about the goals of the firm:

> There was this lady, it is good that she understood to leave, we had totally different views ... there was also another Finn, he was not in the textile business but was a building worker, he understood even less what was necessary. When debts were big, he did not want to pay. I told him that we must pay wages, but he just asked why, let's get rid of the old ones and take new workers.

In this situation, workers often feel that nothing matters and nobody can help or is even interested in their lives. The widespread use of alcohol is one consequence of such social instability, even if it could also be interpreted as an informal means of protesting against those who are well-off or on the opposing end of the situation in which a person finds them self. On an every day level, these problems can be seen in the deep distrust between employer and employees, and in a very instrumental attitude towards work. On the other hand, it highlights elements of paternalism on the part of the employer and 'positional autonomy' on the part of the workers, which for the employer indicates a perceived continuity of the 'kolkhoz mentality.'

It is hardly the whole truth to say, as many entrepreneurs and officials in public administration readily claim, that it is the 'kolkhoz mentality' or the 'Soviet system' that is the ultimate reason for perceived laziness and indifference towards work. The problem is complex, involving social and economic factors.

In the first place, most of the rural enterprises are small and technically under developed. Their 'market' is predominantly local or domestic at best.[3] Therefore rural entrepreneurs cannot compete with the wages found in enterprises in larger urban centres. Secondly, young and qualified labour is constantly drifting away from decelerating rural areas, and therefore local firms must be satisfied with poorly skilled labour. Perhaps the most crucial factor is the deep state of anomie that is characteristic of rural areas in the Baltic countries – decollectivization has destroyed their former means of livelihood, and during that process they have witnessed the destruction not only of the collective farm, but the community,

---

[3]    According to the EMOR survey (2000) in Estonia, 60 per cent of small firms sell their products in their own parish and 13 per cent in their own county. Only 7 per cent of Estonian firms have foreign countries as their main marketing area.

norms and values that existed in that community as well. The Soviet type of collectivism, although it was paternalist and characterized by tight top-down control, along with a certain degree of 'forced autonomy,'[4] was followed by a complete collapse of normative order.[5]

## The Elements of a Successful Firm

Kolodko (2000, 8–9) argues that some reasons for the slow development of the SME sector in many post-socialist countries revolve around the managerial traditions and skills which were relevant during socialism: 'Their management did focus mainly on the procurement of the means of production, raw materials and other assets and, of course, securing a labour.' Therefore, according to Kolodko, the problem is not the lack of managerial skills as such, but the type of management that often does not correspond to the needs of a private capitalistic firm. This view undermines the belief that privatization alone can give rise to a capitalist managerial or enterprise culture. Running a competitive business requires modern managerial skills, and the success of a firm is more dependent on a continuous integration of technical sophistication and specialization.

The entrepreneurs who have succeeded well in their businesses and are exporting or have contacts with foreign firms have a different kind of attitude towards their employees. They acknowledge the rights and value of the labour within the firm and do not 'externalize' the reasons for varying quality of the labour.

### A Furniture Factory in Northern Estonia

The locality in which this particular furniture factory is operating is small a town with slightly less than 4,000 inhabitants. The county of Rapla, in which the town is included, is dominated by the food processing industry and agriculture. There are numerous timber companies, most of which are small firms. The average size of the firms in the county is less than 10 employees. Due to its location near the capital, Rapla county has succeeded in inviting numerous foreign investments.

The furniture factory itself occupies the premises of the former headquarters of the local collective farm, which was privatized in 1993. The current director of the factory was also the last director of the collective farm. His formal training was as a mechanical engineer, but during vocational school he also learned the basics of carpentry work. During the era of decollectivization the farm property was divided into separate units, and farm employees subsequently privatized those units. The

---

[4]    The concept of forced autonomy refers to the fact that workers had to find solutions to a number of problems stemming from the pathologies of the economy of shortages and waste (see for example Clarke et al., 1993; Stark and Bruszt, 1998).

[5]    A vivid description what consequences this collapse had at local level, is given by Ilkka Alanen and his colleagues in their study in a south Estonian rural community (Alanen et al., 2001, 146, 153–156).

director and a friend from the same kolkhoz privatized the sawmill at the collective farm, and along with the sawmill came the mill's employees: 'The one who started with some department, he also took those people who used to work there ... it was compulsory; it was in our reform plan. And so that firm started. In the beginning there was no need to hire workers from outside.'

The number of employees has doubled from 15 to almost 30 since factory's beginning, though in the early days they did not need so many workers. The only clerical worker in the company is the bookkeeper; other workers are engaged in production. The firm mainly produces chairs and tables on order, and the production volume has increased steadily. The quality now meets the requirements of customers, even if the models are simple. There is a gradual movement towards mass production, especially to the German market. The aim of the current entrepreneur is to enlarge their production slowly and gradually, within the limits of financial possibilities.

According to the plant director, the workers have slowly learned the things that are important to success in working life: quality, punctuality, cost-efficiency and higher skills. The learning process has taken place in collaboration with the entrepreneur and he actively acknowledges the importance of a skilled labour force and good working conditions regarding the overall success of the company: 'there are no such good workers who can drink heavily and do good job. If you do lousy work, then nobody will order anything from you any more.' An essential part of decent working conditions is the right of workers to organize and get their voice heard, 'It will come. It is important that they come [the unions], it is not possible that there is only one king, who dictates how things should be and there are no normal negotiations about what the employees might need.'

Anyhow, still the most important issue and biggest obstacle for development of this business is the question of a skilled labour force. There is a labour pool in the municipality, but their level of skills is not what a developed mode of production needs. The possibilities of getting skilled and motivated workers are limited by two phenomena; competition with business in the capital area, and the related issue of decent wage levels. 'For good men you have to pay the average Estonian level, but for very good workers you must pay doubly. Otherwise, they go to Tallinn and can easily find a job.'

Another example of a successful firm where modern workplace relations are a part of success is a woodworking factory from Southern Estonia. The factory operates in the premises of a former collective farm on the outskirts of the municipality. The manager here has no formal training, but after high school he went into the army and afterwards worked at a Swedish farm for a year as a trainee. Thereafter, he gained employment at a Swedish-owned sawmill in a neighbouring county, and participated in the construction and start-up of a sawmill. Thanks to his knowledge of Swedish language, he later gained a job as a quality consultant for the same sawmill company in Latvia. The decisive impulse to start his own business came from a Dutch timber agent, who claimed that there would be a good market for planed timber in Holland. The main portion of the start-up capital came from Sweden, from the farmer for whom he had worked years before.

The rest of the investment capital came from the entrepreneur himself, and his former schoolmate.

Their business has developed rapidly – in the beginning there were only four people working at the firm. Now there are 25. The net revenue has grown from one million crowns to 30 million crowns during the past five-six years. The secret of his success is regarding his entrepreneur is *honesty*, both in respect to his clients and his workers. In relation to his workers, his strategy is simple; he always pays more than others and pays on time. That gives him the leverage to demand diligence in work. This entrepreneur says:

> The problems ... they are probably similar all over the world, these problems that we have, are everlasting and constant. They exist between the employer and employee ... much depends on the employer himself. Those firms that have problems with their workers pay too little – in our firm the average wage is 15 to 20 per cent higher than the national average. In this branch there has not been a single case of wage arrears in our firm.

Another successful entrepreneur from another town in northern Estonia sees the problem of labour as a reflection of the level of entrepreneurial skills:

> I argue that the problem originates with the entrepreneur himself – how he can solve it. Naturally, if things are in confusion, either in the firm or in the management, then these problems are also reflected in the labour force. One can of course argue that the quality of labour is poor, but it stems from the upper level, what kind of discipline there is. The problem is in bad management. All elements have to be in place, economic efficiency, and a view of the future, as well as what has to be done in order to achieve the goal, and this also depends on wages. The worker is not there just for benevolence, but for his or her wage.

In this firm one of the most important success factors has been specialization, the ability to find a niche – in other words to develop a product that is distinctive from the mass of unrefined goods, and which also can connect to markets abroad.

The creation of this entrepreneur's own brand has taken a long time and has come through experimentation. The firm itself began with the production of simple logs, and their next step was production of wooden bases for kettles and pans. Nowadays the firm produces high-quality laminated plates. The company has invested a lot in developing their products and their technical sophistication, which seems to be a kind of a precondition for high quality products: 'Our aim is to buy new equipment and through that we will increase production – the technology grows outdated very quickly. That is a problem, and one has to develop constantly.' The long-term strategy of this firm is to replace human labour with machines as much as possible and for that purpose they now pay more than national average to good workers in order to get the quality of production that is competitive in foreign markets, and thereby brings in money. The entrepreneur in this case does not see himself as particularly obligated to provide employment for local people – for him that is the duty of local or national government. 'We hope that we could engage in more cooperation with local government, whose duty is to create infrastructure. We have had the goal to create an area of cheap municipal services, so that people could stay here and should not be forced to move to Tallinn, for example.'

In this company, there exists the widely esteemed features of traditional entrepreneurship; the director of the company has long-term plans for necessary development and the investments it requires and he also has concrete strategies about the ways to finance them. Contrary to most entrepreneurs, this manager says that 'a bank-loan is a normal thing and the only possibility of financing investments – one must be economical, then loans are not that expensive.'

According to the director the secret for a successful entrepreneurship is simple: 'One must save money, nothing else. Many entrepreneurs use too much money for themselves – one must invest continuously, and therefore we do not travel to the Caribbean.'

In conclusion, it can be said that the ingredients of success (indicated by constant and steady growth of production and turnover along with growth of exports) are not only in the hands of the entrepreneur, but perhaps the most crucial elements are 'external;' existing relations to foreign companies, a clear and consistent business plan and access to foreign markets and financial institutions. These provide an opportunity to follow a personnel policy in which rights and duties of employees are officially acknowledged. Only in this way is it possible to build trust as an ingredient for future success. For employees, it is important to have stability and predictability in their life – to see that wages are paid in time and promises are kept. Workers also expect some kind of respect from their employers, as the above entrepreneur noted 'workers are not working just for benevolence, but for their wages.'

For entrepreneurs, trust is related to the actions of the state. The main issue in this respect is the present institutional performance, in other words, how well institutions are doing their duties. Insufficient legislation, poverty and the low wages of the representatives of public administration often result in semi or patently illegal practices on the part of public servants, business clients and sometimes the entrepreneurs themselves.

## Crime, Corruption and Rural Entrepreneurship

One of the obstacles of the development of 'normal' business in many post-socialist countries is the rampant prevalence of different forms of corruption and criminality. As the European Bank for Reconstruction and Development noted in its 2002 Transition Report: Agriculture and Rural Transition, in 1999 the small private firms in most post-socialist countries paid 7 to 9 per cent of their sales as bribes and tax avoidance was relatively common among small firms. The Baltic countries, particularly Estonia, were among the least corrupt countries, while the position of Latvia and Lithuania has improved during the past few years (EBRD, 2002, 27–30). Crime and corruption are not only typical for mafia-type organizations, but also are engaged in by quite a large segment of local and national bureaucrats. Such bribery and graft is often known as 'greasing' the officials in tax offices, as well as health inspectors, people in charge of licenses and permits, and others.

A study by Mehnaz, Graham and Gonzales-Vega show, for example that Russian micro-enterprises (which engage in manufacturing, are less diversified and have relatively good prospects for growth), are more prone to engage in corruption than other firms are. The relations that enterprises have with mafia-type groups are not unilateral, but instead are such that both sides can profit. These groups are not just ordinary criminals, but include people who have transformed racketeering into the sales of 'security' as a product, and have developed their businesses in the typical capitalist manner – transforming from a loose group of thugs into a company. According to Latvian rural entrepreneurs the influence of this kind of business has not decreased, but it is still controlling some branches of economy. Many of the mafia-type organizations or groups try to legalize their actions, such as establishing personal or business security firms. The most important and profitable branches of business for mafia-type of groups are those that involve bigger money, such as fuel business, and of course, illegal markets such as prostitution, drugs and guns.

The role of social networks and personal relationships that entrepreneurs have with authorities is crucially important; in fact it is more important than formal agreements between e.g. entrepreneurs and authorities. In a society where the legal system and normative structures are underdeveloped, and social relationships are not formalized to the point where there is predictability regarding peoples' actions, informal relations and trust based on friendships or loyalties is the primary basis for organizing relationships between officials and businessmen. The opinion of one Latvian entrepreneur emphasizes the most important elements of 'personalized trust:'

I think that it is very important to know those people well with whom you are collaborating. If you do not know them, then it is difficult to make an agreement without showing money. And even then they wish to go to a bank and to transfer the money to feel surer. But if you know each other, then there is also some mutual trust.

Concerning the role of personal relationships for his business, he notes:

It is important to have contacts not only in your own business, but also in other institutions ... and I have acquaintances everywhere. If some policeman stops me on the road due to exceeding the speed limit, it is easy for me to agree with him because I know people from the police department. If I did not have those contacts, it would be much more difficult to manage my business here. I spent all my life here, and I know many people. There are different institutions whose requirements I should meet: ecological, firemen, the tax office, economic police, etc. I think that it is easy to find some imperfections in every business, but as I have those contacts it is easier for me to ensure that all my documentation is in order. It is impossible to meet the law 100 per cent.

Corruption and criminality have not been limited only to 'ordinary criminals.' The practise of corruption has penetrated deep into the public sphere. As an owner of a construction firm from Lithuania observed: 'Racketeering is finished, all the people who were involved in it, are now in prison or in the government. Now we have state racketeering, and everything goes through taxes.' This statement is, of course, an exaggeration of the existing situation, but for many entrepreneurs unjust taxation and systems of control simply represent more developed forms of 'racketeering.' This kind of attitude is common as well among entrepreneurs in Latvia and Estonia. One such sawmill entrepreneur from Latvia observes:

> It is also the same here [as in Lithuania or Estonia] … It is more difficult for the small firms that nobody knows. Also the bureaucracy here is terrible. Of course, if I would pay all the taxes I would not earn anything, and I would have to close my firm. It is common that taxes are not paid. Those who are honest are not competitors. Therefore, state enterprises are not competitors. And it is very common in the wood business that big part of deals are illegal, without any documentation …We simply do not declare the right sum of the purchases: if we bought a forest for 1000 lats, we declare that we have bought it for 500 lats. We try to legally declare at least the minimum wage for our workers and to pay some taxes for them – social and income tax.

Therefore, in order to survive, one must consciously circumvent certain laws or at least fulfill the regulations while incurring the minimum.

This point is confirmed as well by the bar owner from northern Estonia: 'There are always some problems. And I know very well that the number of those who have all the taxes and payments paid is very small in this town. If the taxes are high, the entrepreneur is also dishonest in relation to the state.'

The other side of this issue is the state of existing corruption and criminality, which still exists, at least to some extent, even though the worst such period was during the early 1990s. A sawmill entrepreneur from Latvia recalls that:

> When I had a small enterprise, I had to pay some money. Very recently, a half-year ago, some gangsters visited me, but they got it in the neck and they have never again been here. Then, there was a case in which some mafia people from Riga arrived, and they also wished to get some money. I told them that I am already paying to local people. Those people met and agreed on something between them, and I have not seen them anymore. If I had some problems in my own business – some debts and so, I asked them to solve it.

In the early days of private entrepreneurship small foreign companies misused the non-existent laws and regulations and abused the trust of many rural entrepreneurs, all for their own profit. The most common form of this practice was a type of fraud, in which foreign company would order goods from local producers without any intention of ever making payment. The sawmill owner from northern Estonia provides a brief description:

> We had a Finnish broker, whose company sold products to Finland, Germany and also to Great Britain. But that man, he was a swindler. We had a lot of problems with him. He had a limited company, and there was big money moving, but he always had problems, about where the money was, nobody could ever find it.

There were very slim chances of initiating litigation against such a businessman, because:

This man, he had a couple of 'armours;' you know what armour is? It is when a man does not have a working permit in Estonia, and does not own anything. He just takes care of bank operations. So how can anybody demand anything from that kind of person?

All in all, there are numerous questions about the general lack of norms (or even trust), which should be the basis for 'normal business;' in which predictability of the actions of one's partners is a crucial element. In a situation where such preconditions are missing, business relations are predominantly *personal or personalized relations*. The processual trust (i.e. trust based on social networks) has a strong position and this plays essentially the same role as it did during the socialist period, opening gates and doors to scarce resources and political influence. In a situation where the operational environment for enterprises and institutions continuously fluctuates and is unpredictable, network relations gain a more important role and eventually become an institutionalized element within the society. The background for this is the ultra-liberal policy of a minimalist state, and a lack of any coherent social policies. The regulative and supportive functions of the state have been stripped to the minimum, thereby ushering in an 'autonomization' of the state sectors – each functions according to rules and laws of its own. It could be argued that the role of network relations will diminish the longer the firms operate within normal market circumstances. But, as such examples as Southern Italy and Russia have shown (Wendell, 2001), highly developed network relations may render market relations unnecessary and there is no ultimate mandate that a normal market should even be the final objective, rather it is the case that institutions and firms can operate very well in numerous kinds of social circumstances.

The causes for a weak state are complex, and their inter-relations vary from country to another. Yet, among the most crucial factors behind the weak state is the way different social groups made their interests heard during the process of change over from socialism to capitalism. This factor combined with the inefficiency of the state (and consequent 'poverty of the state'), to hamper capacity to take care of the very social services that privatized firms have outsourced, either to local communities or for the state.

The consequence of this dysfunctional scenario result in public officials who earn very little, which in turn increases temptations to accept bribes for their services. It also seriously limits the possibilities of developing social services and therefore people are compelled to rely on informal social networks, such as friends and family. This all nurtures a 'robber state' with a widely influential mafia structure and an extensive grey economy. Corrupted and subordinated civil servants then leave room for mafia-type structures and organizations, and the state has no means to act as a third partner for business. The state becomes completely inefficient and there is no support for the state among the people. This further promotes the growth of a grey economy, double bookkeeping, double wages, tax evasion, bribes and other such phenomena.

## Local and National Governments, a Boosting or Hindering Development?

In international discussion it has repeatedly been emphasized that there is a need to develop alternative forms of employment in rural areas, to replace the current losses in employment. Rural unemployment is one of the most serious problems in almost every post-socialist country. There is not just a question just about unemployment per se; rather unemployment involves a complexity of social problems, poverty, alcoholism and mortality, along with the decay of social and productive infrastructure.

Agricultural reform in the Baltic countries was carried out in haste, and mainly with political, as opposed to economic, considerations in mind. In many cases, these reforms were carried out against the will of the local people, who wished to get rid of socialism, but not from large-scale production. This hinges on the fact that the technical and economic basis of agriculture was initially constructed to serve the needs of large-scale production. A significant part of the new private farms and enterprises were thrown into the cold water of the market economy without any of the necessary means of survival. There was a remarkable lack of advisory and training services, the banking system and legislation were inadequate to serve the needs of the emerging private sector. The local administration has also been unable to support or help farmers or small enterprises in almost any way. In recent years, the local administration has been repositioned in social research as an incubator of local development. This means that local administrators could and should have an important role in creating and guaranteeing the preconditions for local development through financial support, training and advisory services (McIntyre, 2002). In practical terms the local government should, according to McIntyre provide local entrepreneurs with subsidies for local goods, as well as channels of wholesale, marketing and local loans. However, in reality, local governments lack the means to support local enterprises, because there is no financial base from which to pursue local development policy. As already noted in rural areas, the income level is significantly lower than in urban areas, and because of high unemployment levels in rural areas, there are no tax incomes for local administrators, who in many cases are dependent on the assistance of a central government.

The cooperation between the 'business community' and 'political community' is necessary in order to promote general well being, and to create the preconditions for future development. Entrepreneurs generally lack both the resources and information they need for development of their businesses. Entrepreneurs should be an asset for local governments; they create jobs and bring financial resources through taxation.

According to rural entrepreneurs, none of the municipalities and states in the Baltic countries has either the interest or resources to support entrepreneurship, except at the level of proclamations. This point is made clear by a mayor of a small

Estonian town that hosts a paper mill and other smaller enterprises. According to him 'as a resting place our town has more possibilities... as the system of municipal financing, it is what it is. We have no interest in being active in the field of entrepreneurship; because the more we have entrepreneurs, the less we get state subsidies.'

Another mayor from southern Estonia notes: 'They write in newspapers that it is the duty of the local government to support and help entrepreneurs, but there are no real possibilities ...here we must sell vacant pieces of land very cheaply, just to get somebody to start something.'

Some of the entrepreneurs understand the fact that cooperation between local administrators and enterprises is a necessity:

> Well, I think there are some deficiencies in the training of municipal authorities. In reality, the municipality is a small enterprise; it has to sell services to inhabitants, which is at the same time the basis of its income. The more the municipality promotes entrepreneurship, the bigger are the incomes it receives, and the better the level of living in the community gets, and the more people are willing to move in to the municipality and the more enterprises are established. The local bosses must understand that this is the way the relations between them and the entrepreneurs should be taken care of (Manager of a heating company, Estonia).

> There should be more cooperation and will to promote common interests; there are a lot of possibilities to improve things if the municipality wants that to happen (A manager of wood works, Estonia).

Certain mayors also see the relationship between the municipality and entrepreneurs in a similar way: 'When a person is able to employ himself and his or her family that is already a big step. He may not bring huge tax incomes to the municipality, but he manages himself, and the city lives, as well as her residents' (Mayor from southern Estonia).

There are entrepreneurs who think that successes or losses are his or her own problem, not the problem of the municipality or the state: 'The municipality can and should not help firms; if it can arrange relations between the people and firms, then it has done enough. The municipality does not give birth to babies, it does not buy toys from us, and it just cannot help us' (Toy-factory owner).

'The state's duty is the maintenance of roads, courts, army, police, making laws and I say once again that the state does not buy anything from me, it does not drink milk, the state has other kinds of duties.'

The quotation above describes the general attitude of entrepreneurs towards local or national civil servants. They wish for authorities simply to leave them alone and not to interfere with their business. On the other hand, this reluctance to cooperate with the local administration (or civil servants in general) produces a situation in which entrepreneurs refuse to take help, even if they would benefit from it. The director of a local business centre in southern Estonia complains: 'Sometimes I think "you stupid entrepreneurs, why do you not understand that this information and all the help we give is useful for you!"' According to her, entrepreneurs only want quick-fix help in situations of immediate urgency, like when it is time to file tax report.

Another side of the coin is the situation in many monocultural towns, where only one large company exists. This company is typically the main employer and source of tax funds. A company like this has a lot of influence on local decision making, as shown in the case of a paper mill from northern Estonia. The paper mill is located in a small town, and it has been the town's main employer since 1920s, when the paper mill was first constructed. After falling into a temporary state of bankruptcy, the paper mill is now functioning again under foreign ownership. The company noted after its fresh start that the attitude of the city administration was unfavourable to them, and just before the elections the local leadership of the company created a factory union. This union sent its own candidates to the local election, and they won a majority in the city council. The vice director of the paper mill was elected as a chairperson of the city council, and she notes: 'Well, if we think it so, that the mill brings most money to local budget, it is only natural that we also control it … the city and the factory, they are one and they will always be together.'

The opinion that local entrepreneurs have about the role of the company in the decision-making and developmental strategies of the town fits well with that of most 'monocultural' cities:

> The problem here is that those people who are in the management of the company also determine the development of the whole city … and nobody knows what will happen if they turn out to be too incompetent to do it. The mayor is in some sense between the bark and the tree – if he wants to pursue his own policy, he has to follow … and they [the people at the factory] decide and determine.

In practice, this means that the major employer has legally 'privatized' the local power within their hands, and that the local administration is dependent on the company, and cannot in this configuration have much influence on the preferences of local development.

The Estonian government acknowledges the vast scope of social problems that exist in its rural areas. In the SAPARD[6] programme for Estonia, for example, it is said that:

> During the last ten years, the unemployment rate has risen during from 1.2 per cent to 11.7 per cent in rural areas. Mostly, this is due to the decreasing number of jobs in the agricultural sector. Regardless of the fact that many people have found new jobs in the cities, the living standards of rural populations have dropped, and the process of social rejection has become more intent. Therefore, the present measure was developed to create new jobs and mitigate the tense labour situation within rural areas of Estonia.[7]

In addition, the regional development plan of Estonia notes the same problems: on the one hand there are bigger urban centres, which have been successful in their

---

6   Special Accession Programme for Agriculture and Rural Development.
7   www.agri.ee/SAPARD/En/index_measures.htm

transformation towards a modern, western economy. Yet, there are still a large number of smaller cities and rural areas, caught in a 'restructuration crisis' (Estonian Regional Development Strategy, 1999). The problem also exists in the regional structure of Latvia; only a hand-full of successful towns exist while there are a number of regions characterized by high rates of unemployment, agricultural and industrial degradation, and poverty.[8]

To overcome the problems of uneven regional development, and its severe consequences, governments in all of the Baltic countries have created national programmes, and with aid from European Union they have also created special SAPARD programmes for agricultural and regional development. These programmes provide funds for the diversification of economic activity; they lend support to investments and help with starting up a new business. The problem with these programmes, according to entrepreneurs is that they involve intensive bureaucratic work and independent investments in order to get them. Therefore, it is no wonder that many of the entrepreneurs such as the construction company owner in Lithuania, maintain that 'these projects are not for us, they are only for rich and large companies who have the time to do all the paper work and the money to invest.'

According to the officer from an Estonian body which represents local governments, the problem is two-fold; on one hand 'all the people who want and can do business are already in Tallinn, and one should train people in those places [rural areas], but people are not able to think of how to do it. Well, there are business support centres in each province, but to my knowledge they only train, nothing else. They say, O.K., if you want to start your own business, you should do this and do that.' Another problem, according to him, is that there is no official industrial policy, except in the beginning, when the official line was: 'now everything is tax-free, just go ahead, develop and search for yourselves.'

The sawmill owner from the eastern part of Latvia thought that entrepreneurs such as himself cannot get such support because: 'I am not sure that without a bribe I could get any support. Anyhow, the bureaucracy is terrible and receiving the money anyway happens through a bank. To receive a loan I should employ some extra man who would work on this project.'

Entrepreneurs argue that there is too much bureaucracy and unpredictability involved in the actions of the state. The laws concerning enterprises are contradictory and constantly changing, which prevents the long-term planning of activities. What the entrepreneurs need, they say, is information about customers, markets and regulations concerning customs and taxes. The entrepreneurs do not think that they need to have very close relationships with either local or state administrators – all they need is support. Now, they think that the state administration is preoccupied with setting up various barriers against their activities and is watching them with suspicion. 'Bureaucracy, tax-policies, today this way, tomorrow in another way. There are changes every day, and amendments. You have to update your knowledge constantly, so that you keep in touch with

---

[8] See SAPARD Programme for Agriculture and Regional Development of Latvia, 2002, 31–32.

what is going on. And that costs a lot of money' (A carpentry entrepreneur from Estonia).

Entrepreneurs tend to think that the most important driving factor and motive for state bureaucrats is their own interests, not the interest of the entrepreneur, community or society. 'In many places, you hear first the question, how do I benefit from this?'(Metal works, Estonia) 'There are some people who defend us until they become members of the Parliament – before that they all fight for our rights, but when they get into power, they start to fight for their own interests. Then, there is no such person, and you can trust only in yourself' (A hotel owner, Lithuania).

## Conclusions

The failures of agricultural reforms have had a direct negative impact on entrepreneurship: the collapse of agricultural production, caused by rapidly rising production costs, the closure of Russian markets (and unprotected domestic markets) have deprived entrepreneurs of numerous customers. Rising unemployment, poverty and an ageing rural population have spelled out weakening purchasing power and a consequently increasing number of bankruptcies among shops, machine stations, and other small businesses.

Even if there are some positive examples of entrepreneurship in the Baltic countries, the overall picture is rather bleak; many firms are operating with outdated technology in a very uncertain and unstable market, and they are desperately short of the kind of labour force they need, as well as solvent customers. On top of that, the set of industrial and regional policies in each country is underdeveloped and insufficient.

The successful firms are those who succeed in establishing export relations with foreign companies, either as a direct producer or as a subcontractor. In this manner, these enterprises have accumulated capital for the modernization of product, the acquisition of new technology, and the payment of higher wages. But as we have already noted, such firms are but exceptions to the rule, and are situated within a very narrow line of business, such as wood processing or carpentry.

For the firms, operating strictly on the local market, and dependent on the local labour market, the prospects are not good. In some cases, entrepreneurship is not simply an alternative source of income and a means for upward social mobility; it is in fact *the only source of income,* and the only means of avoiding poverty.

In the Baltic countries, the levels of development are qualitatively far from each other, particularly between the capital region and rural areas. That also concerns the degree of development of the overall moral order. In the rural areas, the characterizing feature is a state of widespread and deep anomie, while in the capital areas and in some urban areas, business is actually booming and unemployment is a rare phenomenon.

One of the most important ingredients of the solution to the problems presented by the rural development and entrepreneurship is the creation of a clear, consistent and sufficiently resourced agricultural, industrial and regional policy. The necessary institutions and regulative legislation should support this policy. The lack of skilled labour, which is commonly seen as one of the most pressing problems, requires the reform of the educational system to correspond with actual labour needs. Another way to gain (and keep) qualified labour is the payment of decent wages, and a reasonable personnel policy in other respects; labour contracts, working conditions, and training (among other things). One of the necessary elements of the positive development of rural entrepreneurship is a more balanced regional development plan, since up until the present time, the majority of foreign investments and economic reforms have benefited mainly the capital areas and a couple of the larger cities.

Finally many of the problems of establishing a flourishing environment for entrepreneurship revolve around the immaturity of the state. At the present time, the state has no means to provide people with a system of adaptation to a new social order. People who have lost the basic elements of their way of life (their occupation, work collective, norms and values attached to those collectives, their skills, and sometimes even their self-respect), would need at least retraining, various forms of social work and assistance (including physical and mental rehabilitation). All of the above simply emphasizes the need to create the much-needed institutional and legal preconditions for development.

What this spells out is that the regional governments should not, in fact, trust markets to independently create the sufficient preconditions for sustainable growth. Instead, the state should direct and regulate the development of the necessary institutions, because the deficiencies in the institutional preconditions for entrepreneurship are what ultimately represent the most serious obstacle. If there is a continuous deficit of institutions, regulations and laws, there is also the danger of the growth of such unhealthy phenomena as shadow economies, bribery and other mafia types of elements, which in the long run will serve to jeopardize the recovery of the economy (Kolodko, 2000, 22–23).

As the interviews above indicate, the existing legal system ironically represents one of the main obstacles to the long-term development of business for many entrepreneurs. Therefore, legislation and the whole legal system should be streamlined and transformed to serve the market economy. The role of local administrations, (such as municipalities), should be strengthened by decentralizing their base of governance. This is one way to strengthen municipalities financially, and to provide more autonomy to local level (Kolodko, 2000, 24).

A healthy business also requires a healthy community in which to operate. This demands the measures and means to overcome (or at least reduce) anomie and related social problems. For example, Sztompka (1993) argues that the most difficult problem in transformation is not the reform of economy or production, but changing peoples' values, motives and habits. Therefore, the precondition of economic development is the development of the civil society in its all meanings: in economy as a expansion of private sector and the development of the middle classes, in politics in the development of interest organizations and parties, and in the narrowing of the gap between the state and the people along with overcoming the mutual suspicion and mistrust. In this task the development and support of interest and voluntary organizations is crucial, because those organizations play a central role in providing an area of negotiation and cooperation and a channel for interest articulation and representation as well as the formulation of such local and national political goals, which are generally shared. Therefore the emphasis on the third sector necessitates a renewed emphasis on a more developed moral and normative order.

## References

Alanen, I., Nikula, J., Põder, H. and Ruutsoo, R. (2001), *Decollectivisation, Destruction and Disillusionment. Community Study in Southern Estonia*, Ashgate, Aldershot.

Clarke, S. (1994), *Privatisation and the Struggle for Control of the Enterprise in Russia*, A paper presented at the Conference on Russia in Transition, 15–16 December 1994, Cambridge.

Clarke, S., Fairbrother, P., Burawoy, M. and Krotov, P. (1993), *What About the Workers? Workers and the Transition to Capitalism in Russia*, Verso, London.

*Estonian Regional Development Strategy* (1999), Estonian Regional Development Agency. Http://www.erda.ee/english/en_regstrategy.html.

Kolodko, G. (2000), 'Transition to a Market and Entrepreneurship. The Systemic Factors and Policy Options,' *Communist and Post-Communist Studies*, vol. 33, no. 2, pp 271–293.

*Labour Market and Social Policies in the Baltic Countries* (2003), OECD, Paris.

McIntyre, R. (2001), *The Role of Small and Medium Enterprises in Transition: Growth and Entrepreneurship*, Research for Action 49, UNU/WIDER, Helsinki.

McIntyre, R. (2002), *The Community as Actor in and Incubator of Economic and Social Revival: A Local-Level Economic Development Strategy*, Tiger Working Paper Series, no. 22, Warsaw.

Nikula, J. (2001), 'Sour Cream,' in I. Alanen, J. Nikula, H. Põder and R. Ruutsoo (eds), *Decollectivisation, Destruction and Disillusionment – A Community Study in Southern Estonia*, Ashgate, Aldershot.

*OECD Economic Surveys. Baltic States – A Regional Economic Assessment* (2000), OECD, Paris.

Rostow, W, W. (1960), *The Stages of Economic Growth*, Cambridge University Press, Cambridge.

*SAPARD Programme for Agriculture and Regional Development of Latvia* (2002), Latvian Ministry of Agriculture, Riga, pp. 31–32.

Scase, R. (2000), *Entrepreneurship and Proprietorship in Transition: Policy Implications for the Small and Medium-Size Enterprise Sector*, Working Papers no. 139, UNU/WIDER, Helsinki.

*Small-Scale Business in Estonia* (2000), report compiled by A. Tamm, Estonian Market Research Institute (EMOR), Tallinn.

Stark, D. and Bruszt, L. (1998), *Postsocialist Pathways. Transforming Politics and Property in East Central Europe*, Cambridge University Press, Cambridge.

*Statistical Yearbook of Estonia* (2002),Tallinn.

*Statistical Yearbook of Latvia* (2002), Riga.

*Statistical Yearbook of Lithuania* (2002), Vilnius.

Sztompka, P. (1993), 'Civilizational Incompetence: The Trap of Post-Communist Societies,' *Zeitschrift fur Soziologie*, vol. 22, no. 2.

*Transition – First Ten Years*, (2001), The World Bank, Washington.

*Transition Report 2002. Agriculture and Rural Transition* (2002), European Bank for Reconstruction and Development (EBRD), London.

Unwin, T. (1998), 'Rurality and the Construction of Nation in Estonia,' in J. Pickles and A. Smith (eds), *Theorising Transition: the Political Economy of Post-Communist Transformations,* Routledge, London.

Wendell, J. (2001), *Clans, Cliques and Captured States. Rethinking 'Transition' in Central and Eastern Europe and the Former Soviet Union,* Wider Discussion Paper 2001/58, UNU/WIDER, Helsinki.

*Interviews*

A. Estonia
1. Woodworks, Kohila, 16 March 2000.
2. Dressmaker's, Kohila, 17 March 2000.
3. Woodworks, Kohila, 17 March 2000.
4. Brewery, Karksi-Nuia, 27 March 2001.
5. Bar, Väätsa, 10 March 1999.
6. Bar, Kehra, 9 March 1999.
7. Woodworks, Karksi-Nuia, 26 March 2001.
8. Mayor, Kohila, 18 March 2000.
9. Metal works, Karksi-Nuia, 26 March 2001.
10. Consultant, Kehra, 8 September 1999.
11. Woodworks, Karksi-Nuia, 27 March 2001.
12. Paper mill, Kehra, 9 December 1998.
13. Mayor, Räpina, 18 June 2001.
14. Union of Estonian Municipalities, Tallinn, 16 March 2000.

B. Latvia
1. Woodworks, Preili, 12 June 2000.

C. Lithuania
1. Sawmill, Mindunai, 30 Octoner 2000.
2. Hotel, Silute, 1 September 2000.
3. Hairdresser/masseuse, Juknaciai, 6 September 2000.

# Chapter 6

# From Agriculture to Tourism: Constructing New Relations Between Rural Nature and Culture in Lithuania and Finland[1]

Leo Granberg

Since 1945, rural life in Europe has experienced a state of enormous technological progress and a modernization of lifestyle. The way in which production became organized has been called 'fordist' agriculture (Kenney, 1989). Even though it is difficult for one concept to cover the whole range of agricultural systems during this period, the concept of fordist agriculture amply describes the tendency of economic policy particularity as it impacts agriculture, as well as other industries in capitalist market economies. Such a tendency continued in Europe at least until the 1980s, in spite of the increasing environmental and depopulation problems that were its harmful side-effects. During this period, the leading ideology in economic policy was to produce more quantity with smaller costs per product unit, which led to cheaper prices for consumers, increasing profits for industrialists and higher real incomes for workers and farmers. This kind of approach was supported by governments, industrialists, trade unions and farmers' associations. Furthermore, many similar features were prevalent even in the socialist planning economy. The major difference between East and West concerned the way that they organized their economic units – in agriculture, western governments favoured family farming and eastern governments favoured collective farming.[2]

Agriculture seems to be loosing its driving force these days, both in the West and in the former socialist countries. Regional developers often pay attention to

[1] I am grateful for the opportunity to participate in two field research tours with Ilkka Alanen and Jouko Nikula in Lithuania, and to use other data from their project concerning Estonia (Case 4), and Latvia. Furthermore, I had the possibility to collaborate with Kjell Andersson and Erland Eklund during three field research trips in Finland. I have gotten helpful comments on the topic from many people, from Kaija Heikkinen, Janne Hukkinen and Vesa Oittinen among others. I am naturally myself responsible for possible incorrect interpretations in the text.
[2] With some strange exceptions, like conserved peasantism in socialist Poland (see Gorlach and Starosta, 2001).

tourism, claiming that this new and rapidly increasing industry could greatly contribute to rural development in the future. What then is, new in rural tourism? One principal difference between a fordist and a 'touristic' countryside concerns their relationship to nature. In agriculture and fisheries, nature is literally consumed as a raw material source for food production. In forestry, nature is consumed as a raw material source to supply paper factories or the construction industry. In rural tourism, on the other hand, nature is consumed as landscape. Rural tourism does not utilize raw materials from nature, but rather seizes on symbolic values that are coined with each special landscape. These ideas connected with tourism also remark the possibility of combining symbolic meanings of nature and local history in order to support the marketing efforts of food (Ray, 2001). Cultural consumers of the countryside are typically rural tourists and people with second homes in the countryside.

In the following pages, I will first refresh the reader concerning some of the ideological aspects behind the boom of rural tourism. Then, I will present and discuss my data, which was collected mainly from Lithuania and Finland, with some complementary data from Estonia and Latvia as well. My main research question is, whether traces of neo-romanticism can be seen in the speech of rural developers. Two target groups of local actors in this research are local tourism entrepreneurs and authorities. The third actor is the national government, which is responsible for planning and budgeting these said development efforts. The role of tourism in the rural development plans of Lithuania, for example, will be analysed from such plans. Relation to nature, relation to localism and urban-rural relations seem to be undergoing a remarkable change these days. How remarkable this change will be in the Northern European Countries located by the Russian border, I will discuss in the end.

## Romanticism

Romanticism in the 18th and 19th century, according to Isaiah Berlin meant a great turning towards emotionalism, sudden interest in the primitive and the remote – the remote in time, and the remote in place (Berlin, 1999, 14). During the romantic era, new ideas were developed concerning relations of individuals both to nature and to the locality they were born and living in. According to Berlin, one of the main fathers of romanticism is Johann Gottfried von Herder (1744–1803). He was 'the father, the ancestor, of all those travellers, all those amateurs, who go round the world ferreting out all kinds of forgotten forms of life, delighting in everything that is peculiar, everything that is odd, everything that is native, everything that is untouched' (Berlin, 1999, 65). This seems a good list of characteristics, and is certainly typical of rural tourists.

Romanticism was composed of fundamentally different ideas from the rationalism presented by proponents of the enlightenment and the consequent movement of modernism. According to Berlin, romanticism was an anti-modernist movement,[3] and, furthermore, as it seems to me, it would today be an anti-postmodernist movement. One of the main directions of the, criticism of romanticism was the idea of 'cosmopolitan man,' in other words, a man who is equally at home in Paris or Copenhagen, in Iceland or India. On the contrary, according to, for example Herder, there are roots in our mind that cannot be explained, nor made to disappear, roots which are based on personal experiences gained in our childhood and youth. As summarized by Isaiah Berlin:

> A man belongs to where he is, people have roots, they can create only in terms of those symbols in which they were brought up, and they were brought up in terms of some kind of closed society which spoke to them in a uniquely intelligible fashion... This was not a doctrine which could have been understood, nor... approved of, by the rationalist, universalistic, objectivist cosmopolitan thinkers of the French eighteenth century (Berlin, 1999, 63).

The later use of the concept 'Heimat' by German conservatism is related to Herder's thinking, but should not be combined with it. Herder does not use criteria like blood or race when referring to bonds between people, rather he speaks of language and soil.

During the romantic movement, the view of nature became an extremely central factor in European consciousness. In the fine arts, nature was conceived as an independent motive for the painter. In France, during the second half of the 19th century, landscape painting became to the most important branch of painting (Collet, 2002, 13). According to Collet two schools of painting emerged; 'the Barbizon' school and impressionism. The painters of the Barbizon school were seeking for 'loneliness in a forest and feelings given by primitive nature while impressionists were painting populated regions where man was a part of nature.' (Collet, 2002, 35) Berlin seems here to refer more to the former school when he speaks about romanticism in the arts and philosophy.

According to Berlin nature had become reinterpreted as 'something alive, (as) a kind of spiritual self-development' (Berlin, 1999, 97). This was originally the view of Friedrich Wilhelm Joseph von Schelling (1775–1854). Until Immanuel Kant, nature had been regarded in Europe as a harmonious system, or at least as a symmetrical, well-composed system. For Kant, nature was at worst an enemy, at best simply neutral element. For Schelling, there were struggles within nature ... what we admire in nature is some kind of power, force, energy, life, vitality bursting forth. According to Berlin, this doctrine became the basis for aesthetic philosophy in Germany, as well as its philosophy of art; great art had to succeed in catching the pulsations of nature – whether external or human nature (Berlin, 1999, 76 & 97–99). At the turn of the 20th century, this doctrine also had a profound influence upon Nordic national romanticism.

---

[3]   Furthermore, it seems to me, it would be today an anti-postmodernist movement.

## Nature in Environmental Studies

The concept of nature adopted by environmental activists and researchers has been much studied in recent years, by Milton (1996), Macnaghten and Urry (1998) and van Koppen, among others (see Luoma, 2002). Macnaghten and Urry (1998, 35) argue that environmentalism historically originated from the American romanticism of wilderness. A fresh interest in the wilderness, countryside and forests was also connected with the construction of national identities at the turn of the 19th and 20th centuries. On the other hand, the connection of these concepts with romanticism is unclear. On one hand, wilderness (or countryside) could be experienced as primitive nature, which parallels the thinking of the enlightenment that the wilderness should be tamed. On the other hand, the wilderness (or countryside) could be seen, in the light of romanticism, as untouched and pure nature. Kris Van Koppen underlines a similar difference when he analyses these three approaches to nature in environmental sociology: nature as a resource, an arcadian interpretation of nature and the social construction of nature. "The Arcadian view of the relationship between humans and nature is 'devoted to the discovery of intrinsic value and its preservation' in contrast to the imperialist view that lauds the 'creation of an instrumentalized world and its exploitation'" (van Koppen, 2000, 303, including a citation from Worster, 1985).

According to the resource-approach the rational and efficient regulation of nature was in focus for human activities. According to the Arcadian approach, the preservation of nature in its original and untouched form was in focus. As far as agriculture is concerned, only the former alternative is possible – even ecological farming is agriculture, practised by human beings. The countryside also includes other elements other than arable land, however, and therefore rural tourism may utilize a combination of both wild nature and tamed nature. Tourism actually utilizes both arcadian and constructionist approaches in its activities, but basically nature is a resource for the tourist business. The difference as regards to factory industry, however, is remarkable. Nature is not a raw material to be processed in industrial sector, but a raw material for an industry of experiences. On the other hand, alterations in natural environment are needed to make nature accessible for tourists. Material arrangements are needed to build roads, hotels, walking routes, sport centres and airports.

## Tourism as Business

We studied the situation regarding rural tourism in Finland and Lithuania from 1999 to 2001. Information was collected by interviews among two main groups: local entrepreneurs and municipal leaders. Interviews were conducted in three Lithuanian villages and three Finnish municipalities that have a peripheral or semi-peripheral location. They were conducted by two different research groups, in which the writer of this text participated. In Finland and Lithuania I also had the possibility to analyse interviews conducted in Estonia and Latvia by Nikula and Alanen. These interviews took place in a coastal region, and in two interior regions. In the following text, after a remark concerning the role of tourism in regional development from an macro-economic point of view, I will present some pieces of the interview material and summaries of interviews, in order to analyse some plans for rural development both on a local and national level, thus facilitating the topic: in what sense does the present wave of rural tourism include and utilize features of neo-romanticism.

Among the Finnish localities that were studied, the most successful municipality regarding the development of tourism is Nagu, an Island in the Southwest. Andersson calculated changes in employment in Nagu from 1950 to 1996 (Andersson and Eklund, 2000, 8). During this 50-year period, the number of people employed in agriculture, forestry and fisheries fell from 805 persons to 157 persons. At the same time, the combined effect of trade, hotels and restaurants succeeded only in increasing employment from 31 persons to 56 persons. During the most recent 26 years, the overall decrease in employment has ended, thanks to increases in the transportation and communication sectors, as well as in finance, insurance and public services. Some of these increases are the effects of an increase in tourism and summer residences.

In Lithuania from 1993 to 1998, the amount of people working in hotels and restaurants in the entire country increased from 19,000 to 23,000 (NEBI, 2001, 364). This is minor growth, even if one takes into consideration the fact that the tourist industry is the sector with the strongest effect on a black economy, and therefore is partly excluded from statistical indicators. In comparison, Estonia has enjoyed the greatest benefits from the streams of tourists that come from Helsinki region, and also from other northern European regions, via Stockholm.

Anna Pridanova refers to this impact in the daily *The Baltic Times*, but she is sceptical concerning similar possibilities of rapid increase in the tourism potential in Latvia: 'Perhaps Estonia, which recorded the fastest rate of tourism growth in all of Europe for 1999, and where tourism makes up a hefty 17 per cent of the nation's GDP, could teach its southern neighbour how to realize its tourism potential.' Anna Pridanova here underlines the importance of the rural tourism market in Latvia, compared with the tourism of Tallinn in Estonia, but she also highlights a number of the difficulties involved in developing rural tourism (Pridanova, 2000, 9). According to major difficulties are connected to the freshness of entrepreneurial activities; lacking entrepreneurial culture and the infrastructure to facilitate small and medium-sized enterprises in countries where large-scale, state-owned companies dominated the economy for decades. In the countryside, most of the entrepreneurs in the 1990s could be called (according to Nikula) 'accidental entrepreneurs' (Nikula, 2003, in this volume). In other words, they did not start business because of a personal interest in the branch or because of good competence within it, or on the basis of family background – but simply because they had no other alternatives.

**Experiences of Rural Tourism**[4]

We interviewed both municipal leaders and people involved in rural tourism. Let us start with three Lithuanian and one Estonian entrepreneur, who describe how they began supporting activities in tourism.

### Case 1: Lithuania: Rural Tourism Service During and After Socialism

I was a foreman in a boarding house belonging to the kolkhoz. I was responsible for bedding and food in that house. Later [1993] we privatized it ... I am (now) reselling fish and in the summer time we get our incomes from tourists who are staying overnight with us.

We have five rooms and 12 beds ... we are going to build a sauna. Some German families ... want to build a cottage for their holiday time ... because they like our place and our food ... and we could rent out this cottage too. I have some old customers from old (Soviet) times ... and some Germans ... I do not make any advertisements. Therefore, we do not have very many clients. But in the three summer months we have customers with full capacity.

The municipality is not able to help us. They just ask us to buy a licence for tourism business. We have not yet bought the licence, and we are not working legally, but in the spring I will buy the licence, because now I know that we will get some customers. We are going to visit Spain this autumn with my husband. A Lithuanian family has worked there on an orange farm – and they didn't work legally. And all the money, which we earn we are going to invest into our tourism business (A female entrepreneur on the western coast, 2000).

### Case 2: Lithuania: Demand by Foreigners

I started this business because I had a good knowledge of tourism and construction work. And I liked it. And my wife is a professional cook. About eight years ago (1992) I got my grandparents' land in the land restitution. And I built these cottages. During one year I was unemployed, and then I was just building. Later, we sold for a profit some shares in a hotel that belonged to my wife. And we invested this money into the buildings too.

This is not an ideal place for tourism. There is a forest here, but this place is not located by a lake. My first clients were former clients from the leisure centre, where I used to work earlier. I built the sauna later. I borrowed some money from my friends, and built the sauna in the autumn. And that autumn I had many visitors from a town, 40 km from here. There is a beer factory where some Germans worked for three months. They liked my sauna very much, and used to come here every weekend. Therefore, I was able to pay back all the loans during that one autumn (A male owner of a cottage park in the Central Lithuania, 2000).

---

[4]    Interviews in Finland were made by Kjell Andersson, Erland Eklund, Minna Lehtola and Leo Granberg, and in Lithuania by Ilkka Alanen, Jouko Nikula, Lina Krisciunaite (translation) and Leo Granberg.

## Case 3: Lithuania: Demand by Drivers

We did not have any imagination (when planning our hotel business). We built this house not as a hotel, but as a house for ourselves ... we started to rent these rooms... and now we are planning to build an actual hotel.

Local people do not have the money to be customers. The hotel is mainly visited by lorry drivers. In the summertime, there are many tourists, too. Also people with their dogs often stay in the hotel, on their way to Poland to dog exhibitions.

We have a seven hectare park near the hotel, and a river is here too. It is an extraordinary place for recreation. The basis for our business is this road and the customs house. People who succeed in passing the border patrol will stay here and celebrate. And those, on the other hand, who don't have a happy ending, can ease their heavy heart with some drinks in our café (A male hotel owner in northern Lithuania, 2000).

## Case 4: Estonia: Tourists Instead of Cows

This Estonian entrepreneur began his enterprise in rural tourism while losing strategic parts of his field during the Estonian land restitution, with the consequence of having to cancel his plans to begin a family farm. They heard about the Danish type of rural tourism. Thanks to his earlier profession, he was able to build houses and to repair machinery. This proved to be a very useful skill. The beginning of his tourist trade was based on some empty rooms that he rented to travellers. Furthermore, he worked outside of the farm, and participated in reciprocal help groups with neighbours who were repairing their machinery and purchasing food products instead.

Nature does not have special value for him, even if he makes this point: 'this is an interesting landscape, this valley'. In fact, he wonders about the behaviour of his visitors, who seem to enjoy bearing hay in the field, looking at old, ruined agricultural buildings and houses. 'The terrible wild nature, the copse by the side of the road ... that is the attraction which people want to see when they come from the civilized world, with its parks and forests without any animals.'

This entrepreneur had varying experiences from the attitude of local administration. He told that mayors come and go. One of them was eager to promote tourism, but the current one pays only marginal attention to rural tourism as a development effort (A male entrepreneur in southern Estonia, 2000).

## Case 5: Finland: Sport Fishermen Instead of Cows

Brita's farther bought this farm in southwestern Finland, combining a fishery with small-scale agriculture on 3 hectares of arable land, 40 hectares of forest, several small islands, over 100 hectares of land in total, plus assorted pigs, hens and cows. Brita moved out of her home in the 1960s, like all the rest of her sisters and brothers. She lived some years in Turku, Pargas and Helsinki, and moved back to the farm with her husband in the early 1970s, where she took up work as a social worker for the municipality. Up until 1986, she commuted to other villages, but thereafter decided to begin a full time entrepreneurship. A natural harbour, a nearby nature protection park and the long coastal line that was under her ownership helped her to make the decision.

Her family started the business with two cottages, and step-by-step increased the number to seven cottages, all of which were located by the sea or a lake. They used to rent out the cottages for one or two weeks using a self-service approach. Two children helped in the summertime. The season has been prolonged to last into the autumn, because that is a good time for fishing. Many visitors come from abroad, most of whom are Germans. The family has participated in several development projects concerning sport fishing and tourism, which are funded by the EU: PEALS in archipelago, Ring road in archipelago etc. The municipality has thus far been dominated by agrarian politicians, who caused some problems regarding practical questions, but in the mid-1990s a change took place that fostered a more positive attitude towards tourism. In this municipality, 1,400 inhabitants are complemented by 2,000 summer cottages. The island is visited, furthermore, by a number of sailing boats in the summer months. Brita and her husband see the future as positive (RI, c5–6, female small farm owner in Nagu, 1999, interview).

## Case 6: Finland: Conferences on a Rural Island

In 1988, in southwestern Finland a small family enterprise began to rent rooms to tourists, reconstructing old buildings and buying, after some early successes, more capacity for accommodation. The niche became a thriving business in conference tourism. Tourist packages with accommodation, the most popular consultants and elements of recreation were tailored for enterprises needing to arrange training courses and brainstorming workshops. The company is now working in the exotic nature of the southeastern Finnish archipelago, near a wide national park of archipelago. The family enterprise even belongs to the UNESCO biosphere programme. In ten years, it has grown from zero to 80 employees (in the summer season), and has established a functioning network with other local tourist venues, which offers additional accommodation capacity and complementing elements, such as conference packets, that take place in the surrounding nature. This municipality was dominated earlier by agrarians, but the situation is now balanced, and power has partially been taken over by politicians who have a positive attitude about tourism, and even environmental protection as a pre-condition for rural tourism (RI, c1–2, female hotel owner, 1999, interview; Andersson and Eklund, 2000, 9–11, et al.).

## Case 7: Finland: Exercise, Local Food and the Eastern Border

Pyhätunturi, in Salla, is one of several winter sport centres in Finnish Lapland, and lies by the Russian border. It is a relatively small holiday resort. The profitability of this business is continuously in a critical situation, but new innovations are also under way. In the 1990s, three enterprises began to collaborate. A hotel, participating in a multinational hotel chain, a family enterprise with a cottage park and snowmobile services, and Reindeer Park, administered by a co-operative of young, local people. Local food, for example, has been adopted as an element that is included in the programmes of Reindeer Park; reindeer and also dog-sleighs increasingly interest visitors; the snowmobile business is increasing; tourism from Holland to the cottage community is lively and is particularly focussed on marketing strategies and training. In addition, Russians are visiting increasingly and passing through the resort, and collaboration with a neighbouring holiday resort, Ruka, is in progress.

Salla has, however, been far less successful than Ruka, even if they are not located far from each other – both of them are in the border region and nestled in a rather similar landscape. Ruka's success can be measured by its larger number of cottages, and multi-faceted investments (e.g. a tropical bath, and also the quantity of visitors in the region). In Salla it seems, however, that different structures of ownership and complementing services featured by the three above-mentioned enterprises might provide a chance for useful collaboration in a 'win-win' scenario. The cottage community is their most successful enterprise, it singly employs 12 permanent workers, and during one year approx. 66–80 people work in shorter periods for the enterprise. The hotel has an unstable ownership structure, and after our interviews the ownership changed once more (RI, a, several interviews, 1999).

## Case 8: Finland: Military History, Baths and Nature

Suomussalmi is a border-municipality almost 200 hundred kilometres south of Salla. The municipality was trying to develop its image in the 1980s through an eco-municipality project, which did not bring the anticipated results, and has been reduced to minor activities that involve environmental sustainability. The tourist industry in Suomussalmi can be roughly divided into three branches: the bath and healthcare tourist industry, nature and wilderness tourism, and tourism connected to war history. The public bath, 'Kiannon Kuohut,' is a large recreation centre located in the small but modern urbanized central village. Nature and wilderness is featured in some of the remote villages, especially near the large nature park, located on both sides of the Finnish-Russian border. Some private firms, in addition to Villi Pohjola ('The Wild North,' a state-owned unit of the Finnish forest administration) are active agents behind this type of tourism. The third category is built mainly upon the memories of the Finnish Winter War, and the legendary battles of Suomussalmi. The central attraction, a war-museum named 'Raatteen Portti,' is visited by about 30,000 visitors-per-year (Eklund and Andersson 2000, 9–12; several interviews, 1999).

## Traces of Romanticism in Rural Development

*Relation to Nature*

> Everyone who finds the red roofs overgrown
> with moss and having a smell of smoked fish
> attractive and who are tired of noise and willing
> to listen to a song of bird is kindly invited
> to visit our region. Here you will enjoy fishing
> and may have good time in local festivals.
> Mayor of a coastal region, Lithuania (SR, p. 1).

In Western Europe, an environmental awakening has been taking place at least since the 1980s, both inside and outside of agriculture: environmentally-oriented methods have won attention in farming, and the multifunctional use of forests has been accepted on the forestry agenda. Thus, recreation in the countryside has become fashionable. None of these were entirely new practices, (e.g. recreation among relatives in the countryside has always been a part of the town-countryside relationship). A new phenomenon in capitalist countries has been an emerging market for rural tourism. From now on, tourism would become a profession for those living in the countryside, instead of functioning as a hobby or a duty. A new conviction arose regarding the importance of nature as a production factor for the service industries. Even if this remark refers largely to Western experiences, the countryside was nevertheless an important space for leisure activities during the times of socialism. Planning the economy engendered facilities for recreation in the countryside, near nature. Two of the above interviewed Lithuanians (Cases 1 and 2), had both been working for rural tourism during the times of socialism, when Lithuania was a part of the Soviet Union. At that time they learned skills, which they successfully utilized during their transition into starting small-scale tourist businesses.

Nature as an important factor in local development was far more emphasized in our Finnish interviews than in the Lithuanian ones, and the aforementioned text by a Lithuanian mayor seems to be singular exception. We discuss some more information about this question later, when we analyse government documents. During our group interviews, municipal leaders gave a list of the most important factor in regional development (RI, Interviewer 1999, 2). In Salla, the strategic priorities were:

1. Natural resources.
2. Timber.
3. Tourism.
4. Utilising the Russian border.

A similar list of factors in southwest Finland (Nagu) were:

1. Good traffic connections.
2. Pluriactivity.
3. Fine building sites.

The third priority clearly refers to sustaining a beautiful natural environment. In both municipalities, the transition from an agriculturally oriented local policy to a strategy that facilitates tourism was under work during the late 1990s. In Nagu this new orientation was already winning. For example, the Natural Park of Archipelago had been accepted as a positive development, even if it meant a loss of opportunities in fishing and forestry. One reason for this change was that one of Finland's regional stations of forest conservation was located in the municipality, which brought work and educated professionals to the community. Another reason was the number of second homes in the municipality – more than 2,000 second homes in a municipality of 1,500 inhabitants creates a vibrant base for local life. Contradictory views exist, of course, such as fishermen who are disappointed with weak economic preconditions for fishing. These critical voices typically speak out against the tendency to favour sport fishing over professional fishermen. The number of professional fishermen is small, however, and their voice is less powerful than before.

In Salla, the Russian border was not officially open to visitors during the process of the data collection, and travel across the border was limited.[5] Expatriation and a weak municipal economy has caused a severe crisis in the municipality. What complicates the overall relationship of people to nature in Salla is that tourism has mainly been concentrated around only one point in what is a large municipality, that is, a mountain named 'Pyhätunturi.'

Thus, tourism takes place in the middle of nature, but it does not always happen in ecologically sustainable forms: as is the case with most winter sports centres, slalom and snowmobile tracks are literally hooked into hard technology – quite different principles than traditional cross-country skiing or wandering in the wilderness in the summer time.[6] The future of the natural environment since the 1970s has also been under the shadow of a plan to establish an artificial lake named 'Vuotos,' which would cover a good proportion of the land surrounding the municipality, in order to support the energy production that is generated from the Kemijoki river. The debate surrounding this question divided the local population, and their opinions were connected to the different views on nature that were hold by different groups, which Suopajärvi classifies as: 'live in and from the nature,' 'nature as a resource' and 'nature as a value for tourism' (Suopajärvi, 1998, 97). In 2002, the question seems finally to have been solved by Finland's Supreme

---

[5] The border was officially opened in 2002.

[6] Also cross-country skiing is changing to be an increasingly more technology-dependent sport, using machine-made broad tracks, the artificial making of snow and long distant transports of snow.

Administrative Court, resulting in a prohibition to build up the lake, because of the potential irrevocable losses the natural environment (Koivurova, 2002).

More conscious image building by local authorities is being undergone in Nagu (Case 4.1), and in coastal Lithuania. The above-quoted invitation of a Lithuanian mayor is mostly based on the high valuation of natural beauty and peace, supplemented by local cultural festivals and fishing opportunities. In the interviews made by the research team, nature was not often mentioned as an important value. In Estonian Case 4, for example, not one feature of romanticism occurred in the speech of the entrepreneur. Neither was the municipal leader willing to invest on the basis of nature values. The mayor of the municipality in question argued thus: 'Tourism on farms increases slowly, and the activity does not give work for many persons, probably not even after five years. One cannot even wish to reach such an explosive increase of tourism here as in Otepää.' Otepää is a relatively nearby municipality, and boasts the leading winter sports centre in Estonia. In addition, head of a municipality in central Lithuania mentioned agro-tourism as a strategic idea — perhaps the only possible one. This was in a village where kolkhoz was privatized, the cultural house destroyed, and most of the employment opportunities had disappeared except a shrunken state of agriculture and a few small enterprises. Some families had started getting into small-scale tourism (Case 2), while others were planning to do so. In this, as in many other regions in the Baltics tourism seems to be demand-oriented. One of the demand constituencies in Lithuania are Germans, who want to visit their (parents'/grandparents') ancestral home region by the coast. The municipality is located in a region, which belonged to East Prussia, from which both Germans and a large part of the ethnic Lithuanian population emigrated, before the Russian occupation of the Second World War.

The difference in this type of rural tourism is remarkable when you compared it to a developed market economy such as Finland, where supply-orientation is strong: strategic planning, land use arrangements and technological projects facilitate the increase of tourism in the fields of general and rural tourism as well. Since the end of the 1990s, however, all Baltic countries have drawn up large plans to develop their countryside while preparing to become EU members.

*Touristic Nature in Rural Development Plans*

The European Union has developed the SAPARD programme as one of the accession instruments for member candidate countries. SAPARD focuses on rural development, which includes both the agricultural sector and other elements of rural development. One of these elements is the enhancement of diversified economic activities, and rural tourism is often explicitly mentioned as one of the potential new activities. In Estonia, an agreement with SAPARD promises, for the years 2000 to 2006, that 17.6 per cent of their funding will be directed to diversified economic activities, including 'starting or expanding rural tourism, as well as handicraft activities ... food processing in small enterprises' (Press Release 1, 2000). The Latvian SAPARD Programme for Agriculture and Rural Development handles rural tourism as a subchapter, under the title 'Alternative Occupations in Rural Areas.' Their belief in the future of rural tourism is based on the conception that 'attractive rural landscape is a good prerequisite for the development of tourism' (Ministry of Agriculture, Latvia, 2001).

As early as the year 2000 the Lithuanian government had also accepted an agricultural and rural development plan for the years 2000–2006. Eight priority areas have been listed, with agricultural production as the first priority area, and the processing and marketing of agricultural products as the second. Together, these areas receive approx. 70 per cent of the total EU funding for agriculture and rural development. The third priority area is the diversification of economic activities in rural areas, and one of the priority sectors in this area is an explicitly rural state of tourism and recreation, which receives 15 per cent of the total funding channelled to this area, and consequently about 1.5 per cent of the total EU funding for agriculture and rural development. Some activities in other priority sectors may contribute to tourism, (e.g. supporting traditional craft activities (0.5 per cent) and various investments in the local infrastructure).

Lithuania's plan as a whole does not seem to provide a clear concept of rural tourism or rural development. On the one hand, the plan is strongly concentrated on agricultural issues, and mentions agriculture as the special strength of the Lithuanian countryside: 'Summarizing the strengths of rural Lithuania, it is obvious that agriculture is of utmost importance to the future socio-economic development of rural areas in terms of production and employment.' (Ministry of Agriculture, Lithuania, 2000, 7). Seen this angle, natural resources are treated as raw materials, not as image building factors with aesthetic value, or factors which could be marketed to city dwellers and foreigners. Concerning living conditions in rural areas, the plan mentions that 'basic living conditions are comparatively poorer in rural areas than in the cities' (Ibid. 7). This citation is taken from the summary chapter of the plan. When studying the basic text, one finds a more positive attitude towards rural life and non-agricultural rural activities. One central interpretation concerning rural tourism is whether it should be restricted to farm tourism, or if it covers other types of tourism in the countryside. According to Lithuanian statistics only 194 farmsteads were involved in the rural tourism business in 1998, and about 95 per cent of the guests were Lithuanian. Other guests came primarily from former Soviet Union countries (Ibid. 51–52). In this connection, the concept of rural tourism seems to refer only to farmers who receive additional incomes from tourism. The plan, however, clearly promises to support both on-farm and non-farm rural activities (e.g. 99). Furthermore, the complete list of weaknesses also includes an over-dependence on agriculture (Ibid. 73). In the same paragraph, nature is seen to be a strength as such, and takes on a tone of romanticism: 'multiple activities in rural areas ... will exploit existing favourable natural conditions, rural landscape and heritage ...' (Ibid. 73). As we saw earlier, however, the tones of rural neo-romanticism and acceptance of the weaknesses of agriculture were somehow, in the summary part of the text, turned into a belief about the agrarian way to solve development problems in the countryside.

The material used for this study supports the argument that a new concept of nature is being exported to Lithuania and not rising from inside of the country (e.g. on the basis of national history and local traditions). On the one hand, foreign tourists export the concept, and, on the other hand, European Rural Policy and the

EU's funding schemes export the idea. The difference between Lithuanian and Finnish cases is great, and in Finland those interviewed often strongly emphasized the beauty of nature in their home region as a value as such, and as an important reason for tourists to visit the region.

## Localism in Culture and for Tourism

There are structural and cultural factors behind localism. The Finnish peasant agriculture left its land-owning structure to subsequent generations. When a child left his or her parents' farm, the child quite normally inherited a site for a second home. Such a site was often located on invaluable soil, from agricultural point of view, but it had its own value when seen from a recreational point of view, if it was located by a lake or river. The peripheral municipality of Salla, in 1996, had over 1,000 second homes, even if there were only 5,600 permanent inhabitants. Suomussalmi had 1,800 cottages and 11,700 permanent inhabitants. As I mentioned, Nagu had even more cottages than permanent inhabitants. Statistics do not exist about the proportion of cottages that are owned by children and grandchildren of local farmers, but it definitely is a high figure. These summer cottages and second homes form an intermediating structure between old farming generations and new urban generations. They form the structure of rural roots for modern city dwellers.

Localism in everyday life, and localism marketed for tourists, are often two different things, however. Petrinsalo analyses one particular failure in eastern Finland, where a massive tourist centre was built around the reconstruction of a traditional Orthodox-Karelian Culture Centre in a small rural village, 'Hoilola' (Petrisalo, 2001). The image of the centre was promoted as a part of local history; its tradition and its way of life. The distance from the past to present everyday reality was long, however, and traditional culture had to be recreated. The project caused an opposition to spring up on the local and regional level, which at least did not assist the project in the attempts to improve its economic profitability. In five years, the Centre had gone bankrupt.

In rural culture, localism is often combined with cuisine, as well as with art and craft. Usually, this is a small-scale activity, for which identity building is an important function, in addition to income. Both food and handicrafts were visible parts of development strategies, as was seen during our interviews. In Suomussalmi, a local marketing hall was built by a gasoline station in the central village in order to offer an opportunity for local producers to market their products, which consisted of food as well as handicrafts. A plan to develop a wandering pathway between the handicraft centres was presented to us in Salla. In Lithuania we visited an entrepreneur (Case 1), whose husband was a sculptor. His wooden works were visible around the garden, and the wife spoke with pride about this speciality. In the following discussion, we will focus on the role of local food.

Local food is a relatively new concept. Reindeer meals, were earlier considered as a food fit for common people only, and was not accepted on the menus of better restaurants in Lapland until the 1960s, following a breakthrough in a cooking competition in Paris.[7] Food is an excellent product to stir various nostalgic connotations. It sends mediating messages from the past to the present, posing questions about the ways earlier folk collected their food and prepared it, as well as questions about how they lived. Tasty and well-prepared local food may produce heart-warming feelings and a sense of historical nostalgia. In that sense, regional cuisine is about much more than its nutritious content. Complemented with local beverages, of course.

Local food seemed to win recognition in the rural tourism of the 1990s. In Salla's 'Reindeer Park,' local menu items were planned with the help of a set of local vintage receipts, which were collected by a retired teacher in the mid-1980s. Using biological products, a local slaughterhouse, and attempting to be authenticity-oriented was also welcomed by the hotel director during another interview. Reindeer Park simply supplies a complementary programme for the visitors to the hotel. The menu in the hotel, on the other hand, is a typical Lapland combination of international food, with the strong presence of salmon and reindeer meals, including such specialties as 'poropizza' – a variant of pizza that is topped with reindeer meat. In Nagu, one old couple who live on a sheep farm solved their profitability problem by processing all their products into food. They are selling their products in the marketplace by the harbour of the village for sailors, or motoring tourists and summer guests. They also sell them in the market place of the city of Turku, or supply them directly to customers who have learned to know them personally – from a particular TV programme, for example. Other scenarios do not exactly verify the importance of local food for success. In Ruka, which is 100 kilometres south from Salla, local food is not visible. Ruka is a far more economically successful centre, partly because of its history of legendary winter sport competitions and partly because it has become a holiday site for affluent Medical Doctors' Union, as well as a few business companies from the Capital region.

Food as a source of tourist commerce was also mentioned by interviewees in Lithuania, even if it was not a large number of them. In Lithuania, tourism was still in its infancy, and the tourists demand for food was still limited. Furthermore, local demand was lacking, because only foreigners were economically capable of eating in restaurants.

Tourists often want to learn about (and feel) their roots. Forefathers of some Germans have been living in western Lithuania, while the forefathers of certain Austrians are buried in the graveyard across the Russian border in Salla. The forefathers of numerous Americans left the Baltic and Nordic countries in different historical eras. One structural feature concerns the national population: the grandchildren of peasants may spend their summer time exploring their roots in the second home that was built on the prior farm of their grandparents, which is perhaps still cultivated by their cousins. There was more than one person interviewed in central Lithuania who had succeeded in getting back their family's old farm site as well. This kind of search for roots may have, in many regions, a

---

[7] Told in a radio interview by the former cook of the restaurant 'Pohjanhovi' in Rovaniemi.

much stronger effect on rural development than the presence of freely roaming tourists. The question arises about, whether there still exists (or might re-emerge) some kind of locally-oriented community in the 'new countryside' of the 21st century. The answer seems to be that there is no real return to such a community, which has been referred to in classical sociology by the term 'Gemeinschaft.' On the other hand, there is a possibility for new kinds of communities to develop in a network form, consisting of both rural and non-rural actors.

*The New Rural-Urban Relationship*

How is the difference between the town and countryside changing, when farming has become a market-oriented industry, yet the number of farmers has fallen radically? R. Frankenberg underlines the difference in interaction between towns people and people in countryside.

> Social roles in rural society – in contrast to urban society – are complex rather than complicated.
> ... In rural societies people are related in diverse ways and interact frequently, while urbanized societies have an associative nature. Although urban society may offer a greater number of possible relationships, these do not overlap. In rural society a small number of people make up the total social field of an individual. Thus, people in rural society tend to play different roles in regard to the same person, i.e. they have more numerous multiple role relationships. ... People in rural society are treated and expected to behave appropriately to their status, which is outside their control and fixed above all by the family of origin. This status extends from situation to situation and a person's status is the same whatever activity she or he is engaged in.
> ... Now, the general tendency is a move away from the complex (multi-channel and overlapping) social roles typical of rural communities toward complicated (more diffuse and specialized) social roles typical of urban communities. This is also taking place within rural villages (Frankenberg, 1994, 16–21; from Holmila, 2001, 20–21).

Frankenberg's interpretation is based on field research in Britain some decades ago, but seems to fit well regarding the Finnish case. People in villages increasingly work outside their village, relatives of villagers are increasingly living in towns, the frequency of contacts inside villages are falling, and the frequency of contacts between villagers and city people is increasing. Outside the village, certain groups of people are especially interested in the village, keeping the 'community' and 'nature environment' in their heart as a part of their identity. The two most important such groups are relatives (children and grandchildren of former peasants), and cottage owners inside the region of a village. Formerly, these groups used to overlap. Both are outsiders from the farmers' point of view, but perhaps this is not so much the perception in a post-agrarian local society. Tourists who regularly visit a certain village are the third most important group, whose relationship with the villagers may vary, but who remain outsiders: the customers in the rural tourist business scheme. All of these groups have one important thing in common, that is, local nature. It is the focus of much small talk, and an important aspect in almost any communication: 'the lake is marvellous in the summer, isn't it; have you been fishing a lot this summer; was there a lot of algae in the lake during your holiday; has the pollution from the Russian side of the border had some effect on the water quality here.'

Communication takes place between three points: person 1, person 2 and 'an ecological actor' i.e. a lake, river, forest, meadow, hill. Ecological and social actors are not only communicating, but form a kind of network. It is easier for an outsider to settle down in today's community than it is in an old agrarian village. One has to learn, however, to 'personally' know local nature: the lake, the river, the forest, the meadow, the hill. This is not only possible for one born in the village, but can happen to another who has ties to the village through his (grand)parents and relatives. It is far more difficult for a tourist, who only visits the village once or twice. For him or her, there is a performance going on, that is not communication on equal terms. This kind of post-agrarian or neo-rural community could be called a socio-ecological community. Such a community is not a community in the sense of Gemeinschaft (Tönnies), it is a network with stronger or weaker ties connecting actors to the local nature and environment with its cultural and historic features.

The processes of development are, however, uneven. Some villages develop rapidly, and some of them will sprout into towns, while others die out and disappear. This latter trend is not unknown in Finland, in the border regions, in the archipelago. In the countryside of the former Soviet Union, land reforms and the period of collective farming interrupted the continuity of generational changes. During the transition period in the 1990s, the gap between the town and countryside became far greater than it is in Finland today. As concluded from the field research in Estonia by Alanen, et al. 'the line of action ... led to a privatization process that was highly anarchic and even destructive to collective resources, and to the collapse of public morale' (Alanen et al., 2001, 298). This transition process was extremely hard in the post-Soviet countryside, and became a pocket of poverty and social problems, which are discussed further in other articles of this volume.

## From Agriculture to Tourism

As presented in this text, rural tourism has a relatively limited macro-economic impact on local development. Tourism is, however, still under construction in the Lithuanian (as well as the Estonian and Latvian) countryside. The trend to develop rural tourism is similar in Finland and in Lithuania. Resources and the style of accomplishing this work differs dramatically, however. The three cases from Lithuania, which describe the birth of new tourist entrepreneurs in rural Lithuania, speak of 'accidental entrepreneurs' (Nikula, 2003, in this volume). They start a business as a personal reaction to their changing socio-economic environment, as an attempt to survive in one way or another. Often, their strategy is to combine several activities, tourism being one of the few opportunities to gain a source of income. This takes place in a situation with a minimal level of strategic planning from the side of society or the state bureaucracy, and there is more institutional resistance than support – which is indicated by the illegality of certain reasonable economic activities. On the contrary, a small farm in the Finnish archipelago can utilize, in its reorientation to tourism, an infrastructure that is excellent when compared to Lithuanian coastal areas, and the opportunities opened by land-use planning, funding and (after years of struggle) a positive attitude developed by the local administration.

## Culture-Nature Relation

On an individual level, a farmer is a person who reorganizes nature. According to the precepts of the enlightenment, he brings benefit to mankind, while being a dominator and exploiter of nature. Today, however, a tourist seems to be a more valuable person than a farmer. The tourist's relation to nature is quite different than a farmer's. A tourist is an adventurer, a collector of experiences from nature, while at the same time he brings material benefits to those marketing and selling the virtues of nature. Consequently, a tourist actively contributes to the locality and region he or she is visiting.

On the other hand, farmers were historically part of the peasant community. This peasant community has, for the most part, disappeared. What is left is a community, which is less based on local level, which resembles a scattered network with decreasing intensity of interaction at the local level, and an increasing number of interactions between local and external participants. These links can be weak, but they are often important. Not only is the connection between people important, but so are the connections of these people to the landscape, as well as their identification with local nature. This is the basis for any commitment and local political initiatives to safeguard natural conditions and the sense of local community.

## Neo-Romanticism

One can legitimately ask whether we are moving away from the ideas of enlightenment to a new kind of romanticism. Environmental movements, for instance, seem to indicate such a move. Hence, one should develop discussion concerning the role of an overall long-term structural and mental history; one could compare these roles and the mutual differences of increasing new industries (above all tourism and information technology). One might also study the sighs and memories of past history in present time. Furthermore, one should also focus on the European Union's regional and rural policies. Instead of going further with this analysis, I will argue that Nordic national romanticism has left a deep impression on the Nordic interpretation of nature, and that this influence is today utilized by the developers of rural tourism. In Lithuania and Estonia, today's neo-romanticism has the same content for the most part, but its problem is that it focuses on peasant culture before Soviet Union times, without taking into account the western experiences of the evolution of family farming, or the economic crises which hit small-scale farmers in Western Europe during the second half of the 20th century. The revival of family farming is a long and complicated task, and the risk of errors is great, in particular when one is attempting to combine this task with agro-tourism.

The construction of rural tourism takes place side-by-side with the increasing of markets of cultural economy in various forms. The combination of a romantic interpretation of 'wild nature,' ideas of ecological modernization and a materialist approach to the cultural market brings forth a different kind of romanticism than the national romanticism that we saw during the late 19th century. That is why neo-romanticism seems to fit better as a working platform for the ideology of rural tourism.

Neo-romanticism appreciates deep roots as opposed to extreme individualism and the looseness of postmodernism. Roots are sought through examining the local past, and from viewing the aestheticism of nature, but also – following the impressionist stream of Romanism – from traditional farming and life in old-time villages.

The village community is an environment where one naturally expects to find one's roots. The community of grandparents that one may hope to find does not exist anymore, however. The cultural heritage of an entire local community has left only a few remnants of cultural sighs and mental structures. These remnants are the raw materials for neo-romanticism, a precondition for the new rurality of our time.

# References

Alanen, I., Nikula, J., Põder, H. and Ruutsoo, R. (2001), *Decollectivisation, Destruction and Disillusionment. Community Study in Southern Estonia*, Ashgate, Aldershot.

Andersson, K. and Eklund, E. (2000), *From Primary Production to Tourism and Leisure Related Services: The Contemporary History of Two Rural Settings in Finland*, Tenth World Congress of Rural Sociology, Rio de Janeiro.

Berlin, I. (1999), *The Roots of Romanticism*, Princeton University Press, Princeton.

Collet, I. (2002), 'Peindre la nature' and 'La nature dans ses apparences,' in exposition catalogue: *Visions de la Nature, Luonnon näyt*, Taidekeskus Retretti 30.5.–25.8.2002, Punkaharju.

Eklund E. and Andersson K. (2000), *From Peasant State to New Rurality in Finland: Political Change and Discourse at the National Level Illustrated with a Local Case*, COST A 12 Rural Innovation, Environment Working Group, Newcastle Workshop, June 16–17, 2000.

Frankenberg, R. (1994), *Communities in Britain. Social Life in Town and Country*, Ipswich Book Co. Ltd, Ipswich.

Gorlach, K. and Starosta P. (2001), 'De-Peasantisation or Re-Peasantisation?' in L. Granberg, I. Kovách and H. Tovey (eds), *Europe's Green Ring*, Ashgate, Aldershot.

Holmila, M. (2001), *Social Bonds in Rural Life*, Stakes Research Report 113, Stakes, Helsinki.

Kenney, M., Lobao L. M., Curry J. and Goe R. (1989), 'Midwestern Agriculture in U.S. Fordism,' *Sociologia Ruralis*, vol. 29, no. 2, pp. 131–148.

Koivurova, T. (2002), 'Vuotos-päätös tukeutui kuolleeksi luultuun pykälään,' Vieraskynä in *Helsingin Sanomat*, 21.12.2002, Helsinki, p. A4.

van Koppen, C. S. A. (2000), 'Resource, Arcadia, Lifeworld. Nature Concepts in Environmental Sociology,' *Sociologia Ruralis*, vol. 40, no. 3, pp. 300–318.

Luoma, P. (2002), *MTK:n ympäristökäsitykset ja ympäristöpoliittinen toiminta 1980–1999*, Manuscript 23.1.2002, forthcoming.

Macnaghten, P. and Urry J. (1998), *Contested Natures*, Sage Publications, Thousand Oaks.

Milton, K. (1996), *Environmentalism and Cultural Theory. Exploring the Role of Anthropology in Environmental Discourse*, Routledge, London.

Ministry of Agriculture/Latvia (2001), *SAPARD Programme for Agriculture and Rural Development for Latvia*, Revised on 21.11.2001, Vilnius.

Ministry of Agriculture/Lithuania (2000), *Agriculture and Rural Development Plan 2000–2006*, Riga.

NEBI (2000), in L. Hedegaard and B. Lindström (eds), *The NEBI Yearbook 2000. North European and Baltic Sea Integration*, Springer, Berlin.

Petrisalo, K. (2001), *Menneisyys matkakohteena*, SKS, Helsinki.

Press Releases (1) Rapid, DN:IP/00/1211. *Rural Development Programmes for Estonia, Lithuania and Slovakia Endorse*. 25.10.2000. http://europa.eu.int/rapid

Pridanova, A. (2000), 'Rural tourism: Latvia's Potential Gold Mine,' in *The Baltic Times*, April 27 – May 10, 2000, p. 9.

Ray, C. (1998), 'Culture, Intellectual Property and Territorial Rural Development,' *Sociologia Ruralis*, vol. 38, no. 1.

Ray, C. (2001), *Culture Economies*, Centre for Rural Economy, University of Newcastle upon Tyne, Newcastle.

RI (1999), in K. Andersson, E. Eklund and L. Granberg, *Rural Innovation, Intervjuer i Salla, Suomussalmi och Nagu hösten 1999*, Project archive, Unpublished.

SR (no year, about 1999), *A tourist brochure from a coastal region*, Leidykla, Klaipeda.

Suopajärvi, L. (1998), 'Grass-Roots-Level Lapland,' in L. Granberg (ed.), *The Snowbelt*, Kikimora Publications, Aleksanteri Institute, Helsinki.

# Chapter 7

# Trends in Development
# in Lithuanian Agricultural Policy

Donatas Stanikunas,
Irena Krisciukaitiene and
Romualdas Zemeckis

## Introduction

Agriculture is the oldest and an important sector of the Lithuanian economy. Agricultural land covers more than half the country's surface area. In 2001, 16 per cent of the active workforce was engaged in agriculture, generating 6.9 per cent of total value added.

This paper discusses the importance of agriculture to the Lithuanian economy and its development towards a free market economy over the last decade. The main focus is upon resources of agricultural production, structural changes, productivity and preparations for EU membership.

Lithuania is an active member of the United Nations and a member of the Council of Europe. It participates in the activities of the World Trade Organization (WTO), the UN Food and Agriculture Organization (FAO), the International Bank for Reconstruction and Development (IBRD), and the International Monetary Fund (IMF). Lithuania is a candidate country of the European Union and NATO.

Lithuania compares favourably with its northern neighbours in terms of the productivity of its soil and the duration of its growth season, which on the other hand is somewhat shorter than in Poland and southern Scandinavia, for instance. Natural conditions in Lithuania are suitable dairy and meat farming, pig farming, the growing of cereals, potatoes, sugar beets, rape, flax and other crops and livestock characteristic of this latitude. Likewise the conditions are favourable for the development of organic farming.

Lithuania's favourable natural conditions and geopolitical position provide a rather sound foundation for increasing exports of agricultural products to the traditional markets of the CIS, the Baltic States, and Central and East European countries, and indeed to breaking into new markets in the Far East, even in the Americas and Africa. The country's organic farming products could go to the EU and other countries of the world.

On the other hand, Lithuania's geopolitical position may also have adverse, undesirable consequences. If the state borders are insufficiently protected, the inflow of smuggled food products will inevitably distort competitive conditions within the country and undermine the production of equivalent domestic products.

## The Transition Towards Market Economy

Lithuania's transition from a centrally planned economy to a market economy has required radical political decisions that have not always been economically sound or socially acceptable.

Legislation relating to the agricultural reform, which started in 1991, has been amended more than once. The priorities in this reform have gradually shifted towards the principles of restitution. At the present time ownership rights are reinstated up to the third generation, i.e. the grandchildren of the landowner. In an attempt to resolve social problems, plots of land 2–3 ha in size have been allocated to landless rural families for personal needs. Some of these plots have been privatized. In 1991, the maximum amount of agricultural land that any individual was allowed to own was set at 80 hectares, in 1997 it was increased to 150 hectares. Agricultural land may not be sold to legal entities or foreign nationals. All this has facilitated the emergence of small non-competitive farms characterized by low investment capacities. The land reform has taken much longer than expected; its first stage was not completed until 1 October 2000. In 2002 further changes to the legislation gave legal entities and foreign nationals the right to buy agricultural land.

The privatization of the assets of agricultural enterprises was completed in 1992. The separation of the privatization of the assets of collective and state farms from that of land has led to the fragmentation of the productive potential in agriculture. Due to the belated adoption of the Law on Cooperation in 1993, the development of cooperatives, and vertical cooperative relationships in particular, was hampered by the absence of the necessary legal basis.

Over the past decade agricultural policies in Lithuania have shown a marked lack of consistency and continuity. Political views and positions on the land reform and the development of individual agricultural industries have continued to fluctuate. There has been no short-term or medium-term strategy for agricultural development.

Nor has the macroeconomic environment been very favourable for the development of agricultural business. The past decade has been characterized by the differentiation of prices for inputs and agricultural produce. State regulation of the prices for purchasing the main agricultural products (cereals, milk, flax) created a number of problems in agricultural sales and had adverse effects on the performance of both agricultural and processing enterprises as well as on the state budget.

The Law on State Regulation of Economic Relations in Agriculture stipulates that a minimum of 10 per cent of the national budget should go to the implementation of the National Programme for the Development of Agriculture, the implementation of state measures to regulate economic relations in agriculture, to building national reserves of food products, to the reclamation of land and to liming acid soils. However, these provisions have never been put into practice. Considerable amounts allocated from the budget to the agricultural sector have been used to subsidize prices; therefore there has been a constant lack of funds for investment programmes or other state commitments to meet the urgent needs of the land reform or land reclamation.

On the other hand, during the period described individual farmers as well as agricultural enterprises have enjoyed indirect support in the form of reduced electricity tariffs and various tax concessions and reductions with social security contributions, etc.

The whole decade of the 1990s was marred by failure to find an effective system of paying for agricultural produce. The buyers were constantly late with their payments for farmers, while the producers had no protection against the bankruptcy of the buyer organizations.

The credit market offered little help and support. Loans were available for limited periods only, interest rates were high, and difficulties with collateral and guarantees were a recurring problem.

Because of the fixed exchange rate of the litas against the US dollar and the floating exchange rate of the euro, Lithuanian food exporters have found themselves at a distinct disadvantage in comparison to importers. At the beginning of 2002 the litas was pegged to the euro, which very much complicated sales of agricultural produce outside the euro zone. In general, the trade conditions are not favourable for Lithuania. Lithuanian food products exported to the CIS sell at relatively low prices, while the market risk is unacceptably high. Trade with the EU is restricted by small quotas, large import tariffs and certain non-tariff barriers.

As yet there has been no resolution to the problems related to property insurance and social security. State support for agricultural business is not separated from social support to the rural population.

## Importance of Agriculture to the National Economy

The agricultural sector in Lithuania performs important social, economic, environmental and ethno-cultural functions, and it is therefore considered a priority sector of the national economy.

At the beginning of 2001, the rural population accounted for almost 33 per cent of the total population in Lithuania, compared to the EU average of 17.5 per cent. About 16 per cent of the active workforce in Lithuania are engaged in agriculture, whereas the average figure for the EU member states is no more than 5 per cent.

In 2000, the agro-industrial sector accounted for about 20 per cent of GDP, with about 30–35 per cent of all the employed were engaged in this sector (primary agricultural production, processing industry, trade, transport, industry and services related to agriculture).

Agriculture has a crucial role to play in building up Lithuania's national wealth: agriculture and the food sector are among the main economic sectors of the national economy (Table 7.1).

**Table 7.1 Macroeconomic indicators of the agricultural and food sector in 1995–2001**

| Indicators | 1995 | 1996 | 1997 | 1998 | 1999 | 2000 | 2001 |
|---|---|---|---|---|---|---|---|
| Share of agriculture in gross value added*, % | 10.9 | 11.3 | 10.9 | 9.5 | 7.9 | 6.9 | 7.1 |
| Share of agricultural and food products in total exports, % | 18.3 | 17.1 | 16.0 | 14.0 | 12.6 | 11.7 | 12.4 |
| Share of agricultural and food products in total imports, % | 13.4 | 13.1 | 11.1 | 11.0 | 11.3 | 10.2 | 9.4 |
| Share of agriculture in national budget expenditure, % | 9.8 | 9.3 | 8.7 | 7.7 | 6.2 | 5.3 | 4.6 |

\* Including agricultural services, hunting.

*Sources*: Statistical Yearbook of Lithuania 2002, Department of Statistics under the Government of the Republic of Lithuania, 2002; Economic and Social Development of Lithuania 12/2002, Department of Statistics under the Government of the Republic of Lithuania, 2002.

Lithuania is by and large self-sufficient with respect to food production. The need for imports is restricted to certain kinds of grains, oil, vegetables and fruits that cannot be grown at this latitude.

With the reorganization of the national economy, particularly since 1998, the importance of agriculture has gradually declined.

## Agricultural and Food Products

In 2001, the total value of Lithuania's agricultural production amounted to 4.6 billion litas. There have been no major changes in its structure over the past five years. Crop production constitutes more than one-half of total agricultural production, with grain accounting for 23 per cent. The total value of dairy, livestock and poultry production stands at 18 per cent each (Figure 7.1).

The changes in the volumes and structure of agricultural production are primarily explained by cyclical fluctuations. There have also been changes in the marketability of agricultural produce. Compared to 1995, purchases of the main crop products were higher and purchases of the main livestock products were lower than in 1999–2001.

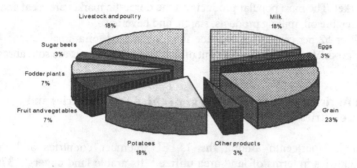

**Figure 7.1 Breakdown of gross agricultural production in 1997–2001**

*Sources:* Production of commodities 2001, Department of Statistics under the Government of the Republic of Lithuania, 2002; Economic and Social Development of Lithuania 12/2002, Department of Statistics under the Government of the Republic of Lithuania, 2003.

In 2001, there were 631 food and beverage industry enterprises in the country (excluding sole proprietorships). Two-thirds or 69 per cent of these companies represented small businesses employing less than 50 people. Sales by these companies represented no more than 13 per cent of total food and beverage sales.

Dairy and meat companies account for almost 37 per cent of the total production of food processing enterprises (sole proprietorships excluded) (Figure 7.2).

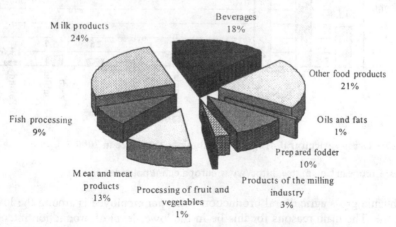

**Figure 7.2 Breakdown of sales in the food and beverage industry in 2001**

*Sources:* Production of commodities 2001, Department of Statistics under the Government of the Republic of Lithuania, 2002; Economic and Social Development of Lithuania (12/2002), Department of Statistics under the Government of the Republic of Lithuania, 2002.

Over 73 per cent of total production by food and beverage companies is sold on the domestic market. The most popular products on the domestic market are meat and meat products, vegetable oil, milling products, starch, and beverages.

Over half or 52 per cent of ready-made fodder and feed for animals, about 40 per cent of dairy products, and over 50 per cent of fish and fish products are sold abroad.

### Lithuanian Agriculture in the Context of EU Member Countries and Candidate Countries

Lithuania ranks fourteenth among the 15 EU member countries and the 10 candidate countries in terms of land area utilized. Its arable land covers 3,370,000 hectares. France has the largest area in this comparison (29,865,000 ha), followed by Spain (25,425,000 ha) and Poland (18,220,000 ha), at the other end of the scale (excluding Luxembourg) are Slovenia (491,000 ha), Estonia (1,001,000 ha) and Belgium (1,396,000 ha).

In Lithuania the number of people engaged in agriculture is relatively high at 16.1 per cent of the workforce. Among EU member countries the largest figure is recorded by Greece (17.0 per cent) and Portugal (12.5 per cent). Among candidate countries this indicator is the highest in Romania (45.2 per cent) and Poland (18.7 per cent). Lithuania ranks fifth in this comparison among EU members and candidate countries.

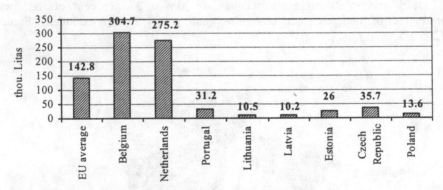

**Figure 7.3 Gross agricultural production value per employee in 2000**

*Sources:* European Committee, http://www.europa.eu.int/pol/index.en.htm.

In Lithuania gross agricultural production value per employee is among the lowest in Europe. The main reasons for this lie in the lower level of production intensity, smaller farms, and lower prices of main products (milk, beef).

Lithuania and its neighbouring countries will be able to reach the level of EU member countries, especially Belgium and Netherlands, within the space of one decade. In Lithuania gross agricultural production value per employee is 14 times lower than the EU average, and 29 times lower than in Belgium.

**Table 7.2 Main indicators of agriculture in Lithuania and EU countries in 2000**

| Country | Share of agriculture in GDP, % | Average farm size, ha of agricultural land | Agricultural land per inhabitant, ha | Share of active work-force engaged in agriculture, %* | Grain yield, t/ha | Milk yield, kg** |
|---|---|---|---|---|---|---|
| Austria | 2.2 | 16.3 | 0.42 | 6.1 | 5.52 | 4,703 |
| Belgium | 1.2 | 20.6 | 0.14 | 1.9 | 7.59 | 5,460 |
| Denmark | 2.7 | 42.6 | 0.50 | 3.7 | 6.04 | 7,516 |
| Estonia | 6.1 | 21.2 | 0.99 | 7.2 | 2.11 | 5,152 |
| Finland | 3.4 | 23.7 | 0.43 | 6.2 | 3.17 | 6,452 |
| France | 3.4 | 41.7 | 0.51 | 4.2 | 6.75 | 5,641 |
| Germany | 1.5 | 32.1 | 0.21 | 2.6 | 7.08 | 5,883 |
| Great Britain | 0.9 | 69.3 | 0.27 | 1.5 | 6.30 | 6,487 |
| Greece | 4.0 | 4.3 | 0.48 | 17.0 | 3.17 | 4,529 |
| Ireland | 4.0 | 29.4 | 1.20 | 7.9 | 7.61 | 4,374 |
| Italy | 4.1 | 6.4 | 0.27 | 5.2 | 4.74 | 5,535 |
| Latvia | 4.5 | 19.9 | 1.00 | 14.7 | 2.20 | 4,055 |
| Lithuania | 7.9 | 14.7 | 0.96 | 18.9 | 2.68 | 3,903 |
| Luxemburg | 1.0 | 42.5 | 0.29 | 2.2 | 5.65 | 5,460 |
| Netherlands | 3.5 | 18.6 | 0.12 | 3.3 | 7.74 | 7,143 |
| Portugal | 2.0 | 9.2 | 0.39 | 12.5 | 2.60 | 5,256 |
| Spain | 7.6 | 21.2 | 0.74 | 6.9 | 2.83 | 4,842 |
| Sweden | 1.8 | 34.7 | 0.35 | 2.9 | 4.67 | 7,674 |
| EU-15 | 2.3 | 18.4 | 0.36 | 4.3 | 5.47 | 5,557*** |

* In the total employed; ** 2001. *** 1998.

*Sources:* European Committee, http://www.europa.eu.int/pol/index-en.htm; Department of Statistics under the Government of the Republic of Lithuania.

## Background of Lithuanian Agriculture

### Land

The main factors, which determine the results of agricultural business are land, labour and capital. Lithuania has rich enough soils and skilled labour force, but suffers a lack of capital.

Over half or 52 per cent of Lithuania's total land area or 3,370,000 hectares is suitable for agricultural purposes. Of this, 85 per cent is arable land, over 13 per cent grasslands and pastures, and 1.3 per cent orchards and berry plantations.

Climate conditions in Lithuania are favourable for dairy and meat farming and the cultivation of wheat, rye, fruit and vegetables, flax, rape, and sugar beets.

In areas that suffer periodic excess of rainfall, 78 per cent of the land is drained. More than 1 million hectares of the land is naturally acid and needs to be limed periodically.

The process of land reform and the restitution of ownership rights to land, forest and water bodies started in 1991. This has increased the amount of private land and decreased the amount of state-owned and state-managed land in the country (Table 7.3).

**Table 7.3 Land used for agricultural activities by owners in 1995 and 2001**

| Land owners | 1995 | | 2001 | |
|---|---|---|---|---|
| | Thousand ha | % | Thousand ha | % |
| Private land used for agricultural activities | 792.6 | 20 | 2,090.5 | 53 |
| State-owned or state-managed land | 3,133.2 | 80 | 1,865.7 | 47 |
| Total | 3,925.8 | 100 | 3,956.2 | 100 |

*Sources:* Land fund of the Republic of Lithuania (as of 1 January 2003), The Ministry of Agriculture of the Republic of Lithuania, National Land Service under the Ministry of Agriculture, 2003.

The first stage of the reform is the restitution of citizens' ownership rights to land, forest and water bodies. The aim of the second stage is the reorganization and consolidation of land plots forming compact farms of rational size.

*Farm Structure*

At the end of 2001, there were about 160,000 farmers, 697 agricultural partnerships, and 244,000 household farms in Lithuania. The average size of family farms was 17.2 hectares, and that of agricultural partnerships 329 hectares. Although the average size of family farms has been increasing and the number of agricultural partnerships decreasing, small family farms still dominate, especially in less favourable farming areas. Among farmers registered in the Farm Register, 38 per cent have a farm sized between 3.1 and 10 hectares, 32 per cent between 10.1 and 20 hectares, and 4.5 per cent a farm over 50 hectares in size.

The financial position of individual farms depends crucially on the price at which they can sell their produce. Those prices are similar across all regions in the country. In 2002, the Lithuanian government introduced a state subsidy system, which takes into account the natural conditions of the farming regions. Direct subsidies are paid out for rye, buckwheat, legume crops, and grasslands and pastures to farmers in less favourable areas.

So far there has been no specialization among Lithuanian farms: mixed farming dominates in the country, combining both crop and livestock production.

Nonetheless there are clear signs of regional differentiation based on farm sizes and the quality of land.

The number of cows on Lithuanian farms is nine times lower than the average for EU countries; for pigs, the figure is 15 times lower. Small, non-specialized farms are less efficient than larger and specialized farms. They also have difficulty finding the funds they would need for modernization. Since 1997, the consolidation of family farms has begun to gather momentum (Table 7.4).

**Table 7.4  Changes in the number of registered family farms in 1997–2002 (as of 1 January)**

| Farm size, ha | 1997 | 1999 | 2001 | 2002 |
|---|---|---|---|---|
| 0.3–3 | 0.8 | 0.9 | 1.3 | 5.2 |
| 3.1–10 | 57.8 | 56.6 | 54.4 | 37.6 |
| 10.1–20 | 29.3 | 29.5 | 29.4 | 31.9 |
| 20.1–30 | 7.8 | 8.1 | 8.6 | 12.7 |
| 30.1–50 | 3.4 | 3.8 | 4.5 | 8.1 |
| > 50 | 0.9 | 1.1 | 1.8 | 4.5 |
| Total | 100 | 100 | 100 | 100 |

*Sources:* Analysis of agricultural enterprises activity in Lithuania and some EU countries, The Ministry of Agriculture of the Republic of Lithuania, Lithuanian Institute of Agrarian Economics, 2003.

EU experience suggests that for reasons of stability the ideal size of a dairy farm is from 30 to 60 cows. As for pig farms, they should have 3–5 pigs per one hectare of agricultural land, with a minimum total of 150–200 pigs or 30 breeding sows. In the EU member states farms typically have from 40 to 100 breeding sows. In less favourable farming areas, sheep farming is recommended.

Financing received from the Special Accession Programme for Agriculture and Rural Development (SAPARD) will further encourage farm consolidation. Farm restructuring will also be influenced by macroeconomic changes in the country and the support it receives from the national budget.

## Labour resources

Agriculture is an important source of employment in Lithuania. In 2001, almost half of the rural population was engaged in agriculture; the figure for the whole country was 16 per cent. Compared to1996, the number of people working in agriculture was down by almost one-third.

In Lithuania the share of the population engaged in agriculture, forestry and fisheries is rather high when compared to other European countries (Figure 7.4).

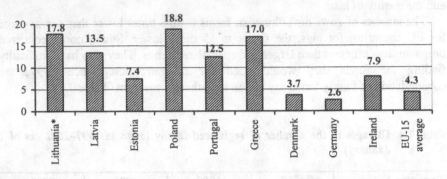

\* 2001

**Figure 7.4 People employed in agriculture, hunting, forestry and fishery as a proportion of the active workforce in 2000, per cent**

*Sources:* Analysis of agricultural enterprises activity in Lithuania and some EU countries, The Ministry of Agriculture of the Republic of Lithuania, Lithuanian Institute of Agrarian Economics, 2003.

On small family farms 27 per cent of family members work less than one-quarter of the total annual standard of working hours (2,033 hours). This means they produce only limited value added, their income does not allow them to save or to invest in modernizing the farm.

On larger farms (50 and over hectares) the technology and machinery is more sophisticated and incomes higher. Labour productivity is also twice as high as on small farms.

Research by the Lithuanian Institute of Agrarian Economics clearly highlights the impacts of age and education upon labour productivity. Farmers under 35 years of age are twice as productive as employees over 65. In Lithuania about 40 per cent of all people working in agriculture are over 65, while young farmers under 40 account for no more than 14 per cent. Many agricultural workers have a high level of vocational education, yet they have only limited management skills and work experience.

The majority of people employed in agriculture are involved in the production of milk, grain and vegetables. The figures for those engaged in industrial crops cultivation, poultry farming and non-traditional agriculture are much smaller (Figure 7.5).

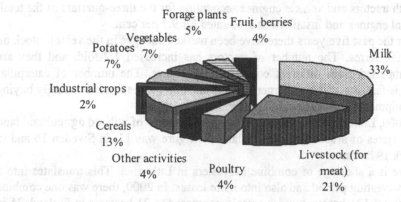

**Figure 7.5    Breakdown of population engaged in agriculture by products in 2001, per cent**

*Sources:* According to calculations by the researchers at the Lithuanian Institute of Agrarian Economics.

## Capital

The composition of fixed assets on family farms and in agricultural enterprises is determined by ownership relations and the nature of farming activities.

The main asset for farmers is represented by the value of their land. Agricultural enterprises, being legal entities, use to have no right to own land; therefore their assets consisted primarily of the value of their machinery and equipment.

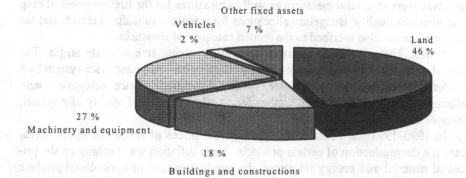

**Figure 7.6    Fixed assets on family farms in 2000**

*Sources:* Agriculture in Lithuania 2001, Lithuanian institute of Agrarian Economics, 2002.

In 2000, the energetic capacity per 1 hectare of utilized agricultural land was 2.3 kW, with tractors and vehicle engines accounting for the three-quarters of the total. Electrical engines and installations accounted for 9.7 per cent.

Over the past five years there have been marked changes in the vehicle stock on Lithuanian farms. The number of tractors has increased 1.2-fold, and they are being imported from different countries than before. The number of caterpillar tractors is fast diminishing. Larger farmers and enterprises are increasingly buying their equipment from western countries.

In 2000, Lithuania had one tractor per 33.4 hectares of utilized agricultural land (28.5 hectares of arable land). In Finland the figure was 11, in Sweden 16 and in Denmark 19 hectares.

There is a shortage of combine harvesters in Lithuania. This translates into a longer harvesting period and also into more losses. In 2000, there was one combine harvester per 121 hectares under cereals, compared to 21 hectares in Finland, 35 in Germany and 52 in Denmark.

The service and maintenance infrastructure for mechanized agriculture is also poorly developed in Lithuania.

### Economic Measures of Market Regulation

Faced with the challenge of transition from administrative to economic measures of market regulation, the Lithuanian government has gradually introduced a series of changes since it gained economic independence. For example, subsidies for agricultural resources have been eliminated, purchasing prices of agricultural produce have been partly liberalized, and food prices have been fully liberalized.

Economic measures of market regulation in agriculture may be divided into two main categories. The first comprises measures of price and income support. The second covers structural measures as well as measures for the improvement of crop and livestock quality. Budgetary allocations for research, subsidized credits and tax concessions are also ascribed to the second category of measures.

*Prices:* The price reform may be seen as involving five separate stages. The first stage resulted in the replacement of a differentiated purchase price system by a uniform agricultural produce price system. Essentially, price categories were eliminated and a uniform price was set for a certain grade and quality of products, irrespective of land productivity.

In 1990–1991 the government approved the prices of resources/inputs necessary for the production of certain products. Since inflation was pushing up the prices of material and energy resources, the purchase prices of agricultural produce were subject to revaluation. Nominal price rates therefore remained stable, while real prices were coming down.

Starting from 1992, the second stage was characterized by support prices for the main agricultural products. The system reflected the situation after free contractual prices of agricultural produce became predominant on the market, their size was determined by supply and demand, and they depended on freely concluded contracts by economic entities selling and buying agricultural produce on the market. The idea behind the price reform was that when prices dropped to the level of the support prices, the government would subsidize the main types of agricultural produce, which would ensure a minimum income to efficient agricultural producers. This system was short-lived, however. It collapsed because processing plants used to contract for low purchase prices and sought state support in order to secure trading margins. On the other hand, the state was not able to ensure a sufficient income to agri-producers as the general price index for goods and services was higher than the purchase price indices for the main types of agricultural produce (Figure 7.7).

In stage three (1993–1994), prices were liberalized. With agricultural produce in abundant supply, prices and profitability dropped.

At the end of 1994, with the adoption of the Law on State Regulation of Economic Relations in Agriculture, minimum purchase prices were set for food grains, milk, livestock, flax, and rape. These prices remained in force with only a few minor changes until 2000.

The year 2000 marked the beginning of a new price stage, which might be called a pre-accession stage, i.e. the stage of adjustment to EU price policy. Intervention and target prices were stipulated. An intervention price is the price at which intervention stocks of agricultural produce or food products are procured. A target (orientation) price is the price stipulated by the government for agricultural production and food products to ensure minimal profitability to efficient farmers.

The dynamics of agricultural production prices and the prices of material resources needed is shown in Figure 7.7.

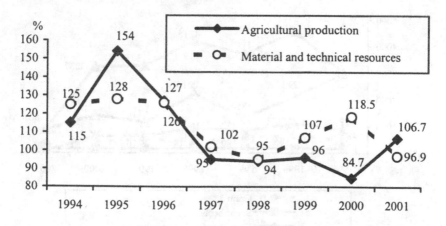

**Figure 7.7** **Dynamics of purchasing prices of agricultural produce and material and technical resources (compared to previous year)**

*Sources*: Statistical Yearbook of Lithuania 2002, Department of Statistics under the Government of the Republic of Lithuania, 2002; Economic and Social Development of Lithuania 12/2002, Department of Statistics under the Government of the Republic of Lithuania, 2002.

The general purchase price index for agricultural produce fluctuated during 1993–1999. The growth in 1994–1995 was far from stable, peaking at 54 per cent in 1995. The period from 1997 to1999 was relatively stable, showing a slow decline at around 4–6 per cent, while the index dropped sharply by 15.3 per cent in 2000 (compared to 1999). In 2001, the agricultural produce price index rose by 26 per cent.

Meanwhile, prices of material and technical resources rose consistently, with the exception of 1998. In 2000 prices shot up by 13.3 per cent, in 2001 prices dropped by 20.6 per cent.

For agricultural producers, 1995 was a good year: that was when the purchase price index for agricultural produce increased by 26 per cent more than the price index for material resources. The situation in 2000 was less favourable to agribusiness, since prices showed the opposite tendency: the prices of material resources increased while those of agricultural products went down. In 2001, though, agricultural products prices rose again.

The government's policy of price regulation has lacked in stability. Grain prices increased by 42 per cent in 1996 compared to 1995, the purpose being to encourage farmers to produce grain, but as it turned out this was a short-lived exercise. The decrease in grain prices in 1997–1998 was followed by an increase (due to the better quality of grain) in 1999 and a rollback of 27 per cent in 2000. In 2001, grain prices fell by 7.4 per cent.

Figures 7.8 and 7.9 describe the changing pattern of the main grain and livestock production prices.

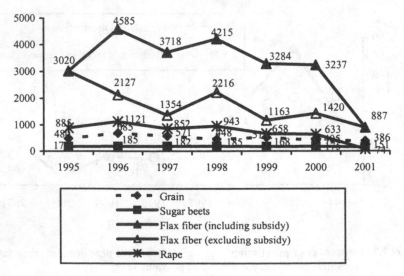

**Figure 7.8    Purchase prices of crop production in 1993–2001, litas per ton**

*Sources:* Agriculture in Lithuania 2001, Lithuanian Institute of Agrarian Economics, 2002; Materials from the Department of Statistics under the Government of the Republic of Lithuania, 2003.

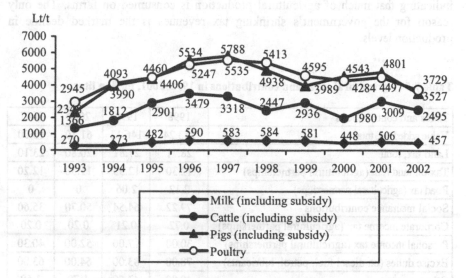

**Figure 7.9    Purchase prices of livestock production in 1993–2002, litas per ton**

*Sources:* Agriculture in Lithuania 2001, Lithuanian Institute of Agrarian Economics, 2002; Materials from the Department of Statistics under the Government of the Republic of Lithuania, 2003.

Cattle prices have been falling since 1996, pig prices dropped from 1997 to 1999. Cattle prices in 2000 were down by 49 per cent on the figure for 1999. The price for pigs, by contrast, increased by 15 per cent.

In 2000 prices for all main products (with the exception of flax, rape and pigs) showed a downward tendency. This was due to the imbalance between supply and demand and to the government's lack of means to intervene.

*Taxes:* With the re-establishment of its independence, Lithuania overhauled its taxation system, including agricultural taxes. In 1992 the government introduced the land tax (at 1.5 per cent) and tax on land lease (3 per cent). Farmers were exempt from land tax for the first three years of farming. Agricultural partnerships have been paying corporate income tax since 1991. In 1994 when value added tax was introduced, agricultural producers were granted special tax allowances. Until 1 January 1997, VAT on agricultural products was 9 per cent as compared to 18 per cent rate on other products. In 1997 this tax allowance was eliminated.

Pursuant to the Law on Tax Administration, agricultural entities shall pay the following taxes: value added tax, excise duty, personal income tax, corporate income tax, land tax, tax on land lease and road tax. In addition, they are required to make mandatory health insurance contributions.

The data in Table 7.5 suggest that the amount of tax paid by agricultural entities has been gradually declining. In 1998 the total amount of tax paid on the value of agricultural production amounted to 8.5 per cent and in 1999 7.8 per cent. The respective figures for agricultural production sold were 17.6 and 16 per cent, indicating that much of agricultural production is consumed on farms. The only reason for the government's shrinking tax revenues is the marked decline in production levels.

**Table 7.5  Agricultural taxes and contributions in 1998–2001, million litas**

| Taxes and contributions | 1998 | 1999 | 2000 | 2001 |
|---|---|---|---|---|
| Value added tax, total | 253.24 | 146.13 | 61.10 | 13.00 |
| Land tax, total | 28.71 | 27.81 | 20.80 | 23.10 |
| Tax on land rent (agricultural partnerships) | 12.30 | 12.35 | 11.50 | 12.20 |
| Road tax (agricultural partnerships) | 2.12 | 2.06 | 0 | 0 |
| Social insurance contributions | 79.22 | 64.54 | 50.70 | 35.60 |
| Corporate income tax (agricultural partnerships) | 0.72 | 0.21 | 0.20 | 0.20 |
| Personal income tax (agricultural partnerships) | 30.00 | 17.00 | 52.00 | 40.30 |
| Excise duties (on diesel fuel, petrol, lubricants) | 78.00 | 83.00 | 64.00 | 63.00 |
| Other taxes | 40.80 | 43.60 | 1.70 | 1.50 |
| Taxes, total | 525.11 | 396.70 | 262.00 | 188.90 |

*Sources:* Agriculture in Lithuania Economic survey 2002, Lithuanian Institute of Agrarian Economics, 2003.

*Foreign Trade Regulation:* The transition from command to market economy also saw the introduction of various quantitative restrictions for purposes of trade regulation. At the beginning of 1993, however, following the liberalization of foreign trade, these restrictions were lifted. Conditions for foreign trade were further improved with the launch of the system of customs duties and with signing of trade agreements with foreign states.

An agreement on economic, commercial and trade cooperation with the EU was signed in 1992, which was followed by a free trade agreement in 1994. Since 1995 free trade agreements have been in force with all EU member states. In June 1995, Lithuania signed the Europe Agreement, granting it the status of an associate country.

Free trade agreements between Lithuania, Latvia and Estonia were enacted in 1994. In addition to the above mentioned free trade partners, FTAs have been signed with Poland, Slovakia, Slovenia, the Czech Republic, Turkey, Hungary, the Ukraine, and EFTA countries.

The main legal documents which lay the foundations for foreign trade in agricultural and food products are as follows: the Law of the Republic of Lithuania on Tariff Rates (19 December 1998 No. VIII-633), the Customs Code of the Republic of Lithuania (18 April 1996, No. I-1292), the Government Resolution 'On the Procedure of Regulation of Export and Import of Goods in the Republic of Lithuania' (24 March 1997 No. 28), the Order of the Minister of Agriculture 'On the Procedure of Issuing Automatic Import Licences for Agricultural and Food Products' (29 January 1998 No. 240), other legislation and tree trade agreements.

Most customs duties in Lithuania are determined on an '*ad valorem*' basis. Depending on the country of origin, autonomous, conventional and preferential customs duties may be imposed on imports to the Lithuanian customs territory.

A preferential duty rate is applicable in cases where goods originate from free trade partners. Where goods originate from countries with which Lithuania has not concluded international agreements providing for a most favoured nation status, an autonomous duty rate is applied. Where goods originate from countries that are parties to such agreements, conventional duty rates are applied.

MFN status has been granted by Lithuania to the Commonwealth of Australia, Belarus, Bulgaria, Georgia, Cyprus, Korea, Cuba, Moldova, India, Uzbekistan, Kazakhstan, the USA, Japan, Romania, the Russian Federation, the People's Republic of China, and the Socialist Republic of Vietnam.

Lithuania's legal acts stipulating the determination of the origin of agricultural and food products are in compliance with the respective EU legislation, presented no barriers to Lithuania's WTO membership and present no barriers to Lithuania's EU accession negotiations.

The list of agricultural and food products which require a licence for importation was extended by Government Resolution No. 1350 (of 20 November 1998), amending the Resolution 'On the Procedure of Regulation of Export and Import of Goods in the Republic of Lithuania' (24 March 1997 No. 28) seeking to protect Lithuania's domestic market from cheap and low quality agricultural and food products. Since 1 February 1999, beef, pork, poultry, meat offal, fish, grain, buckwheat, cereals, starch, wheat gluten, rape and mustard oil, margarine, canned meat and fish, and sugar are imported under an autonomous licensing system, with licences issued by the Ministry of Agriculture.

There are no restrictions in place for agri-food exports, and there are hardly any export duties or quantitative restrictions. Two measures do remain, though, in the form of export subsidies: these are direct subsidies to exporters, and subsidies for agri-food production purchased by the Lithuanian Agency for the Regulation of the Agricultural and Food Product Market.

## State Support for Agriculture

Agricultural reform and modernization requires massive financial investment. However, many agricultural entities in Lithuania lack the necessary capital to make that investment, and borrowing from commercial banks is also difficult.

Concerned about farmers' lack of options and hoping to support their efforts to modernize, the government provides direct support for agriculture in the form of budget allocations. In addition, the government provides indirect support through tax concessions.

The forms and scope of direct and indirect government support for agriculture have seen major changes over the past ten years since independence, which has given rise to much suspicion on the part of farmers in the consistency of state agricultural policy. This has also prevented the implementation of long-term development programmes as well as the effective use of financial support.

*State support 1991–1996:* Seeking to promote the stabilization of farmer's income and productivity in farming, the Lithuanian government approved the Priority Programme for Support to Farmers on 28 January 1991. The document was small in terms of words, but huge in terms of its significance since it was the first time after the re-establishment of Lithuania's independence that the priorities of state support for agriculture were put into words.

On 7 February 1992, the government authorized the Ministry of Finance to set up the Farmers' Support Fund (FSF). This Fund operated for five years. Its sources were the state budget, foreign support and the Agrarian Reform Fund. The funds of the FSF were intended for purposes of supporting farmers who owned at least 15 ha of land used for agricultural purposes and who were engaged in commercial agricultural production.

Alongside the FSF, the government also established the Agricultural Support Fund using proceeds from the sale of grain received in charity from the USA. The Fund was designed as an arrangement for extending medium-term credits to farmers.

On 15 February 1994 the government adopted a resolution 'On the Establishment and Utilization of the Agricultural Support Fund of the Republic of Lithuania.' This fund was used for several different purposes, such as price subsidies, subsidized credits for agri-producers and organizations purchasing agricultural produce and supplying material resources, for extension enterprises, for the development of the cooperative movement in agriculture and support for co-operative agri-trade, for the development of commercial horticulture, and for support to fire stricken agricultural partnerships.

However, in the absence of any specific guidelines and procedures for the utilization of these funds, there have been cases of abuse in extending subsidized credits as well as instances of non-viable projects being financed.

*Rural Support Fund 1997–2001:* On 12 March 1997, the Lithuanian government passed a resolution to set up the Rural Support Fund (RSF). At the same time the Farmers' Support Fund, the Agricultural Support Fund and the Agricultural Support Fund of the Republic of Lithuania were all discontinued.

The RSF funds were to be granted for purposes of the economic regulation of the agricultural market and for the provision of income support to farmers and agricultural partnerships; for financing targeted investment programmes; and for the development of agri-science, consulting, training, and information systems.

The major source of RSF funding is the state budget. Each year about 6 per cent of the state budget is set aside for the purpose: LTL 398 million in 1997, LTL 523 million in 1998, LTL 396 million in 1999 [LTL 434 million in 2000, and LTL 203 million in 2001. Besides, the Fund was supplemented by long-term credits being repaid from the other agricultural support funds. The Privatization Fund became an important source for the RSF during the economic crisis of 1998–1999 when the budget deficit was especially acute. LTL 75.7 million were transferred to the RSF in 1998, and LTL 41.5 million in 1999.

**Table 7.6** **Rural Support Fund in 1997–2002, in million litas**

| Utilization by purpose | 1997 | 1998 | 1999 | 2000 | 2001 | 2002 |
|---|---|---|---|---|---|---|
| Measures for market regulation and income support | 261 | 238 | 184 | 307 | 108 | 144 |
| Priority investment programmes | 94 | 100 | 76 | 68 | 38 | 31 |
| Research, training, consulting, information system | 20 | 23 | 18 | 17 | 13 | 13 |
| Agricultural Credit Guarantee Fund | 8 | 20 | 2 | 5 | 3 | 6 |
| Unforeseen measures | 15 | 142 | 116 | 37 | 41 | 58 |
| Total | 398 | 523 | 396 | 434 | 203 | 252 |

*Sources:* Reports by the Ministry of Agriculture on the Utilization of Funds from the Rural Support Fund in 1997–2002.

*Income Support:* Price subsidies were the major measure of market regulation and income support for agricultural employees before 1997. These subsidies also had a partial impact on the structure and dynamics of agricultural production.

However, as we have seen above, price subsidies were an ineffective tool that merely added to instability and unpredictability in the agricultural market. Their main drawback was the high costs incurred to the state. Therefore the decision was made that subsidies should only be used as a transitional measure and that their scope should be reduced by simultaneously allocating more funds to the co-financing of investment projects.

However the latter provision was halted by the Russian crisis and by sometimes economically unsound political decisions. Therefore rather than decreasing the amount of subsidies increased each year, as did the number of market regulation measures. An ever-growing proportion of RSF funds was used for procuring the surplus of agricultural produce from farmers, for its storage and sales on domestic and foreign markets, and for covering the price difference (Table 7.7).

RSF regulations do not provide for the use of its funds as credit resources. Farmers therefore have to turn to commercial banks and credit unions when they need to borrow capital. To create more favourable borrowing conditions for farmers and agricultural partnerships, support has been made available in the form of 60 per cent subsidies for interest on short-term loans. Since 1999, in cases where farmers, partnerships and cooperatives have financed their investments by borrowing from banks, they have been eligible for an interest subsidy when 30–50 per cent of the interests on their long-term loans are covered from the RSF and the Fund for the Promotion of Small and Medium-sized Enterprises.

*Investment Support:* Both economic theory and business development practice prove that return on investment is the highest when funds are invested in state-of-the-art technologies and intellectual capital. Therefore the priorities of the RSF are defined accordingly.

**Table 7.7 Price subsidies and other short-term measures in 1997–2002, in million litas**

| Measures | 1997 | 1998 | 1999 | 2000 | 2001 | 2002 |
|---|---|---|---|---|---|---|
| Subsidies for procured agricultural products | 205.4 | 190.3 | 128.4 | 137.3 | 52.2 | 198.6 |
| Subsidies for bloodstock and high quality seeds | 16.0 | 14.6 | 7.5 | 13.2 | 16.6 | 9.0 |
| Export subsidies | 16.8 | 45.9 | 60.3 | 14.0 | 0 | 0 |
| Direct payments | 0 | 0 | 0 | 25.6 | 41.0 | 82.7 |
| Expenditure on procurement and storage of agricultural products | 5.6 | 85.3 | 15.3 | 30.0 | 2.9 | 3.7 |
| Interest subsidies | 18.5 | 6.8 | 2.1 | 5.5 | 5.3 | 1.2 |
| Special bank provisions | 7.4 | 14.8 | 4.3 | 0 | 11.4 | 3.0 |
| Cover of sugar beet price difference | 0 | 0 | 77.8 | 80.9 | 81.9 | 97.9 |
| Other measures | 6.8 | 22.6 | 4.5 | 6.3 | 14.4 | 15.9 |
| Total | 276.5 | 380.3 | 300.2 | 312.8 | 225.7 | 412 |

*Sources:* Reports by the Ministry of Agriculture on the Utilization of Funds from the Rural Support Fund in 1997–2002.

At the beginning of 1997 when the RSF was set up, there were in Lithuania 47,000 family farms, the development of which was hindered by lacking infrastructure in rural areas. With this in mind, the RSF provided full financing for infrastructure items such as public roads, power and communication lines, and water supply. During 1997–2002, about 10 per cent of all RSF funds were used for such investment projects.

Investment into advanced technologies was supported through targeted investment programmes: the Modernization of Farmers' Farms, Cooperation, Agricultural Advisory Service, New Technologies, Stock-breeding, Reorganization of Farming Activities on Low Productivity Lands and the Reorganization of Activities in order to Earn Alternative Income, Development of Agricultural Resources, and Development of a Quality Control System. These programmes are aimed at promoting effective and competitive farming.

Between one-quarter and one-half of the real value of investment projects is subsidized by the RSF. The rest has to be financed through one's own resources or borrowed funds. This, in particular, presents a problem in low productivity areas, where payment arrears and unfavourable borrowing conditions prevent farmers from accumulating the necessary share of the anticipated investment. On the other hand, low personal involvement and lack of understanding about the market environment should also be mentioned.

The most effective form of investment is investment into human resources. Each year LTL 23–17 million has been spent on research, consulting and training as well as on the development of agricultural information systems.

**Table 7.8 Financing of targeted programmes in 1997–2002, in million litas**

| Programmes | 1997 | 1998 | 1999 | 2000 | 2001 | 2002 |
|---|---|---|---|---|---|---|
| Programme for settling farmers | 60.7 | 67.9 | 29.5 | 11.7 | 10.8 | 13.4 |
| of which: Infrastructure | 55.4 | 46.7 | 12.9 | 2.2 | 0 | 0.6 |
| Equipment and technology | 5.3 | 21.2 | 16.6 | 9.5 | 10.8 | 12.8 |
| Co-operation | 0.3 | 0.9 | 5.7 | 4.2 | 1.5 | 0.3 |
| Agri-service and technology | 2.8 | 2.5 | 2.3 | 0.8 | 7.6 | 4.9 |
| Reorganization of farming activities on low productivity land and in order to earn alternative income | 2.2 | 1.7 | 1.5 | 0.5 | 1.2 | 11.9 |
| Organic farming | 1.1 | 1.4 | 0.2 | 0.5 | 0.8 | 2.5 |
| Stock-breeding programme | 20.2 | 17.6 | 12.6 | 15.1 | 0.6 | 17.2 |
| Registration and identification of livestock | 0 | 0 | 5.9 | 9.9 | 5.2 | 4.1 |
| Crop declaration | 0 | 0 | 0 | 1.9 | 20.4 | 27.6 |
| Quality testing system | 6.3 | 8.2 | 18.3 | 13.9 | 7.3 | 11.6 |
| Total | 154.3 | 168.1 | 105.5 | 70.2 | 66.2 | 106.9 |

*Sources:* Reports by the Ministry of Agriculture on the Utilization of Funds from the Rural Support Fund in 1997–2002.

*Agricultural Credit Guarantee Fund:* This Fund was set up in 1998, with the necessary monies provided through the RSF. It was designed to issue guarantees to banks in behalf of farmers with long-term investment loans.

The conditions for issuing these guarantees have changed over time. In 2000, aiming at the implementation of the Special Accession Programme for Agriculture and Rural Development (SAPARD), the list of economic entities eligible for guarantees was extended, including all agricultural and economic entities that are engaged in alternative activities.

The Fund undertakes to cover up to 70 per cent of a credit (the bank assumes the remaining risk) extended to farmers and enterprises. Guarantees issued by the Agency for the Regulation of the Agricultural and Food Product Market cover 100 per cent of the credit.

In 1998–2000 the Fund issued 583 guarantees to farmers and cooperatives, with a combined value of LTL 38 million. The majority or 70 per cent of the farmers used the Fund-secured credits to purchase new tractors, 5 per cent used the monies to acquire new milk refrigeration or milking equipment. Several non-traditional and alternative projects stand out among the usual business background: mushroom growing, agricultural tourism, modern vegetable storage facilities, etc. The farmers repaid LTL 13 million to the banks, representing one-third of the guaranteed credits.

The Agency and the enterprises engaged in the purchase and storage of agricultural production made use of LTL 576.7 million guaranteed credits. The largest share (75 per cent) was used for purchasing grain, the rest for purchasing and selling food products.

*Lithuania's Main Agricultural Products: Competitive Edge*

*Milk:* Lithuania's dairy products are of a high quality and competitive on the export market, both in terms of quality and price. There are 17 milk processing enterprises in the country with the right to export products to EU countries. In 2000, total exports of milk amounted to 587,500 tons, accounting for 30.5 per cent of total dairy production. The main markets for Lithuanian dairy products are in the EU, the USA, the Baltic States and Russia. More than one half of dairy products are exported to western countries.

*Meat:* Meat production is in economic terms the second most important sector in Lithuanian agriculture. In spite of their high quality, the export of meat products is problematic. Only three meat-processing enterprises in the country have licenses to export their products to the EU. Besides, beef cattle make up only a small part of the herd. Changes are also needed in pig breeds.

*Grain:* The grain sector is and will remain oriented to domestic needs. Grain consumption for fodder is set to increase while the demand for food grain will decrease.

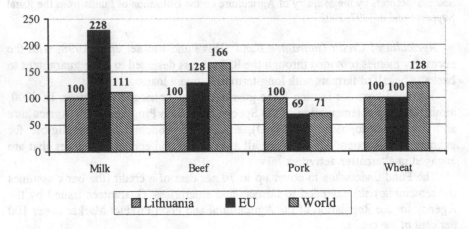

**Figure 7.10  Prices of agricultural products in Lithuania, the EU and the world in 2001 (index = Lithuania 100)**

*Sources:* According to calculations of researchers at the Lithuanian Institute of Agrarian Economics.

Organic farming: Organic agriculture has been advancing in Lithuania in recent years. In 2001, 290 ecological farms, eight enterprises processing organic products, five pickers of natural products and six enterprises supplying auxiliary implements for organic agriculture were certified. The area of certified organic farms increased to 6,500 hectares representing 0.2 per cent of the total area of agricultural land. However, this indicator is 10 times lower than in the EU.

Grains make up the bulk of ecological crops. Organic rye is the main product category. Its popularity is due to the high demand for traditional rye bread. Each year the range of organic products continues to expand. New branches of organic production are also starting up with sheep, goat and rabbit breeding, for instance. In the near future organic fish, sugar and dairy products will also be available in the markets.

At the moment the demand for organic products is greater than the supply.

## Strategic Development Trends

Committed to a balanced development of its agricultural and food sector in economic, ecological and social terms, Lithuania is determined to pursue a rational agrarian policy by means of effective state regulation.

Government agricultural policy is focused on four main areas that are crucial to agricultural and rural development: macroeconomic conditions, market development, land management, and training and education.

The rational development of the agricultural and food sector will be based upon the following principles:

- Political and macroeconomic stability: a long-term agricultural and food strategy supported by democratic government changes.
- The provision of adequate financial resources to agricultural and food entities, including favourable credit terms and state support encouraging positive structural changes.
- The development of domestic and foreign markets, including the creation of an effective market research system, the promotion of competitiveness and exports, the improvement of the intervention purchase model, and protection of the domestic market.
- Completion of the restoration of citizens' ownership rights to land, forest and water bodies, land consolidation and encouragement of land market development.
- Training and education in agricultural science and the improvement of vocational and management competencies.

## Conclusions

Agriculture in Lithuania performs important economic, social, environmental and ethno-cultural functions. It is therefore a priority sector of the economy, whose competitiveness is crucial to the competitiveness of the whole national economy. This will remain the case for some time to come, although it is clear that the share of agricultural production and the numbers employed in agriculture should be reduced.

Lithuania's key competitive advantages both in the domestic market and in the EU common market are its natural resources, production conditions, geographical position, economic policy and the prices of some products (milk, beef, vegetables, fruit, etc.). These competitive strengths will have to be developed further by restructuring agricultural production and processing industries.

## References

Krisciunaite, L. and Uzdavineine V. (2000), 'The Structure of Lithuanian Farms and Their Development in Lithuania,' in *Agricultural Policy and Rural Development in the Baltic States*, Lithuanian Institute for Agrarian Economics, Vilnius.

LIAE (2002), *Agriculture in Lithuania 2001*, Lithuanian Institute of Agrarian Economics, Vilnius.

LIAE (2003), Communications from expert interviews with staff from the Lithuanian Institute of Agrarian Economics.

Stanikunas D. (2002), 'Competitiveness of Agriculture in Lithuania in the Context of EU Accession,' Proceedings on an international seminar entitled *Pre-acession Strategy of Czech Agriculture towards EU* held on the occassion of 90th anniversary of VUZE at Pruhonice on 27–28 September 2002, pp. 177–189.

# Chapter 8

# Means of Subsistence and Welfare in a Small Rural Community in South Estonia

Raija-Liisa Kämäräinen

## Introduction

This article is concerned to examine the impacts of transition from a planned economy to market economy on living standards and welfare among the rural population in south-eastern Estonia following radical changes in the field of agriculture during the 1990s. Special attention is given to income differentiation and to differences in well-being between diverse social groups. The study is based on survey research conducted in the small and remote rural municipality of Kanepi in 1999.[1] The region is one of the country's poorest and least developed. The principal goal of our study was to follow the socio-economic changes taking place following on an agricultural reform during the first half of the 1990s. Kanepi was chosen as the location for our study because it has been the subject of several sociological researches since the 1980s, which means we have had access to useful comparative data. In addition, during the Soviet era the municipality had a sound socio-economic structure.

This case study provides useful insights into the problem of poverty in rural areas in all post-socialist countries. In these countries the risk of poverty is clearly higher among rural people than among urban residents (Assessment of Rural Poverty, 2002, vi). According to Simon Clarke, 'poverty became a leading issue for the World Bank and, following in its wake, the OECD because growing poverty came to be seen as a critical obstacle to the realization of the radical reform programme which the Bank and associated institutions were trying to implement in the transition countries' (Clarke, 1997, 2). Rural economies and societies are under macro and micro scale change in the Western world as well. The rural areas of Europe are characterized by fundamental demographic, social

---

[1] We mailed our questionnaire to a random sample of 300 respondents in Kanepi municipality, and received responses from 157 people. Women accounted for 65 per cent of the responses, men for 35 per cent. By age group, respondents under 40 accounted for 38.9 per cent, those between 40 and 59 for 32.5 per cent and those over 60 for 28.6 per cent. One-quarter or 26.1 per cent of the respondents were pensioners.

and cultural changes. Everywhere the rural standard of living is declining (Shucksmith and Chapman, 1998, 226; Lichter and McLaughlin, 1995). Research has also shown that people in rural societies are poorly understood. We still know very little about how different structural forces affect different people, households and social groups in rural areas (Shucksmith and Chapman, 1998, 239).

Located in south-eastern Estonia, in the western part of Põlva County, the municipality of Kanepi is a rural community with a population of less than 3,000.[2] Until 1993, there were two kolkhozes in the municipality, and half of the population of Kanepi were employed by those kolkhozes (Alanen, 2001, 67). As well as providing for people's material income the kolkhozes had many other functions in the area of social services and culture (Põder, 2001, 41). Their disintegration therefore had broader implications for the whole socio-economic structure of the municipality: this was not just a question of ownership and/or economic performance, but also one, which concerned the organization of everyday life and society. The kolkhoz system was extremely important on account of its communal nature, and 'the decollectivization of kolkhozes and sovkhozes alone had a greater effect on the future development possibilities of rural industries than any other part of the transition process. It also had a wider influence on the restructuring of the entire rural way of life, local administration, and civil society, including other informal social relations' (Alanen et al., 2001, 390).

Until very recently, poverty was a virtually unknown social phenomenon in Estonia. Under the Soviet system, poverty and inequality were officially denied, although there were 'maloobespechenny' (or 'under-provisioned') families (Falkingham, 1999, 15). During the Soviet era, the rural population's standard of living was not comparable to Western standards, yet basic security and continuity was provided to everybody by the traditional system of social assistance with full employment. State subsidies kept the prices of food, industrial goods and services low, and a minimum wage was guaranteed. The differences between the living conditions and salaries of rural and urban Estonian workers were also relatively small. As Helvi Põder describes, 'it was possible for a kolkhoznik to get additional income from his plot[3] and buy food from the kolkhoz at lower prices. According to the sociological polls carried out at the time, the possibilities of obtaining consumer goods in the country or in town did not differ significantly in Estonia' (2001, 40–41). State support for agricultural industrialization thus made farming relatively attractive and reduced urbanization. In short, although the standard of living was relatively low during the Soviet era, poverty did not exist in the same form and to the same degree as today (see Alanen et al., 2001).

---

[2]  In 2000 Kanepi municipality had 2,734 inhabitants; the rural municipalities in Põlva County 23,198 inhabitants; and Põlva County as a whole 32,615 inhabitants. In the whole of Estonia there were 1,367,187 persons (Statistical Office of Estonia).

[3]  The term 'plot' refers here to an area of land that has a legal meaning in the context of the agrarian reform. In the Baltic countries it usually means a family farm with 1–3 hectares of productive agricultural land (Alanen, 1998, 55).

Poverty is a complex, multi-layered phenomenon that can be measured in various different ways.[4] The most common approach is based on the measurement of incomes or consumption levels. Traditionally, economists have chosen income as the basis for measuring and defining poverty, but even that choice allows for a multitude of options. A person is considered poor if his or her income or consumption level falls below a certain minimum level necessary to meet basic needs. This minimum level is usually called the 'poverty line.' The way we calculate the number of people who live in poverty has become very technical and can be off-putting. While no one measure is necessarily correct, experts argue that some are better than others.

The differences between absolute poverty (not having enough to survive) and relative poverty (having less than others) are great, although largely irrelevant to those who know the limitations that inadequate incomes impose. Extensive poverty studies have shown that the social manifestations of poverty are diverse. Poverty is not only characterized by insufficient income, which prevents individuals from covering such basic needs as food, shelter, clothing and health care, but it also affects one's health and ability to work. How people and institutions portray and try to cope with poverty depends essentially on how poverty is measured in the studied society. The definitions of poverty are influenced by value judgements and belief systems and it is difficult to find one that satisfies everyone. 'Poverty is inevitably a political concept' (Alcock, 1993, 3). As a result, the definition of poverty varies by time and place, and each country uses definitions that are appropriate to its level of development, societal norms and values.

Calculations of the absolute poverty line in Estonia are based upon levels of minimal expenditure by household members. The following are taken into consideration: (1) the cost of a minimal food basket; (2) empirically determined housing cost; and (3) basic clothing, education and transport expenditures. The poverty line is thus a dynamic indicator, which is adjusted annually according to changes in the cost of living. During the first half of 1999, Estonia's absolute poverty line was EEK 1,360[5] (€ 87). According to an Estonian poverty study involving the Ministry of Social Affairs:

The poverty line denotes a level of resources below which subjects (individuals, households, social groups), are considered to be in poverty... The poverty line (or the *absolute poverty line*[6]), according to the socio-economic situation, is calculated from real income and consumption... The official poverty line denotes the level of resources, below which the subject is entitled to further measures to ensure economic survival. The official poverty line is a political decision by the legislator of government and it forms the basis for national socio-political regulations' (Poverty Reduction in Estonia, 1999, 10).

---

[4]  For more details, see e.g. Kangas and Ritakallio (1996), Ritakallio (1994) and Alcock (1993).
[5]  On 31 December 1999 the exchange rate of the Estonian krooni (EEK) was fixed permanently to EUR 1= EEK 15.6466.
[6]  Italics added.

Although the official definition of poverty is useful for statistical and regulative purposes, we must remember that those whom the state calls the 'officially poor' might be identified as the 'extremely poor.' In addition, it is important to notice that 'income and expenditure measures cannot grasp the full extent of the impact of the transition on rural poverty, because the rural population has probably also suffered the most from the deterioration in the social infrastructure, in health care and from the sharply increased cost and reduced provision of public transport' (Clarke, 1997, 28). In this article the welfare of the Kanepian population is illustrated mainly by reference to their material income, although some subjective perspectives are also provided by the survey respondents.

## The Agricultural Reform and Its Outcome

Since regaining its independence the Estonian Republic has been working towards the goal of establishing a market-based society. Before 1989 the private sector was immature in agriculture,[7] and following the failure of the Soviet model Estonia embraced neo-liberal strategies that involved the rapid introduction of market relations. As former Prime Minister Laar and his party stressed, 'agriculture should operate under the same market conditions as other branches without any state interference' (Alanen, 2001, 119). The main objective of agricultural reform was to replace the Soviet-type farms with western family ones through the decollectivization and privatization of state owned agricultural assets (Alanen, 1999, 431).

Although the agricultural reform got under way during the last years of the Soviet system, it was not until after independence that it began to gather serious momentum. The closure of kolkhozes and all their socio-economic activities was supposed to release human resources for the new agricultural and other industries, generating greater profits and earnings for people. There are several reasons at both the local and state levels why the reform objectives have not been met. 'The failure of the reform has usually been attributed to a failure in integrating the dissolution of the kolkhozes and sovkhozes with the restitution of land and other assets; and this in turn has been explained either by too much haste or lack of advance planning' (Alanen, 2001, 258). The new nationalist movement's and the neo-

---

[7]   The first small-scale private farms in Estonia were established as early as 1988. From December 1989, the process was based on the Farm Law. However, that law did not concern ownership relations and was not aimed at the liquidation of collective farms. The transition of Estonian agriculture from large-scale socialist production to a system based on private property was guided by two laws: the Law on Land Reform (October 1991) and the Law on Agricultural Reform (March 1992). The aim of the land reform was to create or reestablish relationships and land use based on private ownership. The agricultural reform had two main functions: First, to carry out the property reform of socialist large-scale farms, kolkhozes (collective farms) and sovkhozes (state farms); and secondly, to reorganize or liquidate them and establish a system of private farms. The laws were executed under extremely difficult economic circumstances (see Alanen et al., 2001).

liberalist government's romanticized ideology of family farms rapidly destroyed all large Soviet-type agricultural enterprises, without bringing any small or medium private ones in their place. In short, even though the restitution was a given fact, there might have been a better way to implement the reform (Alanen, 2001, 260).

In Kanepi, the agricultural reform dramatically curtailed the rural population's possibilities of everyday survival.[8] The changes were emphasized after the abolishment of the kolkhoz and sovkhoz system in 1992–93 (Põder, 2001, 15). Large numbers of jobs were lost following the collapse of two kolkhozes because the agricultural reform did not measure up with its promises for better economic structure and performance, and unemployment started rapidly to grow (Alanen, 2001, 114).

**Table 8.1  Labour force participation rate, employment rate and unemployment rate of the population aged 15–69 by county and year (%)**

|  | Labour force participation rate % | Employment rate % | Unemployment rate % |
|---|---|---|---|
| 1989 | – |  | – |
| Tallinn | – | 78.3 | – |
| Põlva county | – | 73.9 | – |
| Whole Estonia | – | 76.4 |  |
| 1994 | – |  |  |
| Tallinn | – | 67.7 | – |
| Põlva county | – | 51.6 | – |
| Whole Estonia | – | 64.6 | 7.6 |
| 1999 |  |  |  |
| Tallinn | 68.3 | 61.1 | 10.6 |
| Põlva county | 54.2 | 42.8 | 21.1 |
| Whole Estonia | 63.0 | 55.3 | 12.2 |
| 2002 |  |  |  |
| Tallinn | 68.8 | 62.4 | 9.3 |
| Põlva county | 49.8 | 42.4 | 14.8 |
| Whole Estonia | 62.3 | 55.9 | 10.3 |

*Source:* Statistical Office of Estonia. Data of the Labour Market Board.

It is easy to agree with Jane Falkingham's comment that 'labour has been the least well protected factor of production during transition' (Falkingham, 1999, 5). According to a World Bank study on poverty in transitional countries, 'the differences in standard of living between those with jobs and those without them are large and growing. Poverty is closely linked to unemployment' (World Bank Report, 2003, 12). However the relationship between poverty and unemployment is

---

[8]  See Alanen et al. 2001 for a detailed analysis of the process, reasons and consequences of agricultural reform and decollectivization in Estonia, and particularly in the Kanepi municipality.

not quite as straightforward as the World Bank report lets us believe. As Clarke points out, 'unemployment has to be explored in relation to the complementary factors of household composition and low wages' (Clarke, 1997, 31). Table 8.1 demonstrates the changes that have happened in Estonia's employment rate since the beginning of transition. The past decade has seen a significant decrease in the numbers who are economically active. In addition, Table 8.1 clearly highlights the depth of regional inequality in employment: unemployment rates are much higher in rural than urban areas. The impacts of economic re-structuring have been the greatest in rural areas where agriculture was the predominant industrial sector, leaving these areas to face enormous job losses in comparison to areas with greater economic diversity. For example, the metropolitan Tallinn area has coped much better with transition than the other areas in the country. In 2002, the employment rate in Tallinn was 62.5 per cent compared to less than one-half of the population in Põlva County. The unemployment figures clearly demonstrate the decline in the amount of work available in agricultural areas after the reform. Põlva County is one of most rural areas in Estonia. In 1999, the unemployment rate in Põlva was 21.1 per cent, while the national average was 12.2 per cent; in 2000 the gap widened further as the figures rose to 22.8 per cent and 13.6 per cent, respectively (Statistical Office of Estonia).

Although official jobless figures have been coming down over the past few years (Table 8.2), the change is mainly explained by current unemployment regulations and the benefit system in place. There is widespread hidden unemployment. Unemployment benefits are paid for no more than six months, which is why it makes more sense for many who are in pre-retirement – age and who have some minor disability to apply for disability pension rather than to register themselves as unemployed (see Hansson and Marin in this volume). In addition, many of those who are out of work do not even bother to register because of the very modest level of benefits. This applies most particularly to rural areas where distances are long and travel costs high. In short, the rural unemployment rate does not tell the whole truth about the labour market situation. In 1999, at the time that our survey was conducted, government statistics indicated that 23.0 per cent of the population in Kanepi municipality were out of work. In our study, only 20.7 per cent of the respondents who indicated they were unemployed were also receiving unemployment benefit.

Although pensioners in many post-socialist states are effectively in the same situation as those who are out of work, in Estonia they enjoy somewhat better financial support than the unemployed. This is true most particularly in rural areas, where they usually own their house that has a small garden and where they can therefore grow food. Although the average pension in Estonia much lower than in advanced western countries, pensions have in fact grown much faster than average wages in the country. Between 1993 and 2001, average wages increased 3.7 times over, the average pension 4.6 times over and the average old-age pension 4.8 times over. Nonetheless pensioners continue to struggle to make ends meet (see Hansson and Marin in this volume).

As was pointed out earlier, rural and urban poverty are qualitatively different kinds of phenomena. The rural poor have less hunger and less money than their urban counterparts (Clarke, 1997, 28). In all post-Soviet countries subsidiary agriculture or plot farming is practised as a major survival strategy. 'It can be argued that plot farming ... usually forms an integral part of the coping strategy of the wage-worker, regardless of whether the rural household in question derives its primary income from agriculture or from some other branch of the economy' (Alanen, 2001, 215). However, in this regard Estonia differs essentially from the other Baltic countries and many other post-socialist states. Ilkka Alanen describes how in Estonia's neighbouring countries Latvia and Lithuania, development is closely similar to that in developing countries where plot farming is characteristic of the proletariat (Alanen, 1998, 58). In developing countries plot farming – both for the family's own use and for sale in the marketplace – is an important source of income in conditions where wages are low and irregular. Plot farming, under these socio-economic conditions, provides people with a stable income and serves as an important strategy of survival. As Alanen points out, 'Latvian and especially Lithuanian agriculture fits in better with... a model for the developing countries. Especially in Lithuania but also in Latvia the number of households for whom plot farming was the primary source of income was much higher than in Estonia,' (Alanen, 1998, 58).

All these findings are supported by our study on Kanepi, where the significance of farming has been dwindling year after year. This is quite surprising in view of the fact that the majority of people here live a meagre existence just above the absolute poverty line, even though they have access to agricultural land for cultivation. Only 6.3 per cent of our respondents indicated that farming was an important means of subsistence for them, while 14.3 per cent described it as an additional source of income. More than half of the people who took part in the survey (53.1 per cent) took the view that farming is an inadequate source of subsistence. In Kanepi only 16.7 per cent of those who practise farming actively have expanded, while more than half (60.4 per cent) have decreased farming since 1993. In the summer of 2000, 23.2 per cent of the respondents were planning to cut back on farming, 10.1 per cent indicated that they were going to discontinue farming for good. The major reason for farming was to produce for private consumption (89.2 per cent). Only 2.7 per cent of those surveyed sold their produce.

Table 8.2 How satisfied are you with your present financial situation and standard of living?

|  | All respondents | Age group | | |
|---|---|---|---|---|
|  |  | 20–39 | 40–59 | Over 60 |
| Very or rather satisfied | 25.8 | 28.3 | 17.6 | 31.8 |
| Not at all satisfied or quite unsatisfied | 70.3 | 68.3 | 78.4 | 63.6 |
| Don't know | 3.9 | 3.3 | 3.9 | 4.5 |
| Total | 100% N=155 | 100% N=60 | 100% N=51 | 100% N=44 |

*Source:* Kanepi Survey, 1999.

The transition to market economy also affected those who still had a job, reducing their real wages and therefore making it harder to make ends meet. In the Kanepi survey in 1999 (see Table 8.2), 70.3 per cent of the respondents said they were not satisfied with their present financial situation and standard of living. Over one-third or 38.7 per cent of the people surveyed said their financial situation had deteriorated. Põder portrays how rapidly the number of people out of work has grown: at the outset of the reform there were no more than four people in the whole community who were out of work, but by 1991 the number had increased to over 100 welfare benefit recipients. There were also increasing numbers of people who suffered from depression, and the number of drunkards started to grow (Põder, 2001, 60). Consequently, the dissolution of two kolkhozes in Kanepi resulted not only in a sharp decline in the standard of living and welfare, but also in increasing psychological problems (see Alanen, 1999; Alanen et al., 2001). As well as restructuring the economy and labour market, the agricultural reform and decollectivization led to changes in the economic structure of the community that had various unintended consequences. 'As the social basis of community life collapsed, not only kolkhoz economy but also many of the norms that regulated people's lives broke down with it' (Alanen, 2001, 146). Meanwhile alcoholism, violence, crime, and social isolation continue to increase among Kanepians (Alanen, 2001, 146).

## Changes in Income and the Growth of Inequality

Conventionally, poverty has been defined in terms of income or expenditure levels, the assumption being that a person's well-being is largely determined by his or her material standard of living. There is thus a close relationship between income inequality and relative poverty, and we can expect to see poverty increase along with income inequality. During the transitional phase, it is useful to examine income inequality and poverty at two different levels: the regional (county) and the local (municipality) levels. There are two reasons for this: first, at the regional level we will be able to gain some insight into uneven regional development, and second, at the local level we will be able to identify the types of social groups that have been the most vulnerable during transition.

### The Differentiation of Income at the Regional Level

There are very marked regional income differences in Estonia: poverty here has a distinct regional character. The areas with the highest risk of poverty are those where income levels are less than 80 per cent of the Estonian average. Most rural districts fall below the average income level, especially in the so-called depression belt in south-eastern and eastern Estonia (Poverty Reduction in Estonia, 1999, 22). Government statistics clearly illustrate the patterns of uneven regional development over the past ten years. Table 8.3 demonstrates the great imbalance of income distributions between urban and rural areas, with poverty clearly more prevalent in the latter. In the most developed urban area of Tallinn, the average net monthly income per household member in 1999 was EEK 2,596.9 (€ 166), whereas in Põlva county the figure was EEK 1,415.6 (€ 90). In the whole of Estonia, the average net income in the same year was EEK 2,015.8 (€ 12) per month. Tallinn and the metropolitan area are among the wealthiest in the country, with strong prospects of economic expansion and development, especially in the service sector. In Põlva County, by contrast, the outlook is far more depressed. As mentioned earlier, the official poverty line in Estonia in 1999 was € 89; in that same year the average net income per person in Põlva was virtually the same, or € 90. Generally speaking, Estonia is one of the highest consuming countries among post-socialist states. One can only wonder how it is possible to keep on consuming with an income of 166 euros a month (as in the capital region), never mind with 90 euros a month?

**Table 8.3  Monthly disposable income per household member in Põlva County, Tallinn capital area and the whole country by source of income in 1996, 1999 and 2002 (%)**

|  | Põlva county | | | Tallinn | | | Whole Estonia | | |
|---|---|---|---|---|---|---|---|---|---|
|  | 1996 | 1999 | 2002 | 1996 | 1999 | 2002 | 1996 | 1999 | 2002 |
| Wage labour | 54.0 | 44.8 | 52.6 | 70.8 | 67.5 | 71.1 | 62.9 | 61.1 | 64.5 |
| Self-employment* | 17.7 | 13.5 | 7.7 | 6.6 | 3.0 | 3.5 | 10.8 | 5.4 | 5.2 |
| Income transfers | 27.6 | 36.6 | 36.3 | 19.4 | 22.4 | 18.8 | 23.3 | 27.0 | 25.0 |
| Other ** | 0.4 | 0.9 | 0.9 | 2.6 | 2.2 | 2.0 | 2.4 | 2.2 | 1.9 |
| Non-monetary | 0.3 | 4.2 | 3.5 | 0.6 | 4.9 | 4.6 | 0. | 4.3 | 3.5 |
| Total in % | 100% | 100% | 100% | 100% | 100% | 100% | 100% | 100% | 100% |
| in EEK | 1,071.5 | 1,415.6 | 1,602.2 | 1,739.9 | 2,596.9 | 3,146.9 | 1,433.1 | 2,015.8 | 2,499.5 |

\* Includes self-produced foodstuffs estimated in monetary value. \*\*Includes property income and other income.

*Source:* Statistical Office of Estonia.

Table 8.3 provides an overview of how the wage distribution and inequality in living standards have changed in Estonia over the past six years. Compared to the average for the whole of Estonia, net income in Põlva County was considerably lower, only 75 per cent of the national average. Whereas in Tallinn average hourly wages in 1999 were EEK 33.11 (€ 2.1), the labourer in Põlva County earned no more than EEK 19.64 (€ 1.3) per hour (Statistical Office of Estonia). From 1996 to 2002 wage labour increased in Estonia as a whole, whereas in Põlva County it declined, indicating an increase in unemployment and poverty in south-eastern parts of Estonia.

The growth of inequality is directly related to changes in the composition of income following independence in Estonia. During the Soviet era the overall income distribution between the various social/professional groups and regions was far more egalitarian than in most market economies, and virtually all income (either income transfers or wages) was received through the intermediation of the state (Põder, 2001, 40–41). In the Soviet Union in the late 1980s, only 14 per cent of total gross income was from private sources, while social transfers accounted for

13 per cent and labour income for 72 per cent; income from property was non-existent (Milanovic, 1998). In contrast to the situation in many Western countries, the principal source of income in most Soviet-type agricultural households was through employment.

In Estonia, the main source income for the rural population consists of paid labour and/or social transfers. Table 8.3 compares two economically and socially dissimilar regions and their disposable income structure compared to the national average. In 1999, the major source of income in Põlva County was wages (44.8 per cent) followed by state transfers (36.6 per cent). The single most important item among all social transfers was represented by pensions (27.5 per cent). During the same year incomes from entrepreneurship[9] accounted for no more than 13.5 per cent, and by 2002 the figure had declined to 7.7 per cent (Statistical Office of Estonia). These figures clearly point at the difficulty of creating successful private enterprises in the rural south-eastern areas. Our survey in Kanepi lends support to these official statistics. While in 1994 wages were the primary source of income for up to 75 per cent of all households, by 1999 only less than half (42.5 per cent) of the survey respondents said that regular wages were their main income source, and 34 per cent of the respondents were completely dependent on social welfare benefits.

In Estonian society another important risk factor for poverty, apart from unemployment, is age. According to the 1999 poverty study, over one-third of older people in Estonia live below the poverty line (Poverty Reduction in Estonia, 1999). The process of economic transition affected the elderly population most significantly in two spheres of life. First, in 1992 the monetary reform was carried out at the expense of the elderly population's life savings. At the same time, the older generation was deprived of its access to the labour market. During the Soviet system, Estonia boasted high levels of labour force participation among workers of pre-retirement and retirement age. Changes in the structure of production and demand for labour put an abrupt end to this activity. In 1989–1995, the employment rate at ages 55–64 dropped by 25 per cent and in age groups over 65 by 60 per cent (Toots, 2000, 5). As they were driven out of employment older people came to rely increasingly on social security transfers.

Following the transition to market economy, older people saw their incomes drop more sharply than any other social group. Today, the vast majority of pensioners are dependent on state transfers or small occupational benefits. Table 8.3 illustrates the importance of transfers as a source of income, especially in rural communities. In Põlva County these transfers (pensions, child benefits, etc.) account for 36.3 per cent of people's total disposable income, while the figure at the national level is 25 per cent; in the more prosperous Tallinn region, it is 18.8 per cent. Therefore, in Estonia as well as in other post-socialist countries, the standard of living of the elderly population dropped dramatically as a result of the economic reforms that some are arguing were carried out at the expense of elderly people. In rural areas where whole communities have witnessed devastating changes and where all social groups are struggling to survive, up to 30 per cent of

---

9   This includes both agricultural and non-agricultural entrepreneurship.

all pensioners may in fact rank among the most well-to-do groups: 'In these mainly rural communities most of the local social funds goes for unemployed people and large families with children' (Toots, 2000, 10; see also Hansson and Marin in this volume).

The patterns of demographic change in Estonia are very similar to those seen elsewhere in Europe: its population is aging. As a consequence the need for income transfers and other social security measures is also increasing. The problem is particularly acute in rural areas. Table 8.4 compares the size of different age groups in Kanepi and the whole of Estonia. In 2002, the population in Kanepi was older than the national average. Kanepi has also seen large number of younger and skilled people move out of the region. During 1989–2002, the population of Kanepi decreased by 13.6 per cent (Statistical Office of Estonia). In general, 'the mainly agricultural regions of southern and south-eastern Estonia are characterized by a higher proportion of people of retirement and pre-retirement age' (Hansson and Marin in this same volume).

Table 8.4  Breakdown of population by age groups in Kanepi and the whole of Estonia in 2002

| Age groups | 0–24 | 25–44 | 45–64 | Over 65 |
|---|---|---|---|---|
| Kanepi | 24.4% | 26.3% | 26.5% | 21.5% |
| Whole Estonia | 32.7% | 27.8% | 26.6% | 14.9% |

*Source:* Statistical Office of Estonia.

The risk of poverty is not shared equally by all groups in society, but some groups are more liable to suffer than others. 'The critical question if poverty is to be reduced is which groups in society are most at risk of being or becoming poor' (Falkingham, 1999, 20). There are strong links between discrimination and poverty, and those who are most vulnerable are often those groups whose access to the labour market is most restricted. Individuals are differently placed in terms of their chances to earn a living and to create welfare. The risk factors for poverty may be either dependent, or independent of the individual. Social factors causing poverty become even more concentrated in certain social groups and regions. Social and macro-economic factors that are independent of the individual's activities, reduce their range of choices. We turn our attention now to those social groups that are more likely to fall into poverty in the small community of Kanepi.

*Income Differentiation at the Local Level*

Kanepi is one of the poorest communities in Estonia, and the poor people in this community have very limited opportunities for gainful employment and by the same token for improving their socio-economic situation. As the national statistics demonstrated earlier, the standard of living in Kanepi is considerably lower than the national average. This is confirmed by our survey data. Table 8.5 illustrates the income distribution[10] among the survey respondents. The figures are very distressing. In 1999, almost half of Kanepian population lived at or around the official poverty line: 47.9 per cent of the respondents earned less than EEK 1,999 (€ 127) a month, while 37 per cent earned between EEK 2,000 and 3,999 (€ 128–196) a month. Only 15.1 per cent of the respondents had a monthly income in excess of EEK 4000 (€ 256).

Several poverty studies in the Baltic counties have shown that the most vulnerable groups with respect to the risk of poverty are large families, single parent households,[11] the rural population, the unemployed, and the elderly.[12] Although Kanepi is poor and the vast majority of its people cannot make ends meet, there are also some indications of a new distribution of 'wealth' and growing inequality among community members. The distribution of incomes appears to be affected not only by unemployment, but also by age, gender and education. Table 8.5 shows the breakdown of different categories of monthly income by age and gender.

**Table 8.5 Monthly income in Kanepi in 1999, including transfers and social benefits (incomes cross-tabulated with age and gender)**

| Income group | All | Respondents under 39 years of age | | Respondents 40–59 years of age | | Respondents over 60 years of age | |
|---|---|---|---|---|---|---|---|
| | | Male | Female | Male | Female | Male | Female |
| I. EEK 0–1,999 (€ 0–127) | 47.9 | 23.5 | 25.7 | 45.4 | 32.0 | 84.6 | 90.0 |
| II. EEK 2,000–3,999 (€ 128–255) | 37.0 | 35.3 | 59.0 | 31.9 | 52.0 | 15.4 | 10.0 |
| III. EEK 4,000–5,999 (€ 256–383) | 8.9 | 23.5 | 12.8 | 13.6 | 4.0 | - | - |
| IV. Over EEK 6,000 (€ 384) | 6.2 | 17.7 | 2.5 | 9.1 | 12.0 | - | - |
| Total | 100% N=151 | 100% N=20 | 100% N=41 | 100% N=22 | 100% N=25 | 100% N=13 | 100% N=30 |

*Source:* Kanepi Survey 1999.

The lowest income category consisted of people who earned less than EEK 2,000 a month; as mentioned earlier, one half of the surveyed population belong to this group. The majority of the respondents were retired (54.3 per cent). Pensioners represented the largest social group among those respondents who earned between EEK 1,000–1,999 a month. Around three-quarters of them (76 per cent) were aged

---

10   Income here includes all transfers and benefits and other sources of income.
11   Our survey material included only two single parents; therefore single parent households are excluded from the analysis.
12   See Kutsar and Trumm (2000); Poverty reduction in Estonia, Background and Guidelines (1999); Toots (2000).

over 60, two-thirds of them were women (66.7 per cent). The main reason for the overrepresentation of women here is obviously their age. Estonian women have a higher life expectancy than men. In Kanepi, the risk of poverty was not significantly higher among pensioners than other social groups. Our findings thus supported earlier results according to which it is not pensioners in rural areas who are the most vulnerable group because they typically are house-owners who are in the position to grow their own food. Almost all persons aged 65 or over (94.1 per cent) owned their house or apartment. There is also a strong tradition in Estonia for adult children to look after of their elderly parents.

The jobless figures have increased significantly over the past few years. According to our findings 20.7 per cent of the Kanepian population said they were out of work at the time of the survey. The true figure was probably even higher than that because it is reasonable to assume that the poorest and most deprived individuals in poor health were severely underrepresented in our survey. People who earned less than EEK 1,000 a month were mainly unemployed (57.1 per cent), farmers (14.3 per cent), housewives, and students. Among those who were out of work, 70 per cent indicated that their monthly income was less than EEK 1,999. Traditionally, poverty and the risk of unemployment are greater among women than for men. Women are more likely to be poor and stay outside the labour market because according to traditional family values and gender roles in rural Estonia, women are still primarily responsible for looking after the home and family. Women were clearly over-represented in the first income group (64.3 per cent), and the majority of unemployed people who earned less than EEK 1,000 a month were also women (58.3 per cent). Men and women were more or less equally represented among all those who were out of work: 48.3 per cent were men, 51.7 per women. The second largest income group (accounting for 37 per cent of the population) comprises respondents earning between EEK 2,000 and 3,999 a month. In this group women were in the majority, accounting for 72.2 per cent of the total as compared to 27.8 per cent males. The proportion of women declines linearly with increasing income levels. In the third and fourth income groups, men outnumbered women (women 44.2 per cent, man 55.6 per cent). All in all our results indicated that women in Kanepi earned less than men.

In Kanepi, unemployment was a common problem in all age groups, but most particularly among respondents aged 30–39. In the present socio-economic environment, everyone has great difficulty finding employment. Most of the unemployed were actively seeking work (81.8 per cent), usually through friends and relatives. State labour agencies were not considered a particularly helpful channel. The majority of the unemployed had tried all avenues available: direct contacts to employers, newspaper ads, state agencies. Since only one-fifth of those who were out of work were receiving unemployment benefits, it is hardly surprising that 17.9 per cent of the unemployed said the income they receive does not even cover the cost of food. Unemployment is a major path to poverty among Kanepians.

**Table 8.6 What can you afford to buy with your present income?**

| | All respondents | Age group | | | |
|---|---|---|---|---|---|
| | | 20–39 | 40–54 | 55–65 | Over 65 |
| All I need including luxury items | 1.9 | 5.0 | – | – | – |
| All I need except luxury items | 28.2 | 30.0 | 27.8 | 19.2 | 32.4 |
| Only for necessary living expenses and food | 50.6 | 48.3 | 44.4 | 57.7 | 55.9 |
| Only for food | 12.8 | 11.7 | 11.1 | 19.2 | 11.8 |
| I cannot even buy food | 6.4 | 5.0 | 16.7 | 3.8 | – |
| Total | 100% N=156 | 100% N=60 | 100% N=36 | 100% N=26 | 100% N=34 |

*Source:* Kanepi Survey 1999.

Table 8.6 clearly demonstrates the depth of deprivation in Kanepi municipality: 6.4 per cent of the local population could not even afford to buy their daily bread. As was just pointed out, the problem of lack of food was significantly more common among the unemployed respondents (17.9 per cent vs. 6.4 per cent). Over one-third or 35.7 per cent of those out of work can afford to buy food but not to pay for their housing. Purchasing power was clearly dependent on age: the middle-aged and pre-retirement generation were financially in the most precarious situation.

Table 8.2 demonstrated earlier how the most discontent groups are to be found in the 'lost generation' of those aged 40–59. In this age group almost four-fifths indicated that they are unsatisfied with their financial situation and standard of living. In particular, the pre-retirement generation is having enormous difficulty finding work and its own place in today's society. During the Soviet era, the people of this generation were in their 40s and had secure jobs; when the economic reforms got under way, they found themselves trapped between two systems: the old one in which they had grown up, had ceased to exist, and the new one no longer needed their education and skills. These people are commonly identified with the old Soviet system and its work culture: the collective farm and agricultural workers in particular are viewed a very negative light. Although the problem of finding work applies to the majority of highly educated middle-aged persons and those in pre-retirement age, the difficulties are greatest of all for women, especially if they are over 35.

In income group I there was only one (female) respondent under 55 with an academic education. The majority of those under 55 had no more than a basic education (see also Table 8.7). All males with a basic education fall into income group I. In general, personal income levels were more or less directly determined by education. Education has become an important socio-economic indicator in Estonia especially in the working age population. During the transitional period the greatest success has been enjoyed by young men with a high level of education. The same trend can be seen in the municipality of Kanepi as well. The most distinctive feature of the highest income bracket (IV) is the large proportion of people with an academic education (77.8 per cent). These were most typically young skilled men or people who in the Soviet era had occupied a managerial position and who during the transition have been able to translate their social capital into financial capital. All young academic men under 40 earned more than EEK 4,000 a month. Evidently, in Kanepi, poverty is associated with being a woman, old, unemployed and/or less educated.

**Table 8.7   Income level of respondents cross-tabulated with education, and in each income group educational level is cross-tabulated with gender**

| Income group | Basic all | Basic education | | Middle all | Middle education | | Voca-tional all | Vocational education | | Aca-demic all | Academic education | |
|---|---|---|---|---|---|---|---|---|---|---|---|---|
| | | Male | Female | | Male | Female | | Male | Female | | Male | Fem. |
| I | 86.1 | 100.0 | 80.0 | 32.1 | 41.6 | 25.0 | 37.3 | 30.0 | 40.0 | 34.8 | 33.3 | 35.2 |
| II | 8.3 | – | 18.0 | 46.5 | 33.4 | 56.2 | 54.2 | 50.0 | 57.4 | 17.3 | 11.1 | 21.3 |
| III | 5.6 | – | 0.8 | 17.9 | 16.7 | 18.8 | 6.8 | 15.0 | 2.6 | 17.4 | 23.2 | 15.0 |
| IV | – | – | – | 3.6 | 8.3 | – | 1.7 | 5.0 | – | 30.3 | 33.3 | 28.5 |
| Total | 100% | 100% | 100% | 100% | 100% | 100% | 100% | 100% | 100% | 100% | 100% | 100% |
| N=146 | N=36 | N=11 | N=25 | N=28 | N=12 | N=16 | N=59 | N=20 | N=39 | N=23 | N=9 | N=14 |

*Source:* Kanepi Survey, 1999

## Conclusions

In Estonia, poverty has a structural and regional nature. Today, poverty is one of the major social problems in south-eastern Estonia, which not only affects the people who are poor but also hampers the development of the whole region. Poverty has wide-ranging and often devastating effects. Many of these effects such as lacking nutrition and physical health problems are the direct results of having too small an income and too few resources. The only way that the people of Kanepi can generate more income, jobs and other resources is through economic growth. Economic growth will both increase wages for those who have work and create new employment opportunities for those who are currently unemployed. The future question for Estonia is, how to promote economic growth in poor rural regions like Kanepi? Is this even possible after the way the transition and agricultural reform took place in the rural communities?

Not all of the long-term effects of poverty are even visible in Estonia today because so little time has passed since its radical socio-economic changes. It is important to remember that the primary effects of poverty often lead to other long-term problems: poverty denies people their basic right as citizens and engenders divisions, which are harmful to democracy. Therefore, as the government is trying to ease the problem of poverty, it is at once creating for poor people the possibility to make a meaningful contribution to society. The reduction of poverty by means of regional policy measures is helping to create equal opportunities in different regions for full participation in the various spheres of community life. The deprivation of one region affects the whole of Estonia and the attempts to resolve problems locally will have a positive impact on society as whole. Just as poverty is a political concept, so too are its solutions.

## References

Alanen, I. (1998), 'Petty Production in Baltic Agriculture: Estonia and Lithuanian Models,' in Kivinen M. (ed.) *The Kalamari Union: Middle Class in East and West*, Ashgate, Aldershot.

Alanen, I. (1999), 'Agricultural Policy and the Struggle over the Destiny of Collective Farms in Estonia,' *European Society for Rural Sociology*, vol. 39, no. 3, pp. 432–458.

Alanen, I., Nikula, J., Põder, H. and Ruutsoo, R. (eds) (2001), *Decollectivisation, Destruction and Disillusionment. A community study in Southern Estonia*, Ashgate, Adershot.

Alcock, P. (1993), *Understanding Poverty*, MacMillian, London.

Clarke, S. (1997), *Poverty in Transition: Final Report*. Http://www.warwick.ac.uk/fac/soc/complabstuds/russia/other.html.

Duncan, C. M. and Lamborghini, N. (1994), 'Poverty and Social Context in Remote Rural Communitites,' *Rural Sociology*, vol. 59, no. 3, pp. 437–164.

Falkingham, J. (1999), *Welfare in Transition: Trends in Poverty and Well-being in Central Asia*. CASE/20. Center of Analysis of Social Exclusion, London School of Economics, London.

International Fund for Agricultural Development, *Assessment of Rural Poverty. Central and Eastern Europe and the Newly Independent States*, Rome. Http://www.ifad.org/poverty/region/pn/index.htm.

Kangas, O. and Ritakallio V.-M. (1996), *Kuka on Köyhä? Köyhyys 1990-luvun puolivälin Suomessa*, Stakes tutkimuksia 65, pp. 231.

Kutsar, D. and Trumm, A. (2000), *Human Rights in Estonia: Poverty as an Obstacle to the Realization of Human Rights*. Http://www.undp.ee/hre/en/3.html.

Lichter, D. T. and McLaughlin, D. K. (1995), 'Changing Economic Opportunities, Family Structure, and Poverty in Rural Areas,' *Rural Sociology*, vol. 60, no. 4, pp. 688–706.

Makeev, S. and Kharchenko, N. (1999), 'The Differentiation of Income and Consumption in Ukraine,' *International Journal of Sociology*, vol. 29, no. 3, pp. 14–30.

Milanovic, B. (1998), *Income, Inequality and Poverty in Transition*, World Bank Regional and Sectoral Study, Washington D.C.

Ministry of Social Affairs of Estonia and United Nations Development Programme (UNDP) (1999), *Poverty Reduction in Estonia, Background and Guidelines*, Tartu University Press, Tartu.

O'Brien, D. J., Patsiorkovski, V. V. and Dershem. L. (1996), 'Household Production and Symptoms of Stress in Post-Soviet Russian Villages,' *Rural Sociology*, vol. 61, no. 4, pp. 674–698.

Ritakallio, V.-M. (1994), *Köyhyys Suomessa 1981–1990. Tutkimus tulonsiirtojen vaikutuksista*, Stakes tutkimuksia 39, pp. 141.

Shucksmith, M. and Chapman, P. (1998), 'Rural Development and Social Exclusion,' *European Society for Rural Sociology*, vol. 38, no. 2, pp. 225–242.

Statistical Office of Estonia. Http://www.stat.ee.

Szelenyi, I. (2001), *Poverty under Post-Communist Capitalism – the Effects of Class and Ethnicity in a Cross-National Comparison*, Center of Comparative Research/Department of Sociology, Yale University. Paper presented at workshop 'Culture and Poverty,' Central European University. 30 November – 2 December 2001, pp. 90.

Toots, A. (2000), *Effects of Estonian Social Security Reform on Old-Age Poverty: Achievements and Problems*. Http://www.nispa.sk/news/toots.rtf.

# Chapter 9

# Rural Community Initiatives in the Latvian Countryside

Talis Tisenkopfs and Sandra Sumane

## Introduction: An Overview of the Rural Community Situation and Rural Policies in Latvia

The countryside is often seen as a loser in the process of post-socialist transformation. Sociological studies have predominantly associated rural areas with the negative consequences of transition and with the limited opportunities available. Latvia's countryside is no exception, showing all the typical hallmarks of declining agricultural production, a scattered enterprise structure, and increasing unemployment and poverty. Studies also point to infrastructure problems, long distances, lack of business skills, low population density, the ageing of the rural population and the weakness of non-governmental organizations. Discourses of this kind usually conclude that modern capitalist development is concentrated in cities, while rural areas are falling ever farther behind. There are few theoretical and empirical studies that indicate any positive development in the Latvian countryside, especially in terms of the stimulating influence that rural ideas and policies can have on overall economic development and social cohesion.

Socio-economic indicators, such as household income, capital investment, entrepreneurial activity, wages, unemployment rate, as well as public opinion and media debate are such that the Latvian countryside can be described as an area with a high concentration of social and economic problems, a kind of disadvantaged zone in contemporary society. During the process of economic transition, rural society has been fragmented into an assembly of weak and partially excluded actors without being replaced by new integrative civil, social, economic and political networks. Rural communities have largely lost their traditional communication patterns and solidarity. Overall, rural society has declined both economically and socially.

The partial social exclusion of rural society goes hand in hand with substantial political exclusion of rural actors from the policy-making process, which takes place predominantly in central government institutions in Riga and within the close circles of governing political parties and leaders of major economic groupings. Local municipalities, small towns, rural communities and rural organizations have only limited impact on the political decision-making process in Latvia. Rural and agricultural policy-making is dominated by central government, state institutions and EU accession negotiations, without sufficient representation of the needs and interests of all groups of the rural population.

Despite the general tendencies of depopulation, the countryside still remains the home and centre for socio-economic activities for a large proportion of Latvia's population. Accordingly to the Central Statistical Bureau of Latvia, 31 per cent of Latvia's population live in rural areas.

However, the Latvian countryside does not constitute a homogeneous space. There are significant regional differences on several social, economic, demographic and natural indicators. For example, the proportion of agricultural land varies from 30 to 60 per cent, and the unemployment rate from 3 to 30 per cent.

Several strata can be identified in the rural society: a minority of wealthy people (big farmers, commercial entrepreneurs), a group of people in a comparatively stable and secure position (permanent employees in state and public service organizations), and a majority struggling in the face of economic and social difficulties (employed in casual jobs, low paid agricultural and forestry sector workers, small subsistence farmers, rural unemployed). There is a marked division of the rural population into strong actors and weak. While the former have more resources and communicate to some extent to establish organizations that promote their interests in government decisions, the latter usually lack resources, they are socially isolated from one another, and excluded from political participation.

Employment opportunities in rural areas are scarce. The most important employers in the Latvian countryside are agriculture and public services (schools, post offices, municipal and state institutions, etc.). Apart from farm labour, some jobs are also provided by the agro-processing industry, the forestry sector, and market service organizations. A small part of the rural population commutes to work in towns.

The study 'Who and where are the poor in Latvia?' (UNDP, 1998, www.undp.riga.lv/undp/programme/poverty) notes that both the frequency and depth of poverty are higher in rural areas. Based on its assessment of factors contributing to poverty, the study concludes: 'What really makes a difference is whether someone is living in an urban or a rural area.' Living in a rural area in post-socialist conditions largely means being a loser.

*Evolution of Agricultural and Rural Development Policies in Latvia*

During the 1990s agricultural policy in Latvia has been predominantly influenced by a neo-liberal discourse that associates development with free market forces and with a decreasing role of the state (Tisenkopfs, 1998, 20). The liberal discourse was modified in the mid-1990s when the state adopted a more active role in regulating and supporting the agricultural sector. An agricultural subsidies programme was introduced, mainly aimed at supporting the modernization of agricultural production and processing enterprises and stimulating high quality production – all geared to Latvia's integration into the European Union (Ministry of Agriculture, 1998).

The shift towards more integrated rural development policies took place in the late 1990s. In 1998 the government adopted the Latvia Rural Development

Programme, which took onboard a wide range of rural development goals and drew attention to the fact that rural development does not only comprise farming, forestry and fishing, but also other types of business. The programme encouraged diversification in the rural economy, environmental protection, and the provision of educational, cultural and health care services for the rural population (UNDP, 2001). It also represented an attempt to coordinate and to integrate rural development policies between different ministries and to promote political communication on rural issues in the country and abroad.

In 2000, responsibility for coordination of the country's Rural Development Programme was delegated to the Ministry of Agriculture, which started to coordinate both national and the EU's pre-accession policies for agriculture and rural development. The Ministry prepared a new Rural Development Plan, identifying five priorities for agricultural and rural development during the pre-accession period: investment in agricultural business, the improvement of food processing and the expansion of marketing, the diversification of farm production, the improvement of the rural infrastructure, and the promotion of environmentally friendly farming methods.

The relationship between agricultural and rural development policies has been a disputed issue in Latvia. Thus far the lobby of agricultural producers has been more influential that the rural development lobby in setting development priorities and allocating public funds. For instance, the Agricultural Development Conception adopted in 1997 clearly states the agricultural priorities: to facilitate agricultural development by making good use of Latvia's natural and socio-economic potential; to develop agriculture as a branch of economy that is capable of integrating and competing in the single European market; to improve living standards among people employed in agriculture, particularly to increase their income; and to facilitate multifunctional agriculture, thus increasing employment opportunities in rural areas. Today, agriculture receives annually LVL 24 million (approximately EUR 40 million) in subsidies in addition to some EUR 20 million of support from EU Pre-Structural Funds. Various community, rural and local development initiatives have also received support from these funds.

The contradiction between rural development and agricultural policies reflects the contradiction between weak and strong rural actors. The policy networks of strong rural actors tend to exclude weak actors. The collaboration that major agricultural producers have with other 'strong' rural actors is more efficient than communication between small farmers and other 'weak' rural actors. Agricultural producers have established various producer associations, which in turn have created an umbrella association for agricultural organizations. A number of producer organizations are active in Latvia's rural areas, including the Farmers' Federation, the Farmers' Saeima, the Young Farmers' Club, the Rural Support Association, and the Association of Agricultural Statute Companies. An Association of Agricultural Organizations was established in 1999 to participate in drafting legislation for the Ministry of Agriculture and to promote openness and participation in agricultural policy issues. Their communication has been strengthened by the policies of the Ministry of Agriculture, which are mainly aimed at supporting the modernization of the productive agricultural sector through different subsidy programmes. As a result, new policy networks among strong actors such as the state and producer associations have emerged, having a major influence on state policies. The socially excluded and poor groups in rural areas are by contrast socially and politically disorganized, and their policy networks develop much more slowly. The policies of the Ministry of Agriculture and the state in general are thus more favourable towards the organizations of strong than they are towards the organizations of weak actors.

Rural policy is still dominated by agricultural policy, and public support for the countryside has in fact accelerated the processes of social and economic stratification in rural areas. As yet rural policies have not sufficiently addressed the rural population's educational, retraining, community development and human resource needs. All in all, the Rural Development Programme has fallen short of expectations. Difficulties have been presented not only by the depth of the country's rural problems, but also by the lack of political will to tackle the rural crisis and by the lack of policy alignment and coordination between different ministries.

## The Rise of the Community Development and Citizen Participation Discourse in Latvia

The development of local initiatives and citizen participation in the Latvian countryside is related to the specific local rural situation and discourses grounded in it. Rural discourses in Latvia are quite different from what we see in other European countries, and they should be analysed in the context of economic and political transition. The consumers' *vis-à-vis* farmers' conflict discourse and the urban *vis-à-vis* rural conflict discourse are not pronounced in Latvia, for various reasons. First, there are no serious agro-environmental problems in Latvia. Second, the rising urban middle class nurtures deep sentiments towards the countryside as a place of their 'social origin' and a 'homeland' of national identity. Third, national independence is closely associated with the history of the peasant-state. The Latvian countryside is not represented in the media as a 'threat' to urban well-being. Production is comparatively low, which means that the damage caused to the environment is also limited. In addition, there have been no major food crises or scandals.

Supporting such ideas as the development of rural tourism, the expansion of non-traditional agriculture and organic farming, and the improvement of rural services, the rural diversification discourse is enjoys increasing popularity in Latvia. This discourse is very similar to that promoting multifunctional agriculture in many Western European countries. However, despite the prominence of the diversification discourse, there are still few examples of genuinely multifunctional rural areas, successful integration of rural and urban territories, effective community development programmes.

The most widespread discourse in the Latvian countryside is the popular social assistance discourse. This represents the language and the political demands of the stratum of small-scale farmers, i.e. those who are primarily engaged in subsistence farming and who are struggling in the face of severe economic and social problems. The ideas of small-scale farmers about rural development are based on their agricultural understanding as well as on their often unjustified expectations that existing farms might see some growth come their way. Small holders are demanding active state involvement, protectionism and support for low-output producers, yet at the same time they are not showing much initiative in searching for alternative types of employment and income (Tisenkopfs, 1999).

It is a European-wide problem that more and more people live in the countryside and rural villages without any contact with the local community. They commute to work in cities, shop in the cities, push up land and property prices, influence local authority decisions, communicate internally, create alienated rural images and constitute an ex-territorial elite. A similar process is gradually unfolding in Latvia, particularly in and around Riga and in coastal regions. The renewed interest of urban dwellers in the countryside is mainly residentially oriented, and it is still difficult to assess the impact that the overflow of the middle-class population from cities may have upon rural life.

The modern attitude towards rural development envisages participatory involvement of communities, non-governmental organizations, and local action groups in the process of designing and implementing various development strategies, plans, projects and interventions. There is a broad consensus of opinion that civil society should have an active role in promoting local development, tackling social problems, alleviating poverty, promoting social integration, and engaging citizens in territorial planning. In Latvia there are currently two parallel discourses of citizen engagement and participation in development: the non-governmental organizations (NGOs) discourse and the partnerships discourse.

NGOs are seen as elements of civil society. They involve and unite people on the basis of common interests, ideas and values. Participation and involvement is voluntary. According to surveys, NGOs promote horizontal relations between members of the community, and they engage in relations with the state and municipal institutions. Most active among rural NGOs are women's, pensioners' and young people's organizations that are concerned with social problems and social assistance. There are also several agricultural producers' associations mainly concerned with advocating their business interests. The new NGOs rely upon voluntary contributions and small grants mainly from international donors. In Latvia the rural poor have not shown very active involvement in NGOs, largely because of their isolation and an unfortunate historical legacy that works against cooperative formation.

In recent years there has emerged a new discourse of partnership for development. Partnerships between civil society groups, local governments, private sector organizations and state institutions are seen as a new mechanism in rural development. The pre-accession assistance instrument SAPARD (Special Action for Pre-Accession for Agricultural and Rural Development) and potentially forthcoming structural funds programmes for regional and rural development entail a partnership approach as one of the main principles. Access to these funds may directly depend on the capacity of local communities to establish efficient partnerships that would prepare and implement local development strategies and projects.

## Partnership Initiatives in Rural Development

A partnership is a subject or agent of collaborative action. Different people are involved in its activities: heads of local governments, leaders of non-governmental organizations, intellectuals, teachers, social workers and others. Partnership participants are not only individuals but also collaborative entities; different institutions and organizations are also involved in partnerships, such as local governments, state authorities (employment authorities, regional higher education authorities, etc.), private sector and business support organizations (banks, credit unions, etc.), as well as different non-governmental organizations and local action groups. Such important institutions as the Church and local media (local press, television, journalists) also are gradually becoming involved in partnerships.

The partnership structure is created through interaction between all these different participants, leading to joint action and to the creation of wider partnership links at district or even regional level. In other words, partnerships are formed as a network of interactions through which its organizational structure is crystallized, joint values and a sense of belonging formed, and cooperation implemented.

Although a wide range of different kind of people and organizations join in partnership activities and although the overall network of interactions changes dynamically, it is still possible to identify a unified basic structure. This structure corresponds to the main areas or levels of partnership activities. The four main areas of partnership activities or partnership levels are as follows: (1) the Partnership Board level, (2) the community level, (3) the wider partnership level, and (4) the policy influence or political institutions level.

## Research Methodology

The present investigation and analysis of partnership initiatives was based on the action research methodology, which is geared to supporting cooperation and interaction between researchers and the actors themselves, i.e. the partnership participants: local governments, state authorities, business support organizations and non-governmental organizations as well as the people who represent these institutions in partnerships.

A partnership has its own life cycle that is followed in action research. It starts with the creation of the partnership and commencement of action, continues with the formation of social links, cooperation with other institutions, strategy development, and project preparation and implementation. The agents involved in the partnership may decide to institutionalize the partnership and register it as a non-governmental organization in order to facilitate its sustainability. Action research depends on these stages in the partnership life cycle and its task is to reflect upon and to analyse processes in the partnership according to the partnership life cycle and to contribute to its further evolution.

Action research has several different functions, including those of analysis, reflection, education, and the submission of recommendations with a view to influencing policy decisions. Experts share their knowledge with partnership participants about the role of social capital, collaborative forms of action, and ways of supporting that action. They promote learning by partnership participants through their own action and draft recommendations on how to improve the partnership, both on the basis of theoretical findings in sociology as well as their observations in practice. Finally, action research experts will try to support partnership members in summarizing their experiences and preparing policy recommendations for partnership support measures. It is also important to note that action researchers themselves learn from the work of partnership members.

In our analysis of partnership processes in the Latvian countryside, we utilized a case of development activities initiated by the Rural Partnerships Project.

## Facts about the Rural Partnerships Project

Supported by the British government, the Latgale Rural Partnerships Project is a rural development initiative in the three districts of Rezekne, Balvi and Daugavpils in the eastern part of Latvia. It is aimed at the creation of rural development partnerships involving several stakeholders: local governments, business support organizations, state institutions, NGOs, and community action groups. The purpose of the partnerships is to elaborate local area-based development strategies, to set priorities and to implement these strategies through the allocation of small grants to community action groups. The project is intended as a test site from which other partnerships in Latvia can draw useful lessons. It is expected that such partnerships will be a key element in the implementation of the LEADER+ Programme in the new Member States after accession.
The Partnership Project has several core objectives:

- To create a partnership structure, test it in practice and to demonstrate the advantages of the partnership approach in local problem solving;
- To achieve tangible results in reducing poverty and social exclusion in rural communities by activating communities and implementing projects;
- To formulate policy proposals in support of rural partnerships for a Single Programming Document, which is Latvia's 'mainstream programme' for the absorption of EU Structural Funds.

Based on the four main partnership levels, our analysis focused on the following areas of activity: (1) activities in the Partnership Board; (2) activities in communities; (3) formation of the wider partnership at district level; and (4) influencing policy-making (see Figure 9.1).

The major processes observed in the partnership were related to the formation of its structure and partnership development processes in communities.

The major stages in the partnership's activities have been as follows: formation of the partnership board, starting up board activities, board training seminars, monthly board meetings, formation of board working groups, meetings of working groups, facilitation of cooperation between the board and communities, selection of active communities and communities which need encouragement, starting concrete cooperation between the board and communities, formation of a group of community facilitators, training community facilitators, developing the first strategy version for the reduction of poverty, and discussing this strategy in communities.

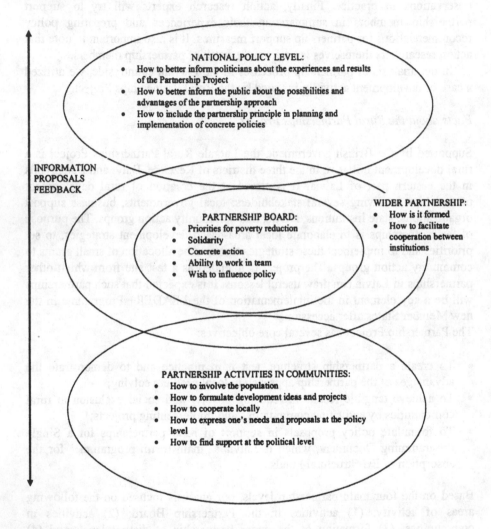

**Figure 9.1 Action research areas and major issues**

In monitoring and analysing these activities, special attention was paid to the formation of the partnership structure, to the motivation of its participants, to perception among partnership participants of the idea of partnership, and to their interests and needs.

With regard to partnership development processes in communities, the main research questions were as follows: how do community groups develop, how do individual strategies converge in collective actions, what development ideas do they set forth, how do those ideas reflect population needs and correspond to the development strategies developed by partnerships? The analyses covered community activities, participation in groups, the variety of the projects submitted, learning and cooperation processes. Attention was paid to communication between and within communities, facilitators and partnership boards.

Several different research methods were applied: participant observation in partnership board meetings, interviews with board members and facilitators, focus group discussions with facilitators, interviews and discussions with representatives of local initiative groups, as well as the analysis of facilitators' diaries.

## Partnership Processes in Rural Communities

### Facts about Rural Partnerships in the Latgale Region

The development of partnerships in rural communities in 2002 was characterized by several important events. A total of 34 rural municipalities joined in the partnerships in the Rezekne, Daugavpils and Balvi districts in the Latgale region. More than 50 local initiative groups were set up.

In the Rezekne district, 31 initiative groups from 18 municipalities became involved, in the Daugavpils district 6 groups from 7 municipalities, and in the Balvi district 16 groups from 9 municipalities.

In all 53 trained community facilitators have been involved in the Rural Partnerships Project.

Local communities have taken an active part in discussing strategies for combating poverty and social exclusion. In the Rezekne district the strategy has been discussed in 17 communities, in the Daugavpils district in 12 communities, and in the Balvi district in 14 communities. The main attention of these discussions has been on the question of whether the strategy sufficiently reflects the problems of the community and offers appropriate solutions.

Communities have been involved in preparing project proposals and have submitted them to partnership boards with a view to obtaining funding. In the Rezekne Partnership 25 proposals were submitted, of which 13 projects were approved; in the Balvi Partnership 34 applications were submitted and 15 projects were approved; and in the Daugavpils Partnership project tenders will be submitted in 2003. Implementation of the projects approved has got under way.

Community facilitators have taken part in training. Three training modules on strategic planning in local communities were organized. Representatives of community facilitators regularly participated in the partnership board meetings and work teams, as well as in the meetings of regional facilitators.

Facilitators maintained regular contact between the partnership boards and communities. They also kept diaries, completed questionnaires, and participated in discussions with Rural Partnerships Project experts who helped to perform work analyses and self-assessments.

**Role of Community Facilitators**

The study of rural community initiatives highlights the crucial need for local leaders in local development activities. In the case of the Rural Partnerships Project, this role is assigned to community facilitators.

Currently 53 community facilitators are involved in the Rural Partnerships Project (34 facilitators work in the Rezekne district, 8 in the Daugavpils district, and 11 in the Balvi district. The vast majority of them are women, who include municipality employees, teachers, farmers, social workers, as well as representatives of other professions. Five of the facilitators are men. All facilitators are engaged in this job on a voluntary basis. Most of them are actively involved, only one facilitator dropped out in 2002 and a further three facilitators did not get involved. In 2002 a total of 13 new facilitators have joined the Rezekne Partnership.

Community facilitators can be characterized as innovative, creative, entrepreneurial personalities who feel responsible for social and economic development in the countryside and who are ready to organize and lead community groups. In their private life, too, facilitators often opt for less traditional solutions, always keen to find new development prospects. 'I work and work all the time, and then I get bored by everything I already know, so then I turn to look for something new. I have to move onto a new stage,' as one community facilitator put it. Facilitators are also creative in their professional life. For example, one community facilitator serves as a manager for local farmers, helps them organise product sales, purchase fuel, do their bookkeeping. Facilitators often say that involvement in the Rural Partnership Project is a personal project for them as well. This synergy of personal and collective motivation significantly increases the prospects of collective projects succeeding.

All community facilitators are devoted to the partnership idea. They feel, update and represent community needs. Some of them have overcome their doubts and disbelief in themselves before becoming facilitators and have learned to speak not only on behalf of themselves, but also on behalf of other people in the same situation and the whole community.

According to the facilitators, the training they have received within the Rural Partnerships Project has been a great asset, providing them with valuable knowledge about partnerships and work with communities. This training and their meetings have strengthened the cohesion of the group of facilitators, created a sense of mutual solidarity and helped them find mutual moral support. Networking and information exchange among facilitators themselves and among facilitators and the partnership board, project experts, lecturers and other partners in cooperation were mentioned as other important benefits. Although the facilitators are in regular contact, some still say they do not know enough about what is going on in other communities.

Within the group of facilitators it is possible to identify several types of leaders with different attitudes to partnership. One type of leader can be characterized by social disposition: these facilitators are mainly oriented to working with communities, to mobilizing groups, to informing, encouraging, involving and organizing people for collective work. Another type of leader is characterized more by a pragmatic approach to project preparation and management. The variety of leader types is also reflected in rural communications. Cooperation among the leaders is important, strengthening as it does partnership relations in the community.

The training of facilitators provided an opportunity for them to set up networks. However, as soon as their training ends, mutual communication tends to be reduced. Nonetheless it is important that these links of contact are maintained: the exchange of experiences is extremely useful, as recognized by the facilitators themselves: 'We should come to Rezekne some time and share our experiences on what we are doing' (Community facilitator).

Working in the partnership, facilitators are keen to see the outcomes of their work: real life improvements in villages, projects in full swing, people showing increased belief in their ability to solve problems. It is also important for the facilitators that the partnership board listens to their views and that their work is evaluated at a political level as well. Some facilitators would also have wanted some small financial remuneration for their work.

## Understanding of Partnership and Attitudes Towards it in the Communities

The Rural Partnership Project has received widespread support in local communities, the clearest indication of this being the number of submitted projects. However, the involvement of the population is a gradual process, from arousing their interest and developing their understanding to support and participation. The development of understanding can be facilitated in different ways. Information was disseminated in several ways in the communities, through pieces in village and district newspapers that directly address the target groups and potential partners in cooperation partners, through discussions with the people who are interested, etc.

Although the notions of 'partnership' and 'community' have not initially met with much understanding among community residents, the partnership principle itself – cooperation among different groups and organizations – is certainly present in the communities. In many places local agents have already recognized that cooperation is essential if their goals are to be achieved:

> What can I do on my own? If not for the municipality that gave the workshops, if not for the women's group where we go together. Then I would have nothing again. All the time you have to look for someone who understands and supports... We too, our women's group, are a kind of partnership (facilitator, representative of women's organization).

The understanding of partnership did not develop in the same way in different groups of rural people. The most sympathetic response was seen in groups who had been active earlier as well as among managers and employees of municipalities. Facilitators admit that it is precisely the poor, socially excluded inhabitants – the main target groups of the Rural Partnership Project – that are the hardest to encourage to get involved. They might take a different view on the way that problems should be resolved, and they may not appreciate the importance of social collaboration. These people would rather receive benefits in kind than get involved in common projects and take the initiative.

Many of the village residents in the Rezekne, Balvi and Daugavpils districts are informed about the partnership and generally approve of it. However, facilitators have also met with negative attitudes: some residents seem to think that the only reason facilitators do their job is because they are getting paid for – and that they are getting 'big money.' However, since facilitators do not in fact get paid at all, such stereotypes based on rumours should be evaluated as symptoms of social passivity, indicating that these rejected individuals often look upon community initiatives with suspicion.

## Community Activities and Involvement in the Partnership

Community activities and projects can be divided into several different groups according to their goals. In all three partnerships, community development initiatives that involve social and culture activities, such as organizing collective undertakings, projects for involving less protected groups, and the development of community centres, are very common.

Plans for setting up community centres in villages warrant special attention in this context: four such projects were approved in the Rezekne Partnership and three in the Balvi Partnership. People regard such community centres as places where they can come together to work and as bases for collective action.

The next most common type of action is represented by initiatives aimed at the improvement of the living environment and community infrastructure. Examples include projects aimed at exploring local resources and using those resources for local development and the delivery of social services to the population. Social involvement is characteristic of these activities, too. Although they may have specific target groups, they are designed to address the whole community.

There is one further widespread community action that deserves separate mention here, i.e. that of organizing training courses. There are fewer projects that deal with economic action. Examples are provided by a group of farmers in Naukseni village who had purchased accounting software as well as craftsmen's and disabled entrepreneurs' groups in the Balvi Partnership.

Collective projects may involve several goals and different lines of action that support one other. As the example of Ozolaine women's group shows, cultural undertakings and training have helped the group diversify its activities: the group has established a sewing workshop, plans to build a hothouse, and wants to engage in biological farming.

Submitted community projects can also be analysed according to their subjects and target groups. It is characteristic of the Rezekne and Balvi Partnerships that community development as well as environment and infrastructure improvement projects are carried out by socially 'mixed' groups: women, young people, farmers, village administrators and other representatives of local society. In other words, they are open to all people in the village. Projects aimed at supporting less protected groups, such as orphans, the disabled and the unemployed also have an inclusive character.

In terms of their goals and composition, the partnerships in Rezekne and Balvi are quite similar. In Rezekne, women and youth groups as well as mixed community groups were particularly active project applicants. In Balvi, too, a number of projects were submitted by mixed community groups and were aimed at improving the quality of the community environment. Several projects were aimed meant at specific target groups: farmers, the disabled, the unemployed, young people out of work, etc. Several applications were submitted that showed an economic orientation.

In the Balvi Partnership many applications were also received from mixed groups and were concerned with the improvement of the community environment. On the other hand there were more projects than in But Rezekne Partnership, which showed an economic orientation, projects submitted by professional groups (firemen) and local history projects.

In spite of their openness and the goal to encourage universal participation, some social strata and demographic groups tend to get involved in partnerships more actively than others. So far the groups that have been the keenest to participate have been women and women's groups, followed by youth groups, farmers' groups, family groups, groups of pensioners, and religious groups.

Several facilitators thought that women got involved in partnerships because they felt responsible not only for the welfare of their children and family, but also for the future of the community.

> It seems to me that women are more successful. They brace themselves up and make demands to the village council. We have Sidrabrasa [Association of Rural Women] in the district and it comprises all women organizations. The people who get involved in new groups are mainly women (Focus group with facilitators).

At the same time, it was said that men get involved in collaborative projects less often; rare examples of men's participation included hunting clubs, farmer cooperation in lending and borrowing vehicles, and farmers' parties.

> Some women are overcome by husbands who don't support any public activities: Why should you go there! What will you do there! This is a major social problem. We organized lectures with a psychologist, but I don't know what to do with men who are sunk in deep apathy – they simply will not step out of their house (Focus group with facilitators).

The rural population in Latgale is ethnically mixed, and in many of the projects financed Russians and other ethnic groups work together with Latvians. In the Rezekne Partnership, one project was submitted in Russian and approved: its aim was to facilitate the employment of young people after they have finished school.

It is groups much harder to encourage some groups to get involved in the partnership than others. According to the facilitators it was most difficult of all to work with the poorest people as well as with the long-term unemployed because they were passive and depressed. It is difficult to encourage them to participate in the groups. It was emphasized that one should try to work individually with these people. 'It is difficult to work in the countryside because of the problem of poverty and passiveness. Alcoholism and unemployment also weighs me down,' one facilitator admitted.

Difficulties were also reported in trying to get people living in the remote parts of the villages to get involved. The group activities usually take place in the village centres, and it is difficult for people who live 10 or even 20 kilometres away to attend. Other obstacles have to do with farm work and lack of time:

> I think that the passiveness is more due to what people do for a living. Because those who do something and who work somewhere are the most active. Those who don't work, it's hard to get them moving (Focus group discussion with facilitators).

Participation in the partnership and collective projects requires active and regular collaboration. As one of the facilitators observed:

> As long as everything is still an idea, there is support, yes. We got together early in the spring and started planning for a sports ground so that children who live farther away could participate in different activities. Yes, everybody agreed: we even have the land rented for the next 3–4 years, everyone promised to participate, to put in some work, to provide technical equipment... But when we got to the stage where the real work was due to start, then suddenly we began to hear: We don't need it.

Both registered non-governmental organizations (NGOs) and unregistered groups, participate in Latgale rural partnerships. Of the 59 projects submitted to the Balvi and Rezekne partnerships, only 10 and 17 per cent, respectively, came from registered organizations. Both registered and unregistered organizations have received support for project implementation. In communities where funding is received by an unregistered group, the municipality operates as a mediator. However, the facilitator of one such community admitted that the idea of establishing an NGO was being considered as it would make fundraising in the future easier. A representative of a women's group expressed a similar opinion: 'That is why we registered. While we were the women's group, we had small projects... But now we have a major project' (Facilitator, representative of a women's organization).

## Communication and Knowledge Processes

Development initiatives in local communities are related to knowledge processes: learning, the exchange of knowledge, etc. A large part of the training activities in the communities are aimed precisely at educating the population, including courses for raising women's self-confidence, courses for landscape designers, etc. Parts of these activities have been organized within the Rural Partnerships Project, while communities implement others by themselves. Within the framework of the training process, local and regional organizations often have close cooperation. Many of the projects submitted to the partnerships include this aspect of knowledge acquisition and consolidation. Facilitators agree that access to information and the accumulation of knowledge are key preconditions for cooperation and development processes at the community level: 'This is all we can give them, information,' as one community facilitator pointed out. However, the local people themselves do not always acknowledge their need for new knowledge: 'If I say there will be a seminar with a specialist giving a talk, what do you think, how many people will come? Five, ten… That's how it is with seminars' (Rural facilitator).

In order to facilitate involvement in development processes, it is crucial to build up social capital at the local level. This is where the skills of facilitators, municipality employees and informal community leaders in raising people's self-confidence and in getting them involved in project preparation are so important. Several facilitators emphasized the necessity of having a leader for both communication and knowledge acquisition: 'We will all help, but we need someone who is in charge' (Facilitator, representative of a women's organization).

## Cooperation in the Community

A number of organizations and interest groups are by now involved in rural partnerships, representing important rural development agents. Among the most active participants identified by facilitators were municipal employees, local amateur groups and farmers. Cooperation among these groups is most typically geared to maintaining links in the local community, but there are also examples of economically minded collaboration. These, however, are rarely growth-oriented but rather concerned to maintain existing economic activities. In the village of Nautreni, farmers had pooled their resources to buy equipment and materials, and they also had a system for sharing their vehicles. The women's group in the village of Ozolaine, for its part, had plans to implement a hothouse project, grow pumpkins and weave blankets.

Facilitators and communities have the closest cooperation with municipalities, schools, museums and cultural institutions. Community facilitators rely most crucially on the support they receive from the municipality and local society. Most typically in the form of premises for group activities, helping with transportation, office equipment, as well as moral support. Support is also received from the business community. The facilitators emphasized that the best response always comes from people who were already involved in cultural and sports activities and in social life generally.

Facilitators have an important role to play in supporting information exchange between the local communities and partnership boards. However, even though the board meetings are open and community facilitators are informed of them, no more than one or two will usually attend. To a large extent communication between the facilitators and the board depends on the initiative of each individual facilitator, as one of them pointed out. 'You have to show an interest in the partnership board. No one will serve you anything on a plate. You have to go there and show an interest all the time and then you will get something out of it.'

Facilitators also have regular and close communication with experts of the Rural Partnerships Project during training seminars and meetings, consulting them about project preparation and implementation. This communication is very important for maintaining the motivation of facilitators, for supporting their work and upgrading their skills. Internal communication between community facilitators is largely realized through the mediation of project experts.

Generally satisfied with the work of the board, facilitators in the Rezekne Partnership expressed the view that the partnership inception period was too long. On the one hand, it was recognized that the start-up of collective initiatives required a sufficiently long period, but if it dragged on for too long, that would undermine people's trust in the project. 'For more than a year, nothing happened in RPP. Many people started to lose their faith. It should have been faster.'

While facilitators accepted that the priorities identified in the programme strategy corresponded to the situation in the communities, they lacked confidence in the processes of evaluating submitted projects, even though it was recognized that the rules and criteria for evaluation had been elaborated with community involvement. Applications were evaluated at several working group meetings. In several cases individual applicants were asked to supply additional information. The working groups reported to the board on its assessment results, and on this basis the board would make its final decision. In spite of careful planning, lack of experience on the part of evaluators meant that it was not always possible to conduct quality evaluation procedures. As one of the working group representatives said: 'The board members did not have a clear understanding of their role in evaluation: you cannot make such choices on the basis of your likes and dislikes, you have to follow the evaluation criteria.'

On the whole board members gained significant experience during the process of project evaluation. On the basis of our analysis as well as the points raised by board members and community facilitators, we have several suggestions and recommendations for the improvement of the evaluation procedure:

- A short training course in project evaluation would be useful for all board members; that could organized in cooperation with some of Latvia's funds.
- The project evaluation process should be organized in the form of as face-to-face discussions with the applicants. That would make it easier for the applicants to explain their project idea and reduce the risk that they are judged on the basis of technical merit rather than content.
- The long-term sustainability of projects should also be weighed in the evaluation process.
- The conditions for the use of the allocated funds should be made clear to the applicants: for example, whether a tender to purchase the planned equipment should be organized, how to deal with project proposals submitted by an unregistered organization.

There remain several obstacles to effective cooperation between local communities and district institutions, including the lack of openness and transparency on the part of those institutions, their bureaucratic public image in society, and limited resources. Although the institutions are aiming to develop more into units that provide public services, in practice they still remain focused more on control, supervision and punishment. This has created the common stereotype that these are not state institutions that serve the people, but it is the people who serve the state. One partnership board member interviewed admitted:

> It seems that in the countryside the bottom sees the top but the top does not see the bottom. There is a lack of understanding as to who is client to whom, an attitude problem.

Indeed there have been recent initiatives aimed at increasing the openness of various state institutions and their regional/district branches, primarily through public discussions and consultations with local residents. Nonetheless the actual working style of these institutions is changing very slowly. Among rural dwellers there is quite a stereotype that it is difficult to get help from these institutions. As one respondent said, 'People are afraid [of institutions], they think of them as places where there are just big bosses.' Another obstacle to cooperation is the lack of resources available for group support. In the words of a facilitator: 'I tried to cooperate with [the organization that helps] the unemployed. I visited the employment service, but they didn't have the funds and no cooperation was established.'

More often than with institutions, communities have cooperation with various specialists – psychologists, doctors, business consultants, lawyers, beauticians, musicians, agricultural specialists – who are invited to attend the cultural events organized in the villages.

> One of the reasons given for this was that partnership members felt it was easier to invite individual experts and indeed easier to pay them than employees of state institutions whose job it is to give advice and assistance to local residents. Whether such individual lectures are more or less effective as strategies or providing advice, is of course debatable. The organizers of these lectures often point out that people 'hear something, get to learn something new.' However the impact of training and education on resolving community problems is more indirect, finding its expression in the social activation of people.

## Partnership Results and Influence in the Communities: Positive Examples

It is still quite difficult to single out the influence of partnership activities and the project upon social and economic life in the communities, for it is only very recently that the projects have begun to receive support. However, several positive trends can already be seen with regard to the local understanding of development ideas and cooperation in the communities:

- The partnership has produced new ideas and understanding of the need for co-operation.
- It has activated some community groups.
- Communication between local residents and institutions has improved.

- Awareness of common interests in local community has been enhanced.
- Community development projects have been set up.

It is difficult at this early stage to anticipate the influence of the projects launched on the battle against exclusion and poverty, even for the project applicants themselves.

The Rural Partnership Programme has gained quite widespread publicity in both the local media (Latgale television, regional newspapers and local radio stations) and national media (newspaper *Diena*, Latvian Radio, *Lauku Avīze*, etc.) and it is therefore recognized outside the immediate target groups as well.

## Success Stories of Rural Partnerships

*Community Centre in Kantinieki Village*  The village of Kantinieki is located far from the district centre. Its infrastructure is poor, it has neither its own school nor cultural centre, and the unemployment rate is high (36 per cent). Various ethnic groups are represented in its population. The Rezekne partnership board appointed an energetic facilitator for the village. She encouraged the local people to be more active, talked about their needs, raised project ideas. Several groups were created which prepared and submitted a proposal for the establishment of a community centre. This project received support from the local municipality and the Rural Partnership Project.

*Farmers' Education and Information Centre in Nautreni Village*  A farmers' group has been set up in the village of Nautreni. They want to establish a community education centre where people from other villages could also get together, acquire information, and realize common ideas. The group manager has the following justification for the need for this centre: 'Economic education is expensive. The knowledge that people have from the days of collective farming is no longer useful, therefore we need a community centre where common sales issues can be discussed, where we can re-qualify.' The farmers' group prepared and submitted to the Rezekne Partnership Board a centre establishment project, which was approved. With the help of project funding, the group will purchase accounting software that can only be used by the village farmers. The group meets once a month, with meetings announced in the local newspaper. The chairman of the local council supports the initiative: 'People will begin to understand that they have to start to earn money and they need knowledge to do that. Ministry of Agriculture advice bureaux have been set up in towns, but they do not work in the countryside.'

*Women's Group in Ozolaine Village*  This group involves 25 women. It was established in 1998 to encourage women, to offer entrepreneurial training and to provide mutual support. Training was an important starting point. Lecturers in psychology, entrepreneurship and economics were invited to give talks. As well as learning, group members gained in self-confidence, helping many of them find a job: 'When we started out, we were all unemployed but now all of us have a job.'

The group has several partners in cooperation, including the municipality, the Rezekne District Partnership, a municipality in Germany, a community from the neighbouring village of Kantinieki, the Employment Agency of Rezekne District, the Society of Biological Agriculture of Preili District, the Society of Rural Women of Latvia, and foreign funds. The Rural Partnership Project met with a sympathetic response among women because they already had experience of cooperation. In 2002 the group received support from several sources: a grant of LVL 800 from the Embassy of Great Britain, a small grant for administrative purposes from the USA mission, and a grant of LVL 10,000 from the Queen Juliana Fund to establish a centre. The group's future plans include growing plants, establishing hothouses, breeding sheep and processing wool. A sewing circle has already been established, and a community centre is being set up in the old workshops. Over time several leaders have emerged in the group and from the initial community several new groups have mushroomed, including a youth club, which publishes its own newspaper and a group of Russian teachers. They submitted two projects plans to the Rezekne District Partnership. A youth business education project was approved.

## Obstacles to Partnership Operation

Cooperative strategies in rural development in Latvia are hampered by bitter memories of totalitarian Soviet collectivity as well as by lack of confidence in the public sphere, lack of leadership in the local population and sense of social exclusion. The recent establishment of rural NGOs and their networking with local governments and the private sector has inspired hope of a more rapid development of rural communication.

There are several potential obstacles to the adoption of the partnership idea in rural communities:

- A prejudiced and cautious attitude towards innovations. People are often reluctant to take risks and get involved in new activities if they cannot be sure of the outcomes. One solution could be the exchange of positive experiences.
- The poor technical infrastructure – roads, telephones, Internet access – makes social communication more difficult.
- Active residents may be held back by the lack of understanding, indifference and passive attitude of others. 'In the beginning, people smirked at us. But I thought nothing of it, because we will prove ourselves over a longer period of time' (Facilitator, representative of a women's organization).
- Although there are several examples of cooperation, certain segments of the population take a sceptical attitude on account of their negative experiences of collective work from the Soviet era.

**Discussion and Conclusions**

The partnership approach is new in Latvia, in several respects. First, it differs from the traditional NGO approach in that it comprises wider cooperation among different groups and agencies. Second, partnership works at all levels: in the partnership board, local community, in the wider partnership and at the political level. It is not organized according to the industry principle, but according to the social, community and territorial principle. Third, partnership aims to address and to include socially excluded groups in the countryside. In Latvia there are now a few programmes aimed specifically at helping these people and there are some examples of success. Fourth, partnership members deepen the understanding in society about rural development that is based on grassroots participation. The Rural Partnership Programme has facilitated the introduction of new notions and terms in the public discourse, such as 'partnership,' 'community,' 'community groups,' 'community centres,' 'community facilitators,' 'initiative groups,' etc.

Partnership can have different roles in community development processes. It can both initiate new development processes by mobilizing resources and encouraging the start-up of new projects/networks/cooperation, but it can also be integrated into existing development projects by offering additional possibilities to enlarge and consolidate their operation.

The Rural Partnership Project assumes a leading role in the initial stage of partnership development. After this stage, it is important that the partnership participants take over. This process can be observed in all three Latgale Partnerships. However, in some cases (for example, the Daugavpils Partnership) the changeover was not a smooth process. In those cases where attendance at board meetings was low and where participants were reluctant to continue working in the partnership, the Daugavpils Partnership Board together with the project experts appointed new, more motivated board members in order to get the strategy development process under way.

The case of the Daugavpils Partnership goes to show that crises can easily result from be caused by poor communication between the board, the communities and facilitators. It is important to maintain regular communication. Currently two representatives of community facilitators are working on the Daugavpils Partnership Board.

Project evaluation has been a critical moment for the communities and facilitators. Although all project applicants were familiar with the project preparation requirements and the evaluation criteria, and although training was provided, many lacked the specific skills they would have needed to compose a project description. Only 28 of the 59 applications submitted (47 per cent) received funding. Many of the rejected projects addressed the social needs of the community and in this sense corresponded to the strategic goals of the partnership, but they fell short of the technical requirements. Rejection caused much disappointment: many communities and their leaders are not yet used to working in competitive situation.

The partnership board and work groups lacked experience in project evaluation, and the evaluation process as a whole was rather hasty and superficial. However, none of the facilitators complained that the evaluators would have lacked in objectivity.

So far the most essential impact of the Rural Partnerships Project at the community level has been the establishment of the group of community facilitators, the provided support to their work, and community activation. The partnership project has generated significant impulses and support for rural communities. In training and in their mutual communication, the facilitators have realized that it is important to work in the partnership, and they promote the idea in their own places of residence. It would be important to continue training for facilitators after the project ends because they have become significant agents of rural development.

Nevertheless the potential of communities to continue activities for resolving their social and economic problems after the Partnership project period, without any external support, may well be insufficient. To some extent they have adopted the partnership project's logic, its objectives, means and methods, but further support is required if lasting effects are to be maintained.

Local development processes are influenced not only by community activity, but also by different external factors and political processes: regional development policy, administrative territorial reform, the liquidation of agriculture organizers' services, the restructuring of agriculture, etc. As the seminars and discussions held with national politicians within the framework of the Rural Partnership Programme indicate, the partnership approach and programmes need support at different management levels – national, regional, and local – in order that the long-term sustainability of local development initiatives can be ensured.

The integration of project work into the logic of community work is a complicated process. Not everything that is perfectly accomplished in the context of a development project can be perfectly accomplished in community and group work. This is the eternal problem of development projects.

## *Partnership Processes and National Characteristics of Rural Communication*

The networking of different agents into partnerships requires the establishment and maintenance of specific links between them. A key precondition for success is thus adequate communication.

In general, rural communication in Latvia is characterized by a limited number of actors involved, the narrowness of communicative practices, weak horizontal ties and the dominance of top-down administrative communication. Although there are some indications now of maturation, civil society in rural areas is still weak and rural communication is very much dominated by state institutions: the Ministry of Agriculture, Regional Development Agencies, the Rural Support Service, regional branches of the State Employment Service and other central government institutions along with local municipalities.

The general weakness of rural communication was indeed the most outstanding research finding. As one of the experts interviewed emphasized:

Rural communication is weak and getting even weaker. One of the reasons is the deep rural recession, the other is the changing lifestyle during the transition from collective to individual farms. In the kolkhoz era, people communicated more. There were various social events and plenty of opportunities to go out, even though these events had an ideological slant to them, now it's much harder to get people out of their homes and there are fewer places where they can meet. Depression and individualism have taken over in rural communication.

Apart from state administration and local municipalities, the strongest actors in rural communication are rural businesses, producers' associations, farmers' organizations, banks, food-processing companies and other economic actors. They engage in political relations with local authorities and regional branches of state institutions with a view to promoting their economic interests.

From an actor perspective we can distinguish two types of rural communication: contacts between economically and politically 'strong' actors and relations between economically and politically 'weak' actors. The latter typically include such segments of the 'ordinary' rural population as small farmers, pensioners, unemployed persons of pre-retirement age, people living in remote areas, rural women, and young people. Some parts of these groups can be characterized as poverty hit and socially excluded.

The main issues of rural communication depend on the respective actors' strategies. Producers' associations and agricultural policy institutions communicate about such issues as the modernization of agricultural production, the development of the forestry sector, state support policies to agriculture, subsidy schemes, production quotas, Latvia's interests and positions in the context of EU enlargement, as well as issues of regional reform. Weak actors are more concerned about other issues, like vitalizing the rural labour market, combating unemployment, creating new jobs, social service provision for the rural population, social benefits for the unemployed, pensioners and families with children. This discourse is grounded in real life conditions and in the attitudes of poor, socially excluded parts of the rural population.

Sectoral and economic communication between rural actors is more active and more advanced than territorially based and community type communication. There is a contradiction between the sectoral and community approaches to rural communication. For instance, there are about 50 active agricultural producers' organizations which operate at the national level (e.g. Latvian Grain Producers' Association, Latvian Dairy Association) and which have established an umbrella Association for Agricultural Organizations. Producers' organizations promote their interests in dialogue with the government. In the meantime, there are a few national umbrella organizations, which represent the interests of rural communities, civic groups, and particularly the interests of the most disadvantaged population. The Association of Latvian Municipalities (*Latvijas Pašvaldību savienība*) has often been criticized for its inability to advocate the interests of the rural population in general.

There is an apparent lack of LEADER type communication between and within rural communities, and a deficit of civic exchange in rural neighbourhoods. The per capita number of registered NGOs in rural areas is lower than in towns. In a typical rural community with 2,000 inhabitants, there would be perhaps a choir, an art group, a youth sports club, a women's club, and a social assistance NGO. Most likely there would not be any community action groups engaged in community planning and undertaking social, economic and environmental projects, nor would there be either a community centre as a meeting place and a focal point for a variety of active citizen groups and local initiatives. Civic ties in rural communities are much weaker than in towns, for economic, social, political and demographic reasons.

Demographic composition, ageing and emigration should be considered as significant obstacles to rural communication. The rural population is ageing faster than the urban population and many young couples with children are leaving rural areas in search of jobs and educational opportunities. Young people who go to study in cities rarely return to their home villages and small towns. As one of the interviewed municipality leaders described the problem: 'The young, the most active and the cleverest people leave for the cities.' The lack of communicative ties creates communicative scepticism. Isolated and weak actors hesitate to engage in groups and associations, and they are sceptical about the gains they will achieve from such communication.

The needs and interests of the ordinary rural population are communicated in local social life predominantly through personal relations. The voice of ordinary people towards local political power is communicated on an individual and ad hoc base. There are few organized mechanisms of regular communication between local power and local civil society. People address municipal power if and when they have problems. There are few regular mechanisms of population engagement in long-term problem-solving and development planning at local level. Weak social actors are not as active in setting up their own representative associations and organizations as partners to local governments. This creates communicative misconceptions and stress on both sides.

Rural communication is hampered by the fact that 'weak' actors who experience economic and social hardships also suffer from low self-confidence and distrust in institutions on which they are dependent. These people have limited access to information and other resources as well as difficulties in mobility. They need communicative assistance and guidance to get access to information, to become involved in community life, to get personally empowered. According to a civil servant from the Ministry of Agriculture:

> There is a need for a community facilitator in every municipality. They would be trustees for farmers, whom they could address for advice. Rural people do not trust official information from the Ministry of Agriculture. They want Rural Support Service employees to explain the information to them in person. Farmers rely more on informal channels.

The case of partnership activities proves that an eventual recipe for efficient communication among 'weak,' socially excluded rural actors would entail such elements as psychological encouragement to break loose from isolation, advice and training, the use of informal rather than formal channels of communication, the presence of people one can trust and animators of communication.

*Diversity and Typology of Rural Communication*

Our overview of community initiatives allowed us to draw some generalizations about rural communication in Latvia and to distinguish some typical traits:

• Communication tends to be localized.
• The number of involved parties is small.
• Communicative action tends to be isolated and event-focused rather than regular and long-term.
• Thematic and issue areas of communication are limited predominantly to social and cultural issues.
• There are few communicative initiatives in the field of economic cooperation.
• There are few success stories from the point of view of communication outcomes.
• Individual leaders inspire communication and it is dependent on them.
• Communication proceeds from one project to another, it is not self-sustained.
• There is a general lack of information, knowledge and communication skills in the rural population.
• There is a general lack of positive experience about rural communication.
• Widespread social and economic depression in rural areas hampers communication.
• Strong actors communicate better than weak actors.

Many communication examples are singular. They mobilize collectivities, explode in action, attract supporters, even achieve some results and then subside. For example, a youth dance group from Mežvidi parish in the Ludza district won the district festival and received the rights to participate in the national Song and Dance festival in Riga. The parish authorities provided support for the dance group, some money, transportation, etc. However, as soon as the festival ended, the youth dance group broke up. This particular case of communication was successful only insofar as the individuals concerned were responsive, determined and able to organize and as long as limited support was available. In the absence of regular financing, self-mobilization can fuel collective action only for some while.

Lack of information is a major obstacle for rural communication. People lack information about services available as well as about the communication process in general. In the words of one interviewee, a mayor of a small town:

> Rural communication is in deadlock; there is an overall lack of information about communication. Communication is mostly associated with solving particular problems and in this context people acquire the necessary information, but there is no communication in terms of broader development. People do not know where to obtain such information.

Individuals find their individual channels for personal problem-solving, but they are unwilling or unable to develop collective solutions.

Financial support for rural communication initiatives is very limited in Latvia. There are practically no support schemes for community initiatives or EU LEADER type projects in the existing state support programmes for rural areas. The only available support comes from NGO support programmes, which are mainly financed by international donors. In 2002 a policy dialogue group was formed to assume responsibility for the elaboration of LEADER type local development measures in Latvia. However politicians believe that in the near future part of current agricultural subsidies will be redirected for support to local initiative groups. Agriculturists and agricultural politicians are more interested in increasing direct subsidies rather than in changing their structure. The most credible financial source for the LEADER+ programme in Latvia could be EU structural funds.

Apart from the areas of rural communication described above, such as territorial planning, community amalgamation, associations of agricultural producers, rural women's movement and development partnerships, a particular area of social interaction is that of cultural communication. Culture is an extremely important component of rural social life. Theatrical collectives, dance groups, choirs, song festivals, national festivities and traditional celebrations all are part of rural life. However these cultural traditions are being fragmented and participation in choirs and folk art groups is declining. On the other hand, traditions are also being upheld and reinvented.

Protest communication is a special type of rural communication. In several cases rural communication has been activated by farmers' protests against the state's agricultural policies. For example, in summer 2000 farmers all over Latvia organized road blocks in protest against low prices and the import of subsidized agricultural products. Farmers achieved high levels of mobilization and political engagement, which however subsided when it came to regular and institutionalized communication with the government. Protest communication is usually short-term and dependent on mobilization.

## Policy Networks for Agricultural and Rural Development

Rural communication is hampered by the lack of policy coordination. One of the main problems is the sectoral division of rural policies. Today rural policy in Latvia does not have an institutional centre, but it is split among several ministries with programmes in their own areas. The Ministry of Agriculture is mainly concerned with strengthening the agricultural production sector, but it also runs a subsidy programme for diversification into non-traditional farming and off-farm activities. The Ministry of Finance encourages small-scale enterprise development in rural areas through various tax concessions. The Ministry of Economy, for its part, operates a Regional Fund that provides grants and interest subsidies for public and private investors in less developed areas, and it is also implementing an SME Development Programme. The Ministry of Environmental Protection and Regional Development is involved in rural development through supervision and through the allocation of funding for capital expenditures to local governments, thus assisting them in the preparation of territorial plans and development strategies. The State

Employment Service runs an active employment policy through its branch offices in rural areas (PHARE, 1998, 26–27). Each ministry operates its own programmes in a rather top-down manner and there is a shortage of coordination. This may prove to be a fundamental obstacle to rural communication.

Recent joint activities between the Ministry of Agriculture and a policy dialogue group which includes representatives of other ministries and government organisations, as well as experts and representatives of NGOs, is an example of a broadened policy network. This new initiative has met with several difficulties. Although the network of political decision-making has been extended, the ministry plays the leading role, and local competence is inadequately incorporated in policy formulation. The convergence of local and governmental competence is a major challenge for this extended policy network. Civil servants are certainly willing to cooperate with representatives of local communities, but they often lack the time to do so. Local actors often lack self-confidence and the skills to formulate their political proposals. The case of the policy dialogue group demonstrates that extended policy networks require mechanisms of regular collaboration between different actors in order to be efficient.

The government of Latvia is currently preparing a Single Programming Document, which defines Latvia's needs and priorities for assistance from EU Structural Funds. The Ministry of Agriculture, which is responsible for agricultural and rural development policies, has established a consultative committee and a policy dialogue group to prepare a LEADER+ type measure for structural fund assistance. Such a measure would support rural partnerships and local action groups in implementing projects in the areas of community planning, social infrastructure development, education, environment and social economy.

The interests of economically strong rural actors have a greater chance to be organized politically and represented in a dialogue with government than do the interests and needs of poor and excluded rural residents. However, some signs of more inclusive policy networks can be seen in the implementation of the Rural Partnerships Project in Latvia, and in the process of developing a new Policy Measure for Local Development, which is designed as a new instrument for rural development after accession. Nevertheless, future rural policy should be even more decentralized, and equal participation of rural organizations from all of the country's regions and groups of rural population should be attained through participatory efforts and consultations with various rural organizations.

## The Logic of Community Action

Partnership provides a new approach to solving local problems by stimulating joint, organized, targeted activities by different groups of society. It facilitates information exchange, cooperation between these groups, and the consolidation of existing available resources for achieving common goals. Collective action develops gradually through the communication process, it has to be learned and requires self-reflection. This process is described in Figure 9.2.

| POLITICS | | | | | | | | |
|---|---|---|---|---|---|---|---|---|
| **INSTRUMENTS OF POLITICAL SUPPORT** | • How should the Government support collaborative action in the countryside? How should social capital be used?<br>• How make a big impact with a small amount of money? How include the partnership principle in policy programmes?<br>• How justify the hopes of people involved in partnerships? | | | | | | | |
| **ACTION PROMOTING FACTORS** | • Entrepreneurship<br>• Innovation<br>• Risk management<br>• Competitiveness | — Communication and creation of social links<br>— Psychological support and enhanced *self-esteem*<br>— *Learning* collaborative action in practice<br>— *Hopes* in local and national political support | | | | | | |

| THE PROCESS OF INDUSTRIAL ACTION | | THE PROCESS OF COLLABORATION IN COMMUNITY | | | | | | |
|---|---|---|---|---|---|---|---|---|
| **ACTION** | Situation | Individual interests and solutions | Situation | Partnership project activities | The role of the community facilitator | Local action groups | Major change agents | Activities and results | Project applications |
| **DEVELOPMENT-ORIENTED ACTION** | | | Majority:<br>Poor people, employed in low-paid jobs, casual labourers, unemployed | Mobilization of people, exploration of opportunities, strategy formation, partnership formation | Development promoters, initiators, encouragers and organizers | Formation of action groups on the basis of democratic features, interests, aims | Women<br>Pensioners<br>Youth<br>Volunteers<br>Intellectuals<br>Men | Overcoming isolation<br>Support<br>Enhanced self-esteem<br>Strategy development<br>Concrete action<br>Project ideas | In social area<br>In environmental area<br>In cultural area<br>In business area |
| **LOCAL NEEDS AND SOLUTIONS** | Minority:<br>Businessmen, self-employed, persons active in the labour market, employed in relatively secure jobs | Business diversity | | | | | | |
| | ECONOMIC DEVELOPMENT | | Socio-Economic Development | | | | | |

**Figure 9.2 How to strengthen collaboration in rural communities**

**Development of a New Institutional Space?**

A critical aspect in rural development is related to harmonizing centrally planned policies and financing with the development visions, needs and participation of the local population. It is important to fuse top-down policies with grass-root initiatives. The paradigm of rural development in the 1990s is based on the assumption that target groups and beneficiaries of development should themselves be participants and authors of the development process. This suggests that successful rural development strategies are impossible without the active and collective participation of local residents, NGOs, business organizations and local governments.

Various initiatives have been undertaken to tackle rural problems and to boost development in the Latvian countryside, including such measures as regional planning, institutional building, revitalizing local action groups and NGOs, establishing regional, environmental and other funds, launching specific rural credit lines and government support schemes, setting up the Rural Development Programme and Support Programme for Disadvantaged Regions, as well as making use of various foreign assistance projects at national, regional and local levels. Many of these activities are claiming to apply an integrated top-down and bottom-up approach.

Rural development 'is about creating a new social field with new rules and new actors. In this new game come new relationships between the actors, and their various forms of capital are given different values' (Almas, 1998, 79). In order to remedy acute social and economic problems in rural Latvia and to achieve a systematic positive impact on rural economic and social life, it is necessary to fit together different actors' interests, to pay proper attention to the development of both human resources and rural support institutions, to complement free market incentives with government stimuli, and to adjust national policies to EU accession opportunities, including broad civil participation at all stages of rural policy-making and implementation.

*Some Recommendations for Strengthening Rural Partnerships*

The research of partnership development in the Latvian countryside allows us to formulate several practical recommendations on how to support this process.

- In order to support local development, Latvia has to introduce rural and regional development programmes that correspond to the EU's LEADER programmes. Experiences gained through the Rural Partnership Project have to be put to good use in working out such programmes;
- Cooperation and coordination networks need to be set up among different management levels to ensure constant information exchange and links among the agents involved in rural development. The communication established among the rural population, communities, facilitators, partnership boards and the Ministry of Agriculture as a result of partnership operation can be

considered such a network. This communication should be regular and more effective;

- It is important that facilitators in the rural communities continue to work closely with one another. The Ministry of Agriculture and other agencies responsible for rural development should consider establishing an institution for community facilitators similar to the former institution of agricultural organizers;
- Training should be continued at different levels;
- At the local level for community facilitators and population groups;
- At the district level for participants of the wider partnership, for specialists and municipality politicians;
- At the regional and national level for purposes of exchanging experiences, preparing political proposals, and coordinating methodology;
- Communication among project participants should be facilitated for better exchange of information and experiences. A system is need for gathering the experiences of supported partnership projects, and; positive examples have to be promoted.

## References

Almas, R. (1998), 'Rural Development in the Norwegian Context,' *Environment and Societe. Innovations Rurales*, no. 20, pp. 79–85.

Central Statistical Bureau of Latvia (2001), *Statistical Yearbook of Latvia*, Central Statistical Bureau of Latvia, Riga.

Davies, S., Slee, B. and Tisenkopfs, T (1997), *Social and Economic Conditions in Rural Latvia*, Scottish Agricultural College.

Keune, M. (1998), *Poverty and Labour Market in Latvia, Households Budget and Labour Force Surveys Data*, Ministry of Welfare and UNDP, Riga.

Kruzmetra, M. and Rivza, B. (1994), 'Some Problems of Rural Development in Latvia,' in B. Rivza (ed.), *Agrarian Reforms in the Baltic States: Experience and Prospects of Development*, Works (LUA) Issue 2, Latvian Agricultural University, Jelgava, pp. 8–12.

Ministry of Agriculture (1998), *Agricultural Strategy Paper*, Riga.

Ministry of Environmental Protection and Regional Development (1998), *Latvia Rural Development Programme*, Riga.

OECD (1996), *Review of Agricultural Policies. Latvia*, OECD, Riga.

PHARE (1998), *Support to the Diversification of Rural Economy*. 2. *Quarterly Report*, ABG GmbH, Riga.

Rungule, R. et al. (1998), *The Social Assessment of Poverty in Latvia*, Institute of Philosophy and Sociology, University of Latvia, Riga.

Tisenkopfs, T. (1998) (ed.), *Latvia Human Development Report 1998*, UNDP, Riga.

Tisenkopfs, T. (1999), 'Rurality as Creative Field: Towards an Integrated Rural Development in Latvia,' *Sociologia Ruralis*, vol. 39, no. 3, pp. 411–430.

UNDP (1998), *Who and Where are the Poor in Latvia? Research Report*, UNDP, Riga. www.undp.riga.lv/undp/programme/poverty

UNDP (2001), *Latvia. Human Development Report*, UNDP, Riga.

# Chapter 10

# The Adjustment of the Elderly to Socio-Economic Change in Rural Estonia

Leeni Hansson and Marjatta Marin

## Introduction

Estonia's socio-economic reforms during the 1990s involved certain costs and consequences and required much adjustment on the part of its citizens. Age, education, gender, ethnic origin and place of residence were among the factors that either helped people adjust to the rapidly changing socio-economic situation, or made that process harder for them.

Several studies (e.g. Laidmäe, 2001) have shown that elderly people in particular had great problems adjusting. Bond and Coleman (1990) have pointed out that even if people with ageing have to face changes in different areas of life, and particularly in relation to work, leisure and social networks, the capacity to adjust to life changes does not diminish in later life but is actually enhanced. However, Bond and Coleman are referring here mainly to age-related changes in a stable society, which is not how Estonia in the 1990s appeared. Older people in Estonia had to adjust not only to age-related but also to a number of socio-economic changes characteristic of a transitional society. And like in most countries in transition, elderly people turned out to be one of the most vulnerable groups. In their case, the cost of the reforms has been a heavy one.

Obviously, older people do not form a homogeneous group. For example, there is a small group of 'millionaire pensioners' in Estonia who have received their family property in restitution and whose main problem today is how to manage that property. Another group is made up of pensioners who have managed to keep their jobs and who accordingly have a double income, i.e. their regular salary plus a full pension, and who even enjoy some additional benefits (e.g. free public transport in Tallinn, subsidized medical treatment). These people are considerably better off than families with dependent children, for instance (Kutsar and Trumm, 1999; Marksoo, 2000). Third, there are those pensioners who do not work and who have to manage with their pension. The financial fitness of this group of pensioners depends largely on the type of household in which they live. High housing costs mean that people who live in one-person households have far more difficulties than those in two-pensioner households or in three-generation households (Marksoo, 2000).

While pensioners represent perhaps the most vulnerable group, it is important not to forget people of pre-retirement age who for one reason or another have lost their jobs. In regions with a difficult labour market situation, it is extremely difficult for these people to find new jobs. The number of such people seems to be increasing, especially in rural areas. Ageing and retirement have thus become synonymous with stress, poverty and exclusion in Estonia (Kutsar and Trumm, 1999).

Is it possible in Estonia to provide normal living standards for the growing number of people in older age groups, including the rural elderly, and to lower their stress level? Hanley and Baikie (1984) have proposed a distinction between four stress factors: *'loss'*, or the removal of external circumstances upon which an individual had relied; *'attack'*, or external forces that produce discomfort; *'restraint'*, or external restrictions on a person's behaviour; and *'threat'*, or any event that warns of possible future loss. Empirical data collected in the 1990s demonstrate that in the first decade after the re-establishment of independence, the majority of elderly people in Estonia were affected by at least some of these four stress factors (Laidmäe, 2001).

At least so far Estonian social policy has does not been able to provide a safety network that fully meets the needs of older age groups. In fact recent legislation has been geared to shifting responsibility for the aged from the public sector to the individuals themselves and their families. Historically it has been the norm in Estonia that families have looked after their own elderly: by pooling their resources older people and the younger generation have managed to pull through the hardest times. Today, the situation is no longer as simple as that. Many families of the younger generation, particularly those with small children and unemployed family members, are in serious financial straits (Marksoo, 2000; Kutsar and Trumm, 1999) and quite simply are not in a position to support their elderly relatives. Psychologically, this is a major source of stress both for those who need support and for those who are expected to provide it. For this reason older people in Estonia will prefer to try and manage on their own as long as possible.

Although Estonia is a small country, regional differences in wealth and opportunities are significant and increasing. Some regions have high potential for investment and an increasing number of jobs, others a very low labour demand and a high level of long-term unemployment (for more details, see *Regional Statistics of Estonia*, 2000). In this article our main aim is to investigate the capacity for adaptation of the rural population of retirement and pre-retirement age (55+) living in the southeastern corner of Estonia. This is the region that together with the northeastern part of the country (Narva, Kohtla-Järve and the surroundings) is struggling most. However, for political reasons this region is still largely neglected at the expense of northeastern Estonia and its problems (Russian-speakers), whereas the southeastern region with its predominantly ethnic Estonian population has remained a 'forgotten periphery.'

Our discussion is based upon qualitative interview data collected in 1995 in a community study of Kanepi and its surrounding villages,[1] data from the 'Estonia 98'[2] population survey as well as official statistics.

## General Problems Related to an Ageing Population

We begin by looking at the recent changes in Estonia that in one way or another have affected older people in general. The problem of older age groups in Estonia derives first and foremost from the sharp decline in birth rates in the early 1990s, when they dropped to exceptionally low levels (Katus et al., 2002). In addition, large numbers of young people moved in from other Soviet regions during the 1960s and 1970s, and now, these immigrants have reached or are reaching retirement age (Ibid.).

The fact that official statistics indicate hardly any increase at all in the number of pensioners as a proportion of the total population during the second half of the 1990s (see Table 10.1) can be explained by changes in the official retirement age. During the Soviet period the retirement age in Estonia was quite low: 55 for women and 60 for men. The pension system was overhauled in the 1990s and since 1994 the retirement age has been increased by six months a year until it reaches 63 years for both men and women. By 2001, the official retirement age had reached 58.5 years for women and 63 years for men. At the end of the 1990s, pensioners accounted for almost 26 per cent of the total population of Estonia (Table 10.1).

**Table 10.1 Number of pensioners (thousands) and proportion of population**

|  | 1980 | 1990 | 1995 | 2000 |
|---|---|---|---|---|
| Population | 1,477.2 | 1,571.6 | 1,491.6 | 1,439.2 |
| Total number of pensioners | 327.0 | 360.5 | 376.2 | 371.4 |
| Pensioners as per cent of population | 22.1 | 22.9 | 25.2 | 25.8 |
| Old-age pensioners | 237.0 | 287.5 | 302.1 | 284,2 |
| Persons on disability pension | 43.0 | 38.9 | 52.3 | 66.8 |
| Disability pensioners as per cent of all pensioners | 13.1 | 10.8 | 13.9 | 18.0 |

*Sources:* Statistika Aastaraamat 1990; Statistical Yearbook of Estonia 1999; Statistical Yearbook of Estonia 2000.

In the 1990s pensions in Estonia were paid out in accordance with the temporary Law on State Allowances, which was adopted in 1993. The law provided for four types of pension: old-age pension, disability pension, loss of provider pension and

---

[1]  For more details, see Alanen et al. (2001).
[2]  A population survey carried out by the Institute of International and Social Studies at Tallinn Pedagogical University in 1998 (2,317 respondents).

national pension. Up until 2000, the payment of old-age pension was not dependent on the individual's previous income. Since 2000, the rate of the pension has been calculated as a sum of three components, one of which is the basic sum as established by *Riigikogu* (the Estonian Parliament). The two other components depend on the individual's contributions, i.e. the number of years in active employment and the insurance component. People who have not been in active employment receive a national pension. It is important to bear in mind that pensions are calculated without any consideration of family size or structure, or other income sources. In Estonia pensions are not taxable income.

Compulsory savings insurance entered into force on 1 January 2001. According to the new laws, the pension insurance system consists of three pillars: state pension insurance, compulsory individual savings insurance and voluntary savings insurance. There are by now several life insurance companies that may sign up customers for voluntary insurance agreements, but for persons already in retirement or pre-retirement age these changes have come too late to be of any help.

As can be seen in Table 10.1, nearly 80 per cent of pensioners in Estonia are entitled to old-age pension. With the introduction of a higher official retirement age, the number of old-age pensioners has slightly decreased in Estonia over the past decade, while that of persons on disability pension has increased. Compared to the situation in 1993, the number of disability pensioners in 2000 was up by 27 per cent. The sharp increase in the number of people on disability pension reflects the changes that have happened in the labour market first of all in rural areas. During the former period of secure jobs and virtually zero unemployment, pre-retirement people with minor disabilities or functional limitations would usually prefer to continue in active employment because the disability pension was (and still is) considerably lower than average wages or salaries. Today, when unemployment benefits are paid for no more than a maximum of six months, it makes sense for a pre-retirement person who has a minor disability and who has difficulties finding a new job to apply for disability pension, otherwise (s)he will remain without any income. According to official statistics the increase in the number of disability pensioners has been the sharpest in regions with the highest unemployment rates (Kõgel, 2001). Given the continuous rise of official retirement age and the insecure labour market situation, the number of disability pensioners looks set to grow in the future as well. This in turn will reduce the economic effect expected from the rise of the retirement age (*Estonian Human Development Report*, 1998).

*Older population groups in the labour market:* As was pointed out above, it was commonplace in the 1970s and 1980s for people who were entitled to old-age pension but still physically fit to continue to work in active employment. During this Soviet period, the employment rate among pensioners was quite high. The structural changes that followed with the re-establishment of independence in Estonia had profound effects on the labour market, and in many companies pensioners were among the first to drop out of active employment. In 1990, for instance, 28 per cent of old-age pensioners were still employed, but by 1996 the figure was down to 15 per cent (*Social Trends*, 1999). At the same time the relative value of the average old-age pension was lower than in the 1980s, and pensioners who were in good health did not usually leave the labour force voluntarily. On the contrary, they preferred to keep their jobs as long as possible, and if they were made redundant, they would try to find a new job, even if it was one of a lower occupational status (Laidmäe, 2001). Therefore, in the 1990s, most people in Estonia did not retire when they wanted to but when they had to. Furthermore, the reorganization of production, high levels of occupational mobility and involuntary job changes, difficulties of re-employment and the growth of unemployment not only reduced pensioners' employment, but also resulted in significantly curtailed job opportunities for people of pre-retirement age (Table 10.2), who according to former Soviet pension legislation would already have been entitled to old-age pension.

**Table 10.2 Employment rates in the population aged 55 (60)+ in Estonia, 1990–2000 %**

|  | Age | 1990 | 1994 | 2000 |
|---|---|---|---|---|
| Men | 60–64 | 64.4 | 39.9 | 43.2 |
|  | 65–69 | 43.3 | 27.1 | 21.8 |
| Women | 55–59 | 55.5 | 41.2 | 48.6 |
|  | 60–64 | 43.8 | 21.8 | 24.2 |
|  | 65–69 | 30.8 | 13.9 | 14.7 |

*Sources:* Statistical Yearbook of Estonia 1999; Labour Force 2000.

*Changes in the economic situation of older age groups:* Compared to advanced western countries, the average pension level in Estonia is low. Even though pensions have risen quite significantly in comparison with wages and salaries – from 1993 to 2001 average wages have increased 3.7 fold, the average pension 4.6 fold, and the average old-age pension 4.8 fold (*Social Trends 2*, 2001) – many pensioners have great difficulty making ends meet. As in the Soviet period, then, it still makes sense to continue to work after retirement, provided one is fit enough to do so.

During the Soviet era the average old-age pension was about 80 per cent of average wages. Pensioners who continued to work were therefore relatively well off: not only did they have enough money to set aside for rainy days, but also to help their children who were setting up their own home. In other words in those days the older generation was often in the role of provider rather than recipient of support (Hansson, 2001).

The situation today is very different. Empirical studies have shown that despite the relative increase in pensions and the introduction of new subsidies that have helped slightly to improve the situation of pensioners, the majority still complain that their income in retirement is inadequate (Marksoo, 2000). The spending of older people is mainly dependent on the costs of food and housing, and on the household type in which they live. Elderly people who live alone can count on one source of income only, and they have no one with whom to share their compulsory costs of living. Lone pensioners who do not work therefore often have to change their consumption habits and make do with less food and a simple diet, to spend less on health and medicines, and possibly give up all 'extra' items of expenditure, such as books and newspapers, visiting friends and relatives and inviting people to see them (Table 10.3).

**Table 10.3 Where do people in different age groups cut costs %**

| I have often given up ...* | < 55 | 55+ |
|---|---|---|
| buying meat | 17 | 28 |
| buying fruit and vegetables | 14 | 20 |
| going to the theatre or pictures | 34 | 53 |
| buying books or magazines | 39 | 54 |
| inviting guests | 16 | 26 |
| visiting relatives and friends | 31 | 40 |

* Answers 'Not relevant' excluded.

*Source:* 'Estonia 98' population survey.

As well as having adverse health effects and increasing isolation, severe shortage can in older people give rise to a growing sense of injustice and by the same token increase alienation from society. Indeed the risk for social exclusion among elderly people has continued to remain high in Estonia (Kutsar and Trumm, 1999).

*Changes in attitudes:* In addition to changes in the labour market and economic well being, Estonia has seen considerable changes over the past few years in public opinion and attitudes towards older people, or the effect of '*attack,*' to use the concept of Hanley and Baikie (1984). During the Soviet era people who retired usually had a welfare party at their workplace in celebration of their 'well-earned leave:' there were flowers and gifts from the trade union and often a certificate of honour as well as a bonus from the employer. Pensioners also continued to be members of the trade union even when they did not continue to work. They were always invited to social events at the workplace and to parties and outings, and they could use the facilities at the workplace like all other workers. In other words, retirement did not make people feel excluded as work-based network ties and connections were kept alive for a long time.

Again, the situation today is very different. As well as losing their job and income, people retiring today also lose contact with the company they have served, many of which have folded in the process of economic restructuring ('*loss*'). The countryside provides a useful example. A kolkhoz was not only a production unit but it also provided certain social and cultural services for its workers (Alanen et al., 2001). Furthermore, it was a closely-knit community of co-workers. The collective farmers often knew one another very well, their children went to the same local school, and they often spent their leisure time together: collective farms were famous for their big New Year parties, harvest feasts, visits to the theatre organized by the trade union, etc. The liquidation of a collective farm resulted not only in the loss of a secure job and a stable income, but very often also in the break-up of all such ties. Increasing economic inequality that was based on restitution and privatization also affected interpersonal relationships in the countryside (Ibid.).

The first decade of independent Estonia has been characterized as a period of success stories for young people. It is still widely thought that older people continue to bear the Soviet ways and habits of their 'red past'[3] and anyone associated with collective farms is viewed as tainted. The new jobs created in the process of economic restructuring require different kinds of skills and competencies than those possessed by older people, especially those who were engaged in agriculture. The prevailing attitude in Estonia today is that people in their fifties or sixties (and sometimes even in their forties) are too old for retraining ('*restraint*'). Furthermore, local media have created an image of '*Auntie Maali*,' i.e. that of an elderly woman in the countryside who fails to understand the changes that are going on in Estonia and who is firmly opposed to Estonia joining the EU. Such stigmatisation of elderly people, who were very active and hopeful when statehood was restored in the early 1990s, has very much disillusioned older people and caused much bitterness and resentment. And as population surveys (e.g. 'Estonia 98' survey) indicate, the social activity of older age groups has been dramatically reduced as they have withdrawn 'into their cocoons,' as put by one of the farmers interviewed.

Popular wisdom has it that adaptation is particularly difficult for older age groups. Younger people who cannot find a new job that matches their qualifications, can usually go back into training and get new qualifications, or they may start up a business of their own. They are free to move to wherever jobs are available, which is rarely the case for those in retirement or pre-retirement age. Older people are also less willing to take risks, and in general major life changes are much harder for them to accept. In part this can be explained by reference to the notion of '*threat*' as described by Hanley and Baikie (1984), but there are also some quite objective reasons why it is hard for them to move: for example, many of them have privatised their flats and they often lack the money to buy a new place for themselves, some have allotments where they grow food crops, some of them keep domestic animals, etc.

### Increasing Regional Differences

Estonia of the 1990s may be characterized as a country of deep regional inequalities, both with respect to the labour market situation and with respect to opportunities more generally. Let us start with the significant differences in the age structure between different regions and counties. According to official statistics,

---

[3]  Even the tragedy of September 2001 when more than 70 people (average age 56) in Pärnu county died after drinking illegal alcohol that turned out to be methanol, was characterized by the political elite as a habit of the Soviet past and typical of elderly people.

the concentration of younger people is higher in the county centres and in the surroundings of major cities, especially Tallinn and the metropolitan area. At the same time, the mainly agricultural regions of southern and southeastern Estonia are characterized by a higher proportion of people of retirement and pre-retirement age (Table 10.4). Following the decollectivization of agriculture, these regions experienced the greatest loss of jobs, and today they are characterized both by a limited range and availability of jobs, as well as by the highest rates of long-term unemployment. This applies most particularly to the rural settlements of these regions (*Regional Statistics of Estonia*, 2000).

**Table 10.4** **Districts with the largest and smallest proportions of pension-age population (men 63+, women 58+) in 2000**

|  |  | Pension-aged people as per cent of total population | | |
|---|---|---|---|---|
|  |  | Total | Men | Women |
| Estonia | Total | 21.0 | 12.9 | 28.0 |
| Counties: |  |  |  |  |
| Harjumaa (incl. Tallinn) | Urban | 20.0 | 12.1 | 26.5 |
|  | Rural | 17.5 | 10.4 | 24.5 |
| ............... |  |  |  |  |
| Võrumaa | Urban | 21.5 | 12.8 | 28.7 |
|  | Rural | 25.3 | 16.0 | 34.0 |
| Põlvamaa | Urban | 17.3 | 11.0 | 22.6 |
|  | Rural | 26.0 | 16.2 | 35.2 |

*Source*: Regional Statistics of Estonia 2000.

It is important to emphasise that due to differences in life expectancy and the remaining differences in the official retirement age between women and men, the proportion of retirement-age women is considerably bigger than that of retirement-age men in virtually all regions.

Looking at the numbers in receipt of different types of pensions, we see that the share of persons on disability pension is relatively stable in Tallinn and in its surroundings, but rapidly increasing in the southeastern counties of Põlva and Võru (Table 10.5), which have very high unemployment levels.

For older people, remaining in active employment would provide a substantial addition to their meagre income. The greater concentration of investments that created new jobs mainly benefited those people who lived in the big cities, especially in Tallinn and its surroundings. In the rural southeast, where large parts of the population were engaged in agriculture or in the food processing industry during the Soviet era, the liquidation of collective farms created an extremely difficult situation (Alanen et al., 2001). The new jobs created in a particular settlement or region in the service sector or in family farming rarely compensated for the loss of jobs in agriculture or related industries. Indeed in the rural areas it was usually the older age groups who had to withdraw, or who were made to withdraw from the labour force.

**Table 10.5 Changes in the number of persons applying for and receiving disability pensions in selected regions**

| Region | Persons applying for the first time or receiving disability pensions | | | | | |
|---|---|---|---|---|---|---|
| | Number of applicants | | Persons declared disabled | | Per 10,000 inhabitants | |
| | 1991 | 1998 | 1991 | 1998 | 1991 | 1998 |
| Harjumaa | 2,892 | 2,863 | 2,056 | 2,367 | 34 | 44 |
| Tallinn | 2,439 | 2,278 | 1,759 | 1,872 | 36 | 45 |
| ......... | | | | | | |
| Võrumaa | 372 | 745 | 282 | 696 | 62 | 161 |
| Põlvamaa | 277 | 644 | 191 | 585 | 53 | 164 |

*Source:* Estonian Statistics (http://www2.ebs.ee/statistika1999/html).

The labour force surveys carried out by the Statistical Office of Estonia have revealed significant differences in the employment and unemployment rates between different regions of Estonia. Tallinn and its surroundings (Harju County) continue to enjoy the best situation, whereas the northeastern (i.e. Narva and its surroundings) and southeastern regions of the country have the most difficult situation (Table 10.6).

**Table 10.6 Employment status of the population aged 15–74 in three counties %**

| | In active employment | | | Unemployed | | |
|---|---|---|---|---|---|---|
| | Harju | Võru | Põlva | Harju | Võru | Põlva |
| 1997 | 63.2 | 49.6 | 50.3 | 8.5 | 11.8 | 12.7 |
| 1998 | 62.9 | 48.3 | 51.3 | 9.1 | 10.7 | 12.2 |
| 1999 | 61.8 | 45.2 | 43.1 | 10.2 | 13.0 | 21,3 |
| 2000 | 60.6 | 45.4 | 40.1 | 11.5 | 15.1 | 23.0 |
| Change 1997–2000 | –2.6 | –4.2 | –10.2 | +2.0 | +3.3 | +10.3 |

*Source:* Labour Force 2000, 2000, 222–223.

Reverting to people of retirement and pre-retirement age, the differences on the urban-rural dimension are also significant. According to official statistics, the employment rate for the urban population aged 50–74 in 2000 was 51 per cent for men and 37 per cent for women. For the rural population, the figures were 40 per cent and 28 per cent, respectively (*Labour Force*, 2000). Unemployment rates were also considerably higher in rural areas. Several studies have pointed out that official statistics in the 1990s did not in fact provide a true reflection of rural unemployment, as the long-term unemployed who were no longer entitled to benefits and who had lost all hope of finding a job, no longer bothered to register as jobseekers (Hansson, 1999).

As was mentioned earlier, the situation of retired pensioners who lived alone, i.e. who had no one with whom to share their costs of living, have had the hardest time making ends meet (for more details, see Marksoo, 2000). As can be seen in Table 10.7, the proportion of elderly women living alone is highest in Estonia's cities, the proportion of men in the countryside.

**Table 10.7    Living arrangements of the population aged 55+ in urban and rural settlements %[4]**

| Living ... | Urban | | Rural | |
|---|---|---|---|---|
| | Men | Women | Men | Women |
| Alone | 7 | 34 | 23 | 25 |
| with spouse/partner | 86 | 49 | 68 | 57 |
| with an adult child or children | 28 | 29 | 18 | 27 |
| with other relatives | 8 | 13 | 12 | 12 |

*Source:* 'Estonia 98' population survey.

Living conditions and spending patterns differ quite widely between urban and rural areas. According to the 'Estonia 98' survey, only about 10–12 per cent of people in cities live in flats that have no modern conveniences, i.e. central heating and running hot water. Almost two-thirds do have central heating. Bearing in mind that in urban areas one in three women aged 55+ lives alone, and that housing costs for a centrally heated flat with running hot water are quite high, it is clear that without an additional income, benefits or community aid these people could hardly manage. In urban areas the living conditions of younger and older age groups are quite similar, whereas in rural areas there are significant differences (Table 10.8).

**Table 10.8   Living conditions in urban and rural areas**

| Flat/house with ... | Urban | | Rural | |
|---|---|---|---|---|
| | < 55 | 55+ | < 55 | 55+ |
| central heating, running water and hot water | 63 | 62 | 39 | 19 |
| stove heating and running water | 27 | 26 | 42 | 45 |
| stove heating and no running water | 10 | 12 | 19 | 36 |

*Source:* 'Estonia 98' population survey.

---

[4]   Some columns do not add up to 100 per cent due to the different composition of households.

On the one hand, living in a stove-heated flat or house, which is more common in the countryside, reduces housing costs. On the other hand, for older people it is quite a major task to get the firewood they need – unless they have a good support network, which provides unpaid assistance whenever they need it. If, however, they have to rely on paid assistance, there is hardly any difference in the costs of living a modern flat in a city and an old house with no conveniences in the countryside.

Elderly people living in rural regions have the additional problem of poor access to services. Public transport is not always available to take them from a remote village to the parish centre or county town. This greatly increases the costs of going to see a doctor, to go to the post office, etc. The city of Tallinn has provided subsidised city transport for all pensioners, and free transport for those over 65. Unemployed persons of pre-retirement age who live in remote villages and who have to visit employment offices, which as a rule are situated in town centres, have often said the reason why they have not registered as unemployed or applied for benefits is precisely the difficulty of getting to the unemployment office.

The policies adopted by local authorities with regard to paying out subsistence benefits or introducing additional support programmes depend upon the monies available to them, i.e. upon local government revenues through personal income tax. The latter in turn is dependent upon the number of people in active employment and upon the average income in the municipality. Official statistics show that county differences in disposable income per household member are increasing. The highest figure is currently recorded for Harju county at EEK 2,771 (about EUR 175), i.e. 27 per cent above the national average. In the southeastern region the figure was about 50 per cent of the disposable income per household member in Harju county, and about two-thirds of the national average (*Regional Statistics of Estonia*, 2000).

As regards the average personal income tax per working-aged person, there are marked differences between the more and the less advanced regions in the country (Table 10.9). Although the less successful local municipalities do get some subsidies from the state budget, these regions are usually unable to provide sufficient support to all elderly people who would need it, whether in the form of direct financial support or services. The growth of poverty and the lack of opportunities in one region is not a local problem, but it affects both the individuals concerned and the whole of society (marginalization of certain population groups, increasing crime, etc.).

**Table 10.9 Advances and less advanced municipalities in Estonia in 1997**

| County/ local government unit | Personal income tax revenues per one working-aged person (Estonian average = 100) | Rate of exclusion from work[5] (1 January 1998) |
|---|---|---|
| Harjumaa | | |
|   Tallinn* | 134.6 | 8.8 |
|   Saku* | 134.6 | 6.5 |
|   ... | | |
|   Kõue** | 72.1 | 11.1 |
|   ...... | | |
| Põlvamaa | | |
|   Põlva (county capital)* | 104.4 | 13.1 |
|   Põlva (parish) | 76.2 | 13.1 |
|   ..... | | |
|   Kanepi | 56 | 24.4 |
|   Mikitamäe** | 49.1 | 23.7 |
|   ..... | | |
| Võrumaa | | |
|   Võru (county capital)* | 85.0 | 14.1 |
|   ..... | | |
|   Võru (parish) | 74.7 | 14.0 |
|   ..... | | |
|   Meremäe** | 43.2 | 25.6 |

\* The most advanced municipality in the county.
\*\* The least advanced municipality in the county.

*Source:* www.undp.ee/poverty/en/VII.html.

## Case Studies in Põlva County

Põlva county, especially its rural areas, is characterized by the relatively large proportion of people of retirement and pre-retirement age. A closer examination of Põlva, which according to official labour statistics is one the most deprived counties in Estonia, also reveals significant differences in the employment rates between urban and rural areas. The employment rate is considerably lower in rural (34 per cent) than in urban settlements (57 per cent), and the unemployment rate considerably higher (27 per cent and 16 per cent, respectively) (*Labour Force*, 2000, 53).

Official statistics show that in Põlva county, both average hourly wages and disposable income per household member are the lowest in Estonia. Although spending on food and housing are also the lowest in Põlva, they still account for

---

5    The index 'exclusion from work' describes the number of non-working and subsistence farming persons of working age within the population capable of work. Students, women on maternity leave and pensioners are not included in the figures.

the lion's share of disposable income (51 per cent). For example, in Tallinn and in Harju county the figure is 44 per cent, while the national average is 47 per cent (*Regional Statistics of Estonia*, 2000). While pensioners in rural areas usually manage to make ends meet by making personal sacrifices in consumption and by growing part of their own food, people of pre-retirement age who are out of work, especially those who are not entitled to unemployment benefit, are in an extremely difficult situation. As was mentioned earlier, most local governments in Põlva county have very low personal income tax revenues, which further complicates the situation. It is obviously extremely hard for any local government in this situation to pay out additional subsidies or to introduce special services or support programmes for those in need.

In the former chapters we have described the financial situation of older people in Estonia as it appears in the light of different statistical sources. In this last part of our article we will be trying to illustrate how elderly people themselves experience their situation. We are particularly interested in how they manage in their everyday life, i.e. in what kinds of cultural and social resources they possess that have helped them pull through the financial problems of the transition period.

For this purpose we draw upon two interviews we conducted in 1995 during the research carried out in the Kanepi area (see Alanen et al., 2001). These interviews represent two different kinds of cases out of a number interviews we carried out at the time. One of the interviews is with a late midlife woman, the other with an elderly man. Both were widows and living on a pension, but they had very different life histories and resources for coping.

Our qualitative excursion is primarily aimed at illustrating the perspective of the individuals concerned, their life histories and the way those histories are connected to their present-day social situation. The life histories of these people form the stories of their life, i.e. how they discursively want to present it to an outsider. At the same time, though, it also forms the basis upon which the researcher can seek to understand how personal histories influence the present day, in the form of social and cultural resources that have taken shape during the life course. We are not suggesting that the structural changes taking place in society have no influence upon the economic position of individuals, but merely trying to illustrate how past personal history – or the way that people experience it – continues to influence their lives.

The social ties and resources that are created during the individual's lifetime may or may not remain unaffected by profound societal change, but it is to these resources that people will most typically resort when faced with problems. However, in the wake of deepgoing societal changes it is often difficult for people to build up new resources that complement or compensate those they have lost (e.g. Hansson, 2001). Cultural resources are also established and accumulated during the lifecourse, but their value as a resource may be drastically affected by societal transitions. It should be stressed, though, that in building up their coping systems people also draw upon the past and that they may experience certain strengths of theirs as important and meaningful on account of their life style and the coping systems they have formed during their lives (in Bourdieu's terms, their habitus; see Bourdieu, 1986). It is this accumulation and de-accumulation of social and cultural resources that we are trying to uncover through these two cases.

## Case 1: A Late Midlife Woman

This woman lived in one of the parish centres of Põlva county. In 1992 she was involved in a road accident and since then has been disabled, having difficulties in walking and generally moving about. She had had an operation to fit an artificial joint, but rejection mechanisms were causing continuing infections.

She was born together with her twin sister into a farmer's family in 1940. Soon after their birth her father had become disabled, leaving the three women in the family – two of them young children – to take care of the farm. When agriculture was collectivized, all the cattle in the kolkhoz were kept in the family's cattleshed, and it was the twins who had to look after all 13–14 heads of cattle from age nine onwards. After completing secondary school, she moved in 1962 to Põlva where she three years later married a tractor driver. In Põlva, she worked as an accountant for a cooperative. She gave birth to two daughters. In 1976 the family moved back to their home parish and she got a cook's job at a canteen. In 1984 in she got a divorce and later married a musician. During this second marriage she did not work outside the home. In 1993 she was widowed.

At the time of the interview her financial situation was not at all sound. She was not entitled to a widow's pension since her husband had not had a job giving entitlement to any pension. She did still have her own old-age pension, which at the time was EEK 640 (about EUR 41) a month. She lived in a two-room apartment that had been privatised by her other daughter, and she did not have to pay any rent. Her refrigerator was broken, but she did not have enough money to have it repaired: that would have cost her about two months' pensions).

Her 'active' social resources comprised her daughter who had privatised her apartment and a few neighbours. This daughter, together with her unemployed husband, brought her firewood and potatoes, and they helped her with various practical matters. She mentioned that her other daughter and her family were doing well, but made no other reference to her at all: the role of this daughter as a potential social resource may therefore not have materialised at all. Her neighbours helped her with the shopping, but she said she did not want to bother them too often. She wanted to manage on her own as far as possible, though because of her poor health she did need at least some outside help.

Her cultural resources came from two sources: her late husband's status and her own handicraft skills. Her second husband had been a nationally known musician, which for the family seemed to bring considerable prestige and thus serve as a continuing cultural (symbolic) resource, even after his death. Her handicraft – tablecloths, scarves, pullovers, mufflers, etc. – seemed to constitute another cultural resource. Primarily it gave her the joy of being able to produce something beautiful with her own hands. These skills she had learned from her mother during her childhood. Financially the work she did was insignificant since there was no organized handicraft sales or resale system in the neighbourhood. These cultural resources gave her a sense of personal and social meaning that helped her overcome her disability, to find her existing capacities, to maintain her personal autonomy, and to find meaning in her life.

The most difficult time of her life had been her childhood and youth. When her father was disabled, that meant all three women in the family had had to work extremely hard. The happiest time of her life, then, was her second marriage: that had provided her with a significant cultural resource that lasted even beyond the death of her husband and even after her own disability. Although her financial situation was somewhat precarious, she had a very optimistic view of the future. All in all her (experienced) life story was a continuous success story, in spite of the various adversities she had encountered.

## Case 2: An Old Man

This old man – he was probably around 75 at the time of the interview – lived in a village six kilometres from the parish centre. He had rented a flat in an old house owned by the municipality (a kitchen and a wall heated room). In the winter he lived with one of his daughters in the parish centre, in a two-room flat.

The life history of this man was colourful indeed. We were unable to establish when exactly he had been born, but during the Second World War he had been called up by the German Army. Refusing to respond, he was sent off to a prisoners' camp. Upon the arrival of the Red Army at the end of the war, he escaped from the camp. After the war he worked as a lumberman in northeastern Estonia, where he met his future wife. He was injured in an accident and had to change jobs. The family therefore moved to the Võru region where he started to work as a herdsman. He describes this period as the happiest time of his life, since the work he did was 'in the open air, the shepherd's dogs did the job and he felt like a bird.' He also thought the Soviet era had been the easiest part of his life. Everything was cheaper in those days (he even remembered the prices of individual items), which meant he had no difficulty managing on his pension.

His wife had been a milkmaid and she had died of cancer just three months before our interview. At present his family comprised five adult children, two daughters and three sons. They all had their own families. His daughters lived close by, his sons had moved farther away (the tractor driver to Põlva, the policeman to Tartu and the third of whom the old man did not know much, to Pärnu).

This man lived a meagre existence indeed: he hardly managed to make ends meet with his monthly pension of EEK 490 (about EUR 31). The money was not enough to pay the rent, but he did pay for his electricity. He had a plot where he grew potatoes, and since his neighbour had a cow, he was able to buy 'better milk cheaper' than from a shop. His daily diet usually consisted of potatoes with salt, milk and bread, which had to be good – even though this was an expensive 'indulgence'. He also smoked, and when he got his monthly pension he always set aside a certain sum of money for cigarettes. He spent almost all of the the rest of his money on food.

His social resources consisted mainly of his two daughters, two of his sons and the neighbour who had a cow. Together, they made it possible for him to cope with his daily routines and the maintenance of a summer cottage in Karte – especially after his wife's death – and to keep his monthly budget in a manageable balance. His younger daughter (a former milkmaid in the kolkhoz, now unemployed) lived in Karste near him, and earned extra money by picking berries for export to Finland. His third son was a potential social resource, though at the time of the interview he did not know much about what he was doing ('He lives with some woman, I don't know whether they have any children or not. He has not been here for a long time'). All his other social ties seemed to be rather weak because of the kind of people around him: for instance, all the women he knew were either 'too young or drank too much.'

His cultural resources seemed to derive from his past, from his adventures as a young man (a period which seemed to give him something exciting and 'heroic' to remember) and from the financially trouble-free years during the Soviet era. The present day was the worst time of his life, possibly partly because of the recent death of his wife, partly because of his poor income. Although the cultural heritage (or past) he had did not apparently have much symbolic value in the society in which he lived at present, he did have something in his family past of which he could be proud: one of his daughters had graduated from college ('She did it all by herself'.) In family generational terms: his nearest kin was moving upwards in terms of cultural resources and was facing the challenges of changing society in a better position than he ever had. This was something he could be proud of.

One type of cultural resource that he also seemed to possess and that was based on his habitus, was his ability to organize his everyday life. Especially his use of money, his diet and smoking arrangements were such that his daily routines were very well organized in the summertime. This kind of systematism makes daily life easier, but it also tends to be rather vulnerable to outside changes and to changes in one's health and mental condition. It may also be read as a search for personal autonomy, as it also appeared in his adventurous life story (he was a man who always wanted to go his own way).

## Conclusions

Studies of ageing have shown how quality of life in old age is dependent on several factors: the quality of the social environment, standard of living, personal autonomy and privacy, family and other social ties, psychological well-being and daily activities (e.g. Hughes, 1990). In this article we have examined some of these factors in the light of both statistical data and two qualitative interviews.

Our statistical data have clearly shown how the quality of the environment and the standard of living among rural elderly people have changed during and after the transitional period in Estonia. Older people in Estonia have encountered both *'losses'* and *'attacks'* (see Hanley and Baikie, 1984). The external circumstances on which people traditionally relied, have changed for the worse (losses) and produced discomfort (attacks). Financially the elderly were in quite a precarious situation, given their small pensions and the difficulty to mixing paid work and retirement. The former kolkhoz system that had evened out the effects of low pensions by providing daily support, had collapsed and left elderly people to cope on their own or with the help of their children. Older people also had difficulties with the changing housing situation and rising housing costs.

Many elderly people in Estonia were dependent on the help of their children or other family members.Some of our interviewees, for instance, said they had to live at least part of the year with their children (usually during the winters), which obviously was a major threat to – their personal autonomy and privacy. The elderly man we interviewed seemed to maintain his privacy in his memories and partly in the system of his daily routines; the elderly woman was more fortunate in this respect in that she could still live on her own. Nonetheless both were quite vulnerable because of their disabilities.

The family had clearly gained in importance as a source of financial help and support with daily routines and household chores (see also Alanen et al., 2001). For elderly people these were essential resources, but we were unable to establish to what extent this burdened their children and how long they can continue to offer this support. However, the family networks rarely comprise all the children: some of them live farther away, possibly as potential resources. The elderly have to rely and trust on the keenness of kin ties.

Older people in Estonia seemed to had found different ways of coping with their difficulties. As our two cases show, they very much lean on their former life resources. Some older people are capable of carving out a meaningful existence on the strength of their cultural resources (as in the case of the elderly woman we interviewed), but also those who are deprived and feel lonely (the case of the old man). Our argument then is that it is the interplay of all the various life quality factors that determines the level of satisfaction and general life management during transitional times.

Table 10.10 presents a summary of our analysis and illustrates those *threats* that in our diagnosis represent the most important factors in the welfare of present day elderly people in Estonia. The summary comprises factors at both the societal and individual level. These factors are closely interwoven, having a mutual effect upon one another. The summary shows how difficult a task it will be for Estonian society to respond appropriately to the needs of elderly people. Difficult, but not impossible.

**Table 10.10 Summary: Threats against the welfare of elderly people in present-day Estonia**

1) Government and politics:
   - Small old-age pensions
   - Pensions are not linked to the size or structure of the pensioner's family
   - Scarce opportunities for paid work after retirement
   - Responsibility for old age welfare has been shifted from the public sector to individuals and their families
   - Access to services is poor in rural areas which is where many old people live

2) Civil society and institutional structure:
   - Civil society is still rather underdeveloped after the liquidation of collective farms
   - The liquidation of collective farms created a social and cultural vacuum since civil society did not develop at the same pace
   - Kolkhozes provided social and cultural services and maintained contact with people who retired – decollectivization destroyed such relations

3) Economy:
   - Cash has gained increasing importance: pensions are no longer enough to cover costs of living
   - There are marked regional differences in economic welfare: it is difficult for elderly people who have privatized their flats or who own land, domestic animals, etc. to move

4) Housing:
   - Housing is costly, especially in cities
   - Inconvenient heating systems and water supply

5) Family:
   - Family members (children's families) do not live in the same area
   - Family members themselves have financial problems because of unemployment, the need to support dependent children, etc.

6) Neighbours:
   - Neighbours do not have the time, money or other means (e.g. suitable vehicles) to support others
   - Absence of neighbours one knows well enough to ask for help

7) Life-course factors:
   - Cultural resources formed during the life course are losing their value in changing society
   - Social ties (to others than family members) tend to lessen and become more dependent on physical distance with advancing age

8) Elderly people themselves:
   - Poor health and disabilities make older people more dependent on others, undermining their personal autonomy
   - Older people often want to manage on their own for as long as possible, without leaning on their relatives or neighbours even when that would be necessary (need to maintain personal privacy and pride)

# References

Alanen, I., Nikula, J., Põder, H. and Ruutsoo, R. (2001), *Decollectivisation, Destruction and Disillusionment. Community Study in Southern Estonia*, Ashgate, Aldershot.

Bond, J. and Coleman, P. (eds) (1990), *Ageing in Society. An Introduction to Social Gerontology*, Sage, London.

Bourdieu, P. (1986), 'The Forms of Capital,' in J. G. Richardson (ed.), *Handbook of Theory and Research for the Sociology of Education*, Greenwood Press, New York.

Coleman, P. (1990), 'Ageing into the Twenty-First Century,' in J. Bond and P. Coleman (eds), *Ageing in Society. An Introduction to Social Gerontology*, Sage, London.

*Estonian Human Development Report 1998* (1998), UNDP, Tallinn.

Hanley, I. and Baikie, F. (1984), 'Understanding and Treating Depression in the Elderly,' in I. Hanley and J. Hodge (eds), *Psychological Approach to the Care of the Elderly*, Croom Helm, London, pp. 213–236.

Hansson, L. (1999), 'Töö,' in A. Narusk (ed.), *Argielu Eestis 1990ndatel aastatel*, RASI, Tallinn, pp. 11–27.

Hansson, L. (2001), *Networks Matter. The Role of Informal Social Networks in the Period of Socio-Economic Reforms of the 1990s in Estonia*, Jyväskylä Studies No 181 in Education, Psychology and Social Research. University of Jyväskylä, Jyväskylä.

Hughes, B. (1990), 'Quality of Life,' in S. M. Peace (ed.), *Researching Social Gerontology. Concepts, Methods and Issues*, Sage, London.

Katus, K., Puur, A. and Põldma, A. (2002), *Eesti põlvkondlik rahvastikuareng*, Eesti Kõrgkoolidevaheline Demouuringute Keskus, Tallinn.

Kõgel, H. (2001), 'Pension Insurance,' in *Social Trends 2*, Statistical Office of Estonia, Tallinn.

Kutsar, D. and Trumm, A. (1999), *Vaesuse leevendamisest Eestis*, Tartu Ülikooli Kirjastus, Tartu.

*Labour Force 2000* (2000), Statistical Office of Estonia, Tallinn.

Laidmäe, V.-I. (2001), 'Vanemaealised ja eluga toimetulek [Older People and Coping with Life],' in L. Hansson (ed.), *Mitte ainult võitjatest*, Teaduste Akadeemia Kirjastus, Tallinn, pp. 67–91.

Marksoo, Ü. (2000), *Living Conditions Study in Estonia 1999. Baseline Report*, Tartu Ülikooli Kirjastus, Tartu.

*Regional Statistics of Estonia 2000* (2000), Statistical Office of Estonia, Tallinn.

*Social Trends.* (1999), Statistical Office of Estonia, Tallinn.

*Social Trends 2* (2001), Statistical Office of Estonia, Tallinn.

*Statistika Aastaraamat 1990* (1990), Tallinn.

*Statistical Yearbook of Estonia 1999* (1999), Statistical Office of Estonia, Tallinn.

*Statistical Yearbook of Estonia 2000* (2000), Statistical Office of Estonia, Tallinn.

# Chapter 11

# The Significance
# of the Research Results

### Ilkka Alanen

The transformation of the post-socialist system is a comprehensive structural changeover from a Soviet-type society to a capitalist market economy. This process involves different problems at different stages. Furthermore, none of the constituent processes in this changeover, such as the privatization of companies and their integration into the capitalist system, can be taken too far from their broader context, i.e. the formal and informal institutions supporting their privatization and integration (see Chapter 2 by Alanen, as well as Chapters 4 and 5 by Nikula). Nikula (Chapter 4) very much emphasizes the importance of governmental regulation systems as highlighted in Smith's regulation theory, adapting this to the conditions of post-socialist transition. the existing structures of institutional regulation (and their immaturity or absence) impose a certain mode of adaptation and development for the privatized companies as the external conditions of their management.

The stages in the process of transformation also involve various contradictions. This means that the nature of original accumulation at the first stage – that of privatization – frequently undermines the moral foundation (Nikula, Chapters 4 and 5) that the companies nonetheless need (e.g. in the form of generalized trust) in order to operate efficiently (Nikula, Chapter 4). The development (or inadequacy) of this moral foundation is reflected in the everyday problems of all the actors involved as well as in the ability of the government to act as a neutral third party (Alanen, Chapter 2; Ruutsoo, Chapter 3; Nikula, Chapters 4 and 5; Granberg, Chapter 6).

From a rural perspective, the conditions of adaptation are somewhat different for agricultural and non-agricultural enterprises (in this volume mainly companies in the industrial sector). For both types of industries, the support provided by formal institutions is essential. They have also found that managing a successful business in a capitalist society requires different skills to those that were important in Soviet society. In addition, the products of these industrial companies must meet consumer needs, and entrepreneurs need to know how to market them. None of these skills were included in the legacy inherited from the Soviet era. As far as the major non-agricultural companies in countryside are concerned, their adaptation to the capitalist market economy has depended mainly on foreign investments and the new technology thus acquired, as well as on their ability to rèstructure their production and work organization. However, foreign companies have only been interested in certain types of businesses, especially those with large market shares or in monopoly positions (Nikula, Chapter 4). On the other hand, weak consumer demand on the domestic marketplace during the early stages of transition highlighted the importance of western markets for small industrial companies. Companies that have managed to establish direct contacts with clients in the west or become subcontractors for western companies have generally done very well and are technologically the most advanced. However, the number of such companies is quite limited, and in rural areas they are mainly found in the wood processing and carpentry businesses (Nikula, Chapter 5). Most of the remaining small-scale industries and other enterprises have followed a subsistence strategy, either consciously or because people have had no other source of livelihood: lacking the ability to reorganize themselves technologically, these companies easily find themselves in a vicious circle of poor wages and incompetent labour (Nikula, Chapter 5). The industrial transition has in fact led to de-industrialization, redundancies and wage arrears far more often than to technological restructuring, increased effectiveness, new jobs and higher wages (Nikula, Chapter 4 and 5). Since agricultural companies were unable to find new resources from anywhere else during decollectivization, the key question with regard to the successful transformation of the agricultural system was how far the privatized companies were able to make use, at least during the start-up phase, of the technology and know-how they had inherited from the so-called socialist economy (Alanen, Chapter 2). Since the resources of the Soviet system had been developed through large-scale production, successful transition would require adapting large agricultural companies to the conditions of capitalist market economy. However, not even the big new companies would succeed in it, if their management prerequisites were inadequate for some other reason (Alanen, Chapter 2). Although the sphere of governmental policies has expanded and new institutions have been established, this has not necessarily led to better conditions for company management (Stanikunas, Kriciukaitiene and Zemeckis, Chapter 7). The majority of the agricultural labour force are continuing the private household plot tradition by producing food mainly for the needs of the immediate family. The problems associated with household plot farming have more to do with social policy than with business economy (Stanikunas, Kriciukaitiene and Zemeckis, Chapter 7; Alanen, Chapter 2).

There is indeed a great need for new types of companies. Much hope has been pinned upon rural tourism, even though it has still not taken off in any big way. In this sector too, companies are often established quite casually, without any rigorous planning (Granberg, Chapter 6). Finnish entrepreneurs in rural tourism, for instance, can rely on much stronger support from formal institutions than their Baltic counterparts (Granberg, Chapter 6). On the whole, the establishment of adequate institutional regulation and support systems and the development of special regional policy programmes is the most central and topical task in all the Baltic countries (Nikula, Chapters 4 and 5).

The importance of the initial conditions as compared to actual transition politics has been extensively discussed in the transition debate. The research results in this volume indicate that the decollectivization policy pursued has clearly been decisive from the point of view of the final outcome – even though that outcome, even when successful, does not necessarily meet the political goals originally set out. The success of Estonian decollectivization policy in rural areas, when compared to the two other Baltic countries, was not due to the adoption of the family farming system as its target, but above all to the fact that the local people who put the reform into practice were able to preserve large-scale production. Considering the level of development achieved in Soviet agriculture and the initial conditions in each of the Baltic countries, all of them should have been able to do better. This applies to Latvia and Lithuania in particular (Alanen, Chapter 2). However, other aspects of the initial conditions also stand out in the final outcome of decollectivization, especially the relationship between small-scale and large-scale production (Alanen, Chapter 2). The outcome probably also reflects differences in the social preparedness of the three countries (Alanen, Chapter 1 and Ruutsoo, Chapter 3), as well as special features that go even further back in time (Ruutsoo, Chapter 3). The local conditions that vary from one country or region to another should have been given closer attention in developing the decollectivization policy of rural areas (Alanen, Chapter 2) and the business economy in general (Nikula, Chapters 4 and 5).

The failure of transition policies is also reflected in the research findings in the shape of anomie among the rural population, poor work ethics, administrative inefficiency and corruption (Nikula, Chapter 5), the exclusion of large numbers from working life (Alanen, Chapter 2; Kämäräinen, Chapter 8; Hansson and Marin, Chapter 10), subsistence production (Stanikunas, Kriciukaitiene and Zemeckis, Chapter 7; Alanen, Chapter 2), and unemployment and poverty (Kämäräinen, Chapter 8, Tisenkopfs and Sumane, Chapter 9; Alanen, Chapter 2). Many of these problems, and long-term unemployment and poverty in particular, are extremely difficult to eradicate once they have set in (Tisenkopfs and Sumane, Chapter 9).

Older people in the Baltic countries have been forced to adapt to various radical social changes without having any say on the course of events (Hansson and Marin, Chapter 10; Ruutsoo, Chapter 3). The social structure of rural society has been reshaped both by collectivization during Stalin's regime and decollectivization that followed with independence. People experienced both of these processes primarily as external constraints. The research results, however, suggest that by letting people participate in decision-making and implementation, the final outcome will be economically sounder than any outcome based on top-down politics (Alanen, Chapter 2). Activating civil society by supporting NGOs is beneficial as well. Relatively small amounts of money have enabled people to get to grips with social problems at the local level (Tisenkopfs and Sumane, Chapter 9) and, as a by-product, they may also have strengthened the normative foundation of the local community that was weakened during the transition.

# Annex Tables and Figures

## Annex Table A-1 Some transition time series in agriculture

### A. Estonia

|      | 1. GAO | 2. GDP | 3. Per cent of GDP | 4. Employees in Agriculture | 5. Employees in Agriculture (%) | 6. Corporate as % of GAO | 7. PSE | 8. Inf. |
|------|--------|--------|--------|--------|--------|--------|--------|--------|
| 1990 | 100 | 100 | – | 100 | 16.6 | 76 | +71 | 23 |
| 1991 | 94 | 86 | – | 94 | 16.0 | 71 | +59' | 211 |
| 1992 | 76 | 87 | 10.8 | 84 | 13.0 | 65 | –97 | 1,076 |
| 1993 | 67 | 61 | 9.3 | 67 | 11.0 | 54 | –30 | 90 |
| 1994 | 58 | 60 | 9.4 | 56 | 8.5 | 49 | –10 | 48 |
| 1995 | 58 | 63 | 7.8 | 41 | 8.1 | 46 | 0 | 29 |
| 1996 | 55 | 66 | 7.5 | 38 | 8.1 | 46 | +7 | 23 |
| 1997 | 54 | 72 | 7.0 | 33 | 6.9 | 46 | +5 | 12 |
| 1998 | 51 | 75 | 6.5 | 32 | 6.8 | 47 | +16 | 8 |
| 1999 | 47 | 75 | 6.1 | 28 | 6.2 | 46 | – | 3 |
| 2000 | 46 | 80 | 5.5 | – | – | 47 | – | 4 |
| 2001 | – | 84 | 5.2 | – | – | – | – | 6 |

### B. Latvia

|      | 1. GAO | 2. GDP | 3. Per cent of GDP | 4. Employees in Agriculture | 5. Employees in Agriculture (%) | 6. Corporate as % of GAO | 7. PSE | 8. Inf. |
|------|--------|--------|--------|--------|--------|--------|--------|--------|
| 1990 | 100 | 100 | 21.1 | 100 | 15.5 | 72 | +76 | 11 |
| 1991 | 96 | 90 | 21.9 | 101 | 15.7 | 66 | +83 | 172 |
| 1992 | 81 | 58 | 17.2 | 114 | 18.5 | 48 | –110 | 951 |
| 1993 | 63 | 50 | 11.7 | 102 | 17.9 | 35 | –39 | 109 |
| 1994 | 50 | 50 | 7.5 | 96 | 17.3 | 27 | +7 | 36 |
| 1995 | 47 | 50 | 9.9 | 80 | 16.6 | 24 | +4 | 25 |
| 1996 | 44 | 51 | 8.2 | 77 | 13.5 | 24 | +3 | 18 |
| 1997 | 51 | 56 | 5.6 | 79 | 16.6 | 20 | +4 | 8 |
| 1998 | 45 | 58 | 4.4 | 75 | 15.7 | 21 | +10 | 5 |
| 1999 | 40 | 58 | 4.3 | 72 | 15.0 | 23 | – | 2 |
| 2000 | 41 | 62 | 4.9 | – | – | – | – | 3 |
| 2001 | 41 | 67 | 4.7 | – | – | – | – | 3 |

C. Lithuania

|      | 1. GAO | 2. GDP | 3. Per cent of GDP | 4. Employees in Agriculture | 5. Employees in Agriculture (%) | 6. Corporate as % of GAO | 7. PSE | 8. Inf. |
|------|--------|--------|--------------------|-----------------------------|--------------------------------|--------------------------|--------|---------|
| 1990 | 100    | 100    | –                  | 100                         | 17.6                           | 68                       | +70    | 4       |
| 1991 | 95     | 94     | –                  | –                           | –                              | 59                       | -262   | 225     |
| 1992 | 73     | 74     | 11.6               | 109                         | 19.5                           | 42                       | -124   | 1,021   |
| 1993 | 68     | 62     | 11.0               | 121                         | 22.4                           | 31                       | -37    | 410     |
| 1994 | 54     | 56     | 10.1               | 119                         | 23.3                           | 36                       | -15    | 72      |
| 1995 | 61     | 58     | 10.7               | 118                         | 23.7                           | 31                       | 0      | 40      |
| 1996 | 69     | 61     | 11.1               | 121                         | 24.1                           | 25                       | +5     | 25      |
| 1997 | 76     | 65     | 10.7               | 110                         | 21.7                           | 24                       | +7     | 9       |
| 1998 | 72     | 69     | 9.1                | 108                         | 21.4                           | 21                       | +14    | 5       |
| 1999 | 62     | 66     | 7.5                | 100                         | 20.1                           | 19                       | –      | 1       |
| 2000 | 65     | 68     | 7.0                | –                           | –                              | 21                       | –      | 1       |
| 2001 | 60     | 72     | 6.7                | –                           | –                              | 21                       | –      | 1       |

* Lithuanian statistics include the population working in forestry, but their share is marginal.

Sources and explanations:
1. GAO = Gross Agricultural Output (Index 1990=100)
*Sources:* SYEs.
2. GDP = Gross Domestic Product (Index 1990=100)
*Source:* Lerman, 2001c.
3. Per cent of GDP = Share of agriculture in GDP
*Sources:* SYEs; SYLas; SYLis; EBRD, 2002.
4. Employees in agriculture (Index 1990=100)
*Sources:* SYEs; SYLas; SYLis; WB, 1992; WB, 1993b; WB, 1993c.
5. Employees in Agriculture (%) = Agricultural employees as a proportion of total employed population
*Sources:* SYEs; SYLas; SYLis; WB, 1993d.
6. Corporate in GAO (%) = Corporate farming as a proportion of GAO (%)
*Sources:* SYEs; SYLas; Poviliunas and Batuleviciute, 1997 (years 1990–1995); SYLis (years 1996–1999).
7. PSE=Producer Support Estimate (percentage of value transfers to agricultural producers)
*Source:* OECD, 2000.
8. Inf. = Inflation
*Source:* EBRD, 2002.

## Annex Table A-2 Time series on livestock farming and milk production

A. Estonia

|      | 1. Livestock as % of GAO | 2. Milk index | 3. Cow index | 4. Cows on corporate farms index | 5. Livestock production on corporate farms (%) | 6. Cows on individual farms index | 7. Cows on corporate farms (%) | 8. Livestock production on individual farms (%) |
|------|------|------|------|------|------|------|------|------|
| 1990 | 66 | 100 | 100 | 100 | 71 | 100 | 84 | 48 |
| 1991 | 62 | 91 | – | – | 69 | – | 84 | 43 |
| 1992 | 60 | 76 | 94 | 90 | 67 | 114 | 80 | 37 |
| 1993 | 57 | 67 | 90 | 80 | 59 | 145 | 75 | 37 |
| 1994 | 57 | 64 | 81 | 61 | 70 | 186 | 63 | 37 |
| 1995 | 52 | 59 | 75 | 56 | 70 | 176 | 63 | 38 |
| 1996 | 51 | 56 | 66 | 48 | 68 | 162 | 61 | 39 |
| 1997 | 52 | 59 | 61 | 43 | 68 | 155 | 60 | 40 |
| 1998 | 56 | 60 | 60 | 42 | 72 | 156 | 58 | 41 |
| 1999 | 56 | 52 | 57 | 41 | 76 | 141 | 60 | 39 |
| 2000 | 57 | 52 | 49 | 35 | 75 | 127 | 59 | 44 |
| 2001 | 57 | 57 | 46 | 32 | – | 49 | 59 | – |

B. Latvia

|      | 1. Livestock as % of GAO | 2. Milk index | 3. Cow index | 4. Cows on corporate farms index | 5. Livestock production on corporate farms (%) | 6. Cows on individual farms index | 7. Cows on corporate farms (%) | 8. Livestock production on individual farms (%) |
|------|------|------|------|------|------|------|------|------|
| 1990 | 66 | 100 | 100 | 100 | – | 100 | 70 | – |
| 1991 | 62 | 78 | 99 | 92 | – | 118 | 65 | – |
| 1992 | 56 | 61 | 90 | 66 | – | 147 | 51 | – |
| 1993 | 56 | 53 | 66 | 29 | – | 152 | 31 | – |
| 1994 | 50 | 50 | 58 | 20 | – | 149 | 24 | – |
| 1995 | 54 | 55 | 55 | 16 | – | 140 | 20 | – |
| 1996 | 45 | 49 | 52 | 14 | – | 140 | 29 | – |
| 1997 | 46 | 50 | 51 | – | – | – | – | – |
| 1998 | 50 | 42 | 45 | – | – | – | – | – |
| 1999 | 45 | 39 | 39 | 21 | – | 111 | 15 | – |
| 2000 | – | – | – | – | – | – | – | – |
| 2001 | – | 45 | 36 | 18 | – | 106 | 13 | – |

C. Lithuania

| | 1. Livestock as % of GAO | 2. Milk index | 3. Cow index | 4. Cows on corporate farms index | 5. Livestock production on corporate farms (%) | 6. Cows on individual farms index | 7. Cows. on corporate farms (%) | 8. Livestock production on individual farms (%) |
|---|---|---|---|---|---|---|---|---|
| 1990 | 54 | 100 | 100 | 100 | 64 | 100 | 62 | 32 |
| 1991 | 55 | 92 | 99 | 97 | 63 | 104 | 60 | 46 |
| 1992 | 43 | 77 | 98 | 89 | 52 | 112 | 56 | 38 |
| 1993 | 53 | 66 | 87 | 60 | 61 | 131 | 42 | 48 |
| 1994 | 46 | 60 | 80 | 45 | 64 | 137 | 34 | 36 |
| 1995 | 47 | 58 | 73 | 33 | 66 | 137 | 27 | 38 |
| 1996 | 45 | 58 | 69 | 27 | 61 | 138 | 24 | 40 |
| 1997 | 42 | 58 | 70 | 22 | 54 | 147 | 19 | 38 |
| 1998 | 41 | 62 | 63 | 17 | 60 | 153 | 15 | 36 |
| 1999 | 42 | 54 | 57 | 10 | 59 | 144 | 14 | 37 |
| 2000 | 40 | 55 | 58 | 10 | 58 | 136 | 11 | 35 |
| 2001 | 45 | 55 | 52 | 7 | 67 | 124 | 9 | 39 |

Sources and explanations:

1. Livestock as % of GAO = Livestock production as percentage of Gross Agricultural Output (GAO)
Sources: Lerman, Kislev et al., 2003, 1003 (years 1950–1989); SYEs; SYLas; SYLis.
2. Milk index = Milk index (Index 1980–1989 average=100)
Sources: SYEs; SYLas; SYLis.
3. Cow index = Dairy cow index (Index 1980–1989 average=100)
Sources: SYEs; SYLas; SYLis; ACE, 2002; AgrE, 2003.
4. Cows on corporate farms index = Corporate farm dairy cow index (Index 1980–1989 average=100)
Sources: SYEs; SYLas; SYLis; Pirksts and Rozenberga, 1997; AFE, 1994, 1997 and 2000; ACE, 2002; ACLa, 2003.
5. Livestock production on corporate farms (%) = Livestock production as a proportion of total production on corporate farms
Sources: SYEs; SYLas; SYLis.
6. Cows on individual farms index = Plot farm and family farm dairy cow index (Index 1990=100)
Sources: SYEs; SYLas; SYLis; Pirksts and Rozenberga, 1997; AFE, 1994, 1997 and 2000; ACE, 2002; ACLa, 2003.
7. Cows on corporate farms (%) = Share (%) dairy cows on corporate farms
Sources: SYEs; SYLas; SYLis; Pirksts and Rozenberga, 1997; AFE, 1994, 1997 and 2000; ACE, 2002; ACLa, 2003; AgrE, 2003.
8. Livestock production on individual farms (%) = Livestock production as a proportion of total production on individual farms
Sources: SYEs; SYLas; SYLis.

**Annex Table A-3  Population employed in agriculture and forestry and total employed population 1990–1999**

A. Estonia

| | 1990 | 1991 | 1992 | 1993 | 1994 | 1995 | 1996 | 1997 | 1998 | 1999 | Change |
|---|---|---|---|---|---|---|---|---|---|---|---|
| 1. Agriculture (1,000s) | 149 | 141 | 125 | 102 | 88 | 63 | 60 | 54 | 53 | 48 | –68% |
| – Change (1,000s) | – | –8 | –16 | –23 | –14 | –25 | –3 | –6 | –1 | –5 | –101 |
| 2. Total employed (1,000s) | 826 | 808 | 766 | 708 | 693 | 656 | 646 | 648 | 640 | 614 | – |
| – Change (1,000s) | – | –18 | –42 | –58 | –15 | –37 | –10 | +2 | –8 | –26 | –212 |
| – Change (%) | – | –2.2 | –5.2 | –7.6 | –2.1 | –5.3 | –1.5 | +0.03 | –1.2 | –4.1 | –26 |
| 3. Employment index (1990=100) | 100 | 98 | 93 | 86 | 84 | 79 | 77 | 78 | 77 | 74 | – |

B. Latvia

| | 1990 | 1991 | 1992 | 1993 | 1994 | 1995 | 1996 | 1997 | 1998 | 1999 | Change |
|---|---|---|---|---|---|---|---|---|---|---|---|
| 1. Agriculture (1,000s) | 233 | 236 | 262 | 235 | 224 | 188 | 181 | 187 | 178 | 171 | –27% |
| – Change (1,000s) | – | +6 | +26 | –27 | –11 | –36 | –7 | +6 | –9 | –7 | –63 |
| 2. Total employed (1,000s) | 1,409 | 1,397 | 1,345 | 1,245 | 1,205 | 1,046 | 1,018 | 1,037 | 1,043 | 1,038 | – |
| – Change (1,000s) | – | –12 | –52 | –100 | –40 | –159 | –28 | +19 | +6 | –5 | –371 |
| – Change (%) | – | –0.9 | –3.7 | –7.4 | –3.2 | –13.2 | –2.7 | +1.9 | +0.6 | –0.5 | –26 |
| 3. Employment index (1990=100) | 100 | 99 | 95 | 88 | 86 | 74 | 72 | 74 | 74 | 74 | – |

C. Lithuania

| | 1990 | 1991 | 1992 | 1993 | 1994 | 1995 | 1996 | 1997 | 1998 | 1999 | Change |
|---|---|---|---|---|---|---|---|---|---|---|---|
| 1. Agriculture (1,000s) | 335 | – | 362 | 399 | 390 | 390 | 399 | 363 | 355 | 331 | –1% |
| – Change (1,000s) | – | – | +27 | +37 | –9 | 0 | +9 | –36 | –8 | –24 | –4 |
| 2. Total employed (1,000s) | 1,837 | – | 1,855 | 1,778 | 1,675 | 1,644 | 1,659 | 1,669 | 1,656 | 1,648 | – |
| – Change (1,000s) | – | – | +18 | –77 | –103 | –31 | +15 | +10 | –13 | –8 | –189 |
| – Change (%) | – | – | +1.0 | –4.2 | –5.8 | –1.9 | +1.0 | +0.6 | –0.8 | –0.5 | –10 |
| 3. Employment index (1990=100) | 100 | – | 101 | 97 | 91 | 89 | 90 | 91 | 90 | 90 | – |

*Sources:* SYEs; SYLas; SYLis; WB, 1993d.

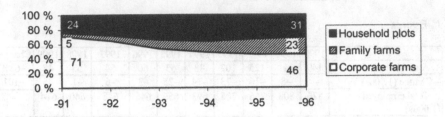

A. Estonia

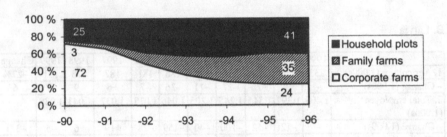

B. Latvia

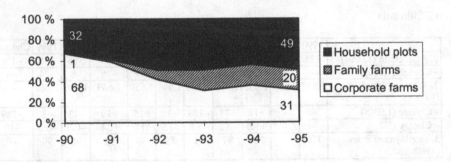

C. Lithuania

**Annex Figure A-1     Breakdown of Gross Agricultural Output (GAO) between corporate, family and household plot farms 1990–1996 (%)**

*Sources:* AFE, 1994 and 1997; Bratka, 1998; Pirksts and Rozenberga, 1997; Poviliunas and Batuleviciute, 1997 (years 1990–1995); SYEs; SYLas.

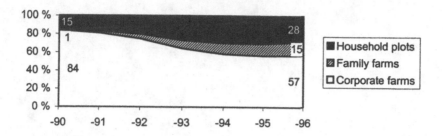

A. Estonia

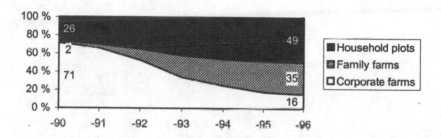

B. Latvia

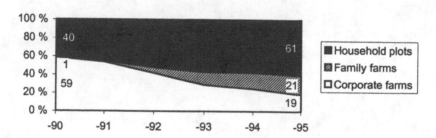

C. Lithuania

**Annex Figure A-2**   **Breakdown of total milk production by household plot, family and corporate farms in 1990–1996 (%)**

*Sources:* AFE, 1994 and 1997; Bratka, 1998; Pirksts and Rozenberga, 1997; Poviliunas and Batuleviciute, 1997; SYEs; SYLas; SYLis.

# Index

Printed in the United States
by Baker & Taylor Publisher Services.

Printed in the United States
by Baker & Taylor Publisher Services